T0132999

The Fleas of the Pacific Northwest

The Fleas of the Pacific Northwest

Robert E. Lewis
Joanne H. Lewis
Chris Maser

Oregon State University Press
Corvallis, Oregon

Journal Paper No. J-12183 of the Iowa Agriculture and Home Economics Experiment Station, Ames, Iowa. Project No. 2581.

The paper in this book meets the guidelines for permanence and durability of the Committee on Production Guidelines for Book Longevity of the Council on Library Resources and the minimum requirements of the American National Standard for Permanence of Paper for Printed Library Materials 239.48-1984.

Library of Congress Cataloging-in-Publication Data
Lewis, Robert Earl, 1929-
The fleas of the Pacific Northwest / Robert E. Lewis, Joanne H. Lewis, Chris Maser.
p. cm.
Bibliography: p.
Includes index.
ISBN 0-87071-355-8 (alk. paper)
1. Fleas—Northwest, Pacific—Classification. 2. Insects—Classification. 3. Insects—Northwest, Pacific—Classification. I. Lewis, Joanne H., 1932- . II. Maser, Chris. III. Title.
QL599.6.P17L49 1988
595.77'5'0795—dc19 88-1612
CIP

Printed in the United States of America

DEDICATION

This work is dedicated to the memory of

Dr. GEORGE PEARSON HOLLAND

27-VIII-1911 10-XI-1985

whose studies of the Siphonaptera of Canada over the past 40 years have clarified the faunal complexities of northern North America and adjacent Eurasia.

CONTENTS

Introduction

Since 1967, Maser has been conducting various faunistic studies in Oregon. Much of this research has involved the mammals and their interrelationships within the diverse ecosystems found there. As a part of these studies, an effort has been made to accumulate as much information as possible about the subjects, whether it had a direct bearing on the study or not. One of the fringe benefits has been the accumulation of a large collection of ectoparasitic arthropods. The senior author has been the grateful recipient of the fleas that form the basis for this report.

Nothing of a comprehensive nature has been published on the fleas of the Pacific Northwest since the volume by Hubbard (1947). During the intervening period, a number of cases of human plague have been reported from the area and there has been a concomitant increase in human contact with fleas and their hosts. Public health workers are exposed to fleas and other disease vectors during the course of their work. The increased popularity of recreational areas in the western states has also exposed a larger segment of the lay public to areas where vector-borne diseases are endemic. This aspect of the subject is treated in more detail elsewhere in this study.

Any thorough study of the epidemiology of a vector-borne disease requires the accurate identification of the vector species. The availability of good keys and other types of literature used in identification largely depends on the availability of specialists in that particular group of organisms. At one time Oregon had one of the few recognized authorities on the fleas of the western United States in C. A. Hubbard, but by the mid-1950s his interests had turned to other areas of the world. Since then additional species have been described, and the known ranges of the endemic taxa have been further elucidated. The literature has not kept pace with these advances, and the following is an attempt to rectify this deficiency.

The need for well documented, properly preserved material continues, and information on collection, preservation and mounting is treated in a later section. But first, it seems appropriate to acquaint the reader with some basic information about this fascinating group of insects.

Since humans began to live in caves and other permanent or semipermanent abodes, one of their constant companions has been the flea. Although seldom a problem in developed countries, other than occasional outbreaks on pets and other domesticated animals, much of the world's population still lives in intimate contact with these pests.

Fleas belong to the insect order Siphonaptera. As adults, they are small (0.5-10 mm), laterally compressed, blood-sucking, external parasites of some birds and mammals. Hosts that nest on or near the ground or in burrows are preferred by most bird fleas. Mammals that occupy more or less permanent nests or lairs are preferred by most mammal fleas. Mammals that do not build nests, such as primates and ungulates, and aquatic mammals, such as beavers, otters, and seals, rarely have fleas. Most fleas show a fair amount of host specificity, but a few species feed on a broad range of hosts.

There currently are slightly in excess of 2,375 described species and subspecies of fleas in the world and they are distributed over all of the continental land masses, including Antarctica. The bulk of the species occur in the North Temperate zone, or at higher elevations in the tropics and semitropics. About 75 percent of these are parasites of rodents and insectivores.

In spite of the large number of species, fleas have a remarkably uniform life cycle. Being holometabolous insects, they have a "complete" life cycle that consists of four stages: egg, larva, pupa, and adult.

Life History

EGGS

The eggs of most fleas have never been observed, except as they occur in the abdomen of gravid females. Among those species in which the eggs are known, they range from pearly white to black in color, and spherical to oblong-oval in shape. They usually are laid a few at a time while the flea is on the host or in the nest of the host. In the latter instance, the eggs may be firmly glued to the nesting material; in the former case, the eggs are not adhesive and quickly fall to the ground. Usually only a few eggs develop at a time but in neosomic species, such as *Vermipsylla alakurt* and *Dorcadia ioffi,* all of the eggs evidently mature at the same time. The only species known to attach its eggs to the pelage of the host is *Uropsylla tasmanica,* the larvae of which are dermal parasites of dasyurid marsupials in Tasmania and Victoria, Australia. The incubation period ranges from 2 to 12 or more days, depending on the temperature. Relatively high temperatures of 20-30 degrees C and 75 percent relative humidity, or above, are optimum for those species that have been reared in the laboratory.

LARVAE

Unlike the immatures of hemimetabolous ectoparasites such as lice, flea larvae are free-living organisms and are not nutritionally dependent on the adult habitat. The larvae have chewing mouthparts and feed on various organic materials, both plant and animal, in the larval habitat. Some are suspected to be predaceous and many consume the small dried blood pellets produced by the adults during feeding. Larvae are vermiform and vary in color from white to creamy yellow. The head is usually well developed with serrated mandibles, 2-segmented antennae, and no eyes. In the first larval instar the dorsum of the head is provided with a keeled oviruptor that is lost at the first moult. There are 13 body segments, frequently with many rigid setae. The thorax is not differentiated from the abdomen, and there are no thoracic legs. The terminal abdominal segment bears a pair of finger-like projections known as anal struts. There are 3 larval instars, the third terminating with a cessation of feeding activity, the spinning of a cocoon, and a moult to the pupal stage.

Since the larvae are free-living organisms in the nest, burrow, or lair of their host, it follows that the suitability of the larval habitat exerts a profound influence on the survival of the larva. This has led some observers to conclude that ecological factors in the larval habitat do more to determine host specificity in fleas than do the physiological requirements of the adults. Obviously both are important, and it is known that there is a broad range of tolerance among species. The long list of host species and broad geographical distribution of *Xenopsylla cheopis,* for example, would suggest that this is one of the more ecologically tolerant taxa.

PUPAE

Flea pupae are adecticous and exarate. They are enclosed in a silken cocoon produced from secretions of the labial glands and usually incorporate sand grains and detritus from the larval habitat. When the adult stage is reached, the pupal cuticle is shed but the adult flea remains in its cocoon until stimulated to leave. It then emerges as a perfectly formed, non-teneral adult capable of instant feeding. Transient wing buds have been reported in the pupae of some species, a testimony to the theory that fleas evolved from winged ancestors.

ADULTS

Except for the neosomic forms, adult fleas tend to be remarkably similar in structure. They range in size from the tiny males in the genus *Tunga* which may be less than 0.8 mm to the giant females of *Hystrichopsylla schefferi* which may exceed 8 mm in length. Adults vary in color from pale amber yellow to black; the majority are golden brown. They become much darker when engorged with blood.

While the tagmata of adult fleas are distinct, they are so thoroughly integrated that they are not distinct as they are in many insect orders. The ectoparasitic mode of life has encouraged the development of specializations with survival value for an organism that depends on the vital fluids of a large host animal for its own existence. As a result, the strong lateral compression, great lateral flexibility, the tough, smooth cuticle and caudally directed combs, false combs, spinelets and setae all combine to allow the flea to course through the pelage or plumage of its host with the least resistance. To be sure, fleas in some genera have evolved toward a more sedentary life, and in so doing have lost most of the traits mentioned above, if they ever had them. However, they would still never be mistaken for some other type of insect. Most of the known species conform to the stereotyped behavior of the cat flea or the human flea, spending much of their time in almost constant motion on the body of their host. Because of this great similarity in form, characters on the nongenitalic segments are of little or no value in identification at the species level. Because they are still useful in delimiting the higher categories, the following review of these characters is given.

Morphology

THE HEAD

The head capsule of fleas (Fig. 1-4) is divided obliquely into preantennal and postantennal portions by the antennal fossa. The preantennal portion is usually called the frons, while that behind is termed the occiput. The frons usually bears a small, tooth-like structure on its anterior margin. This is the frontal tubercle and it is deciduous in some fleas. There also may be a variable number of rows of bristles. A row of two or three bristles is commonly encountered below the eye, or where the eye would be if it is vestigial. This is the ocular row, and may be the only setation on the frons. Directly in front of the ocular row may be another row of a variable number of bristles, the frontal row. In addition, there may be another row of bristles extending along the anterior margin of the frons. This is the submarginal row. In some genera, some or all of these may be

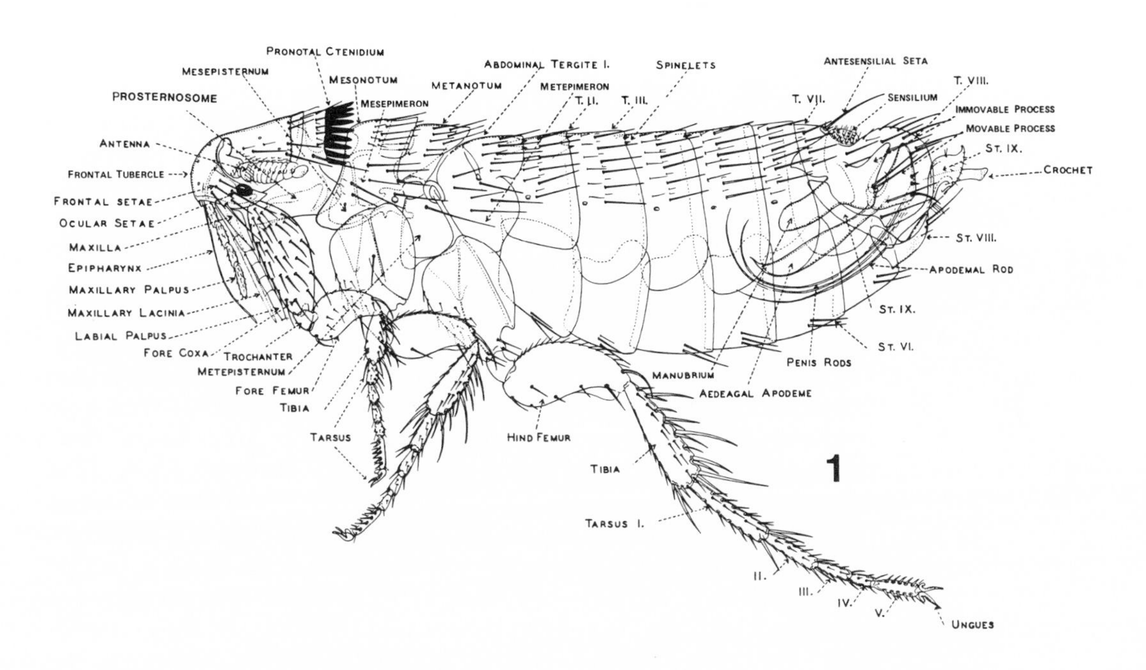

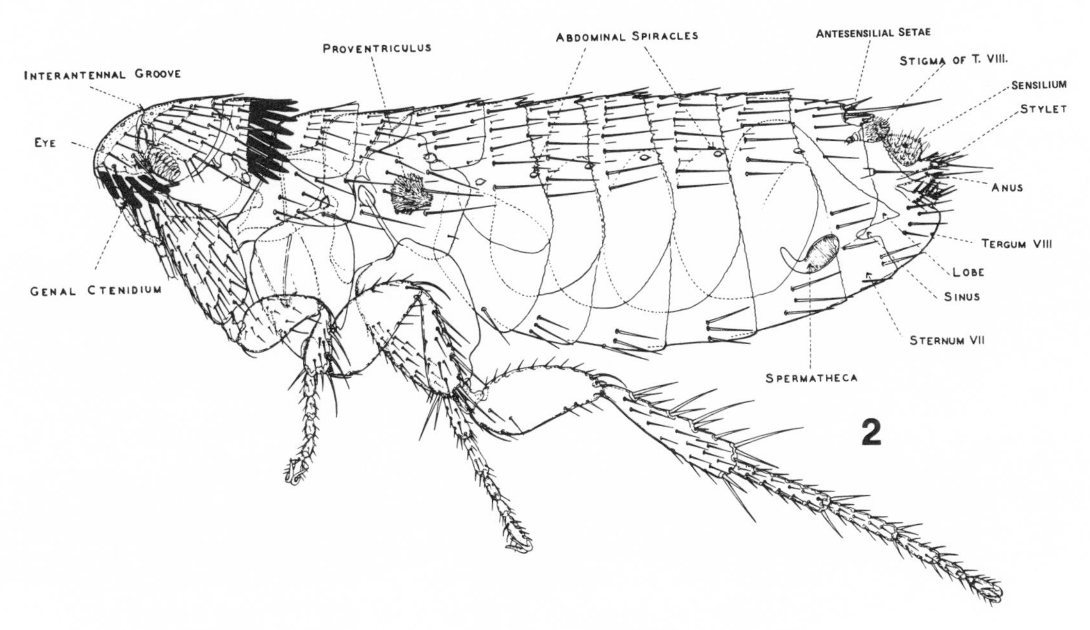

1 Male *Aetheca thamba* (Jordan) showing general anatomy. After Holland, 1985.
2 Female *Corrodopsylla curvata obtusata* (Wagner) showing general anatomy. After Holland, 1985.

thickened and spiniform. The postantennal portion of the head capsule may bear as many as four diagonal rows of bristles, but the caudal or occipital row is always present.

In many genera, the inner walls of the antennal fossae are connected by a rod-like sclerite visible as a round or oval darkened area about midway on the anterior margin of the fossa. This is the trabecula centralis. It reinforces the head capsule. In addition, some genera show a narrow, sclerotized structure extending anteriorly in front of the eye. This is the tentorium, and the presence or absence of it and the trabecula centralis are used for separating some genera. The antennal fossa itself may be open almost to its dorsal margin, as it is in most male fleas, or it may be closed dorsally, as it is in females. The antennae of male fleas are erectile and are used to grasp the basal abdominal sternite of the female during copulation. The antenna is composed of a basal scape, a median pedicel and an apical flagellum or clava. The length of the setae on the pedicel is used as a taxonomic character in some groups.

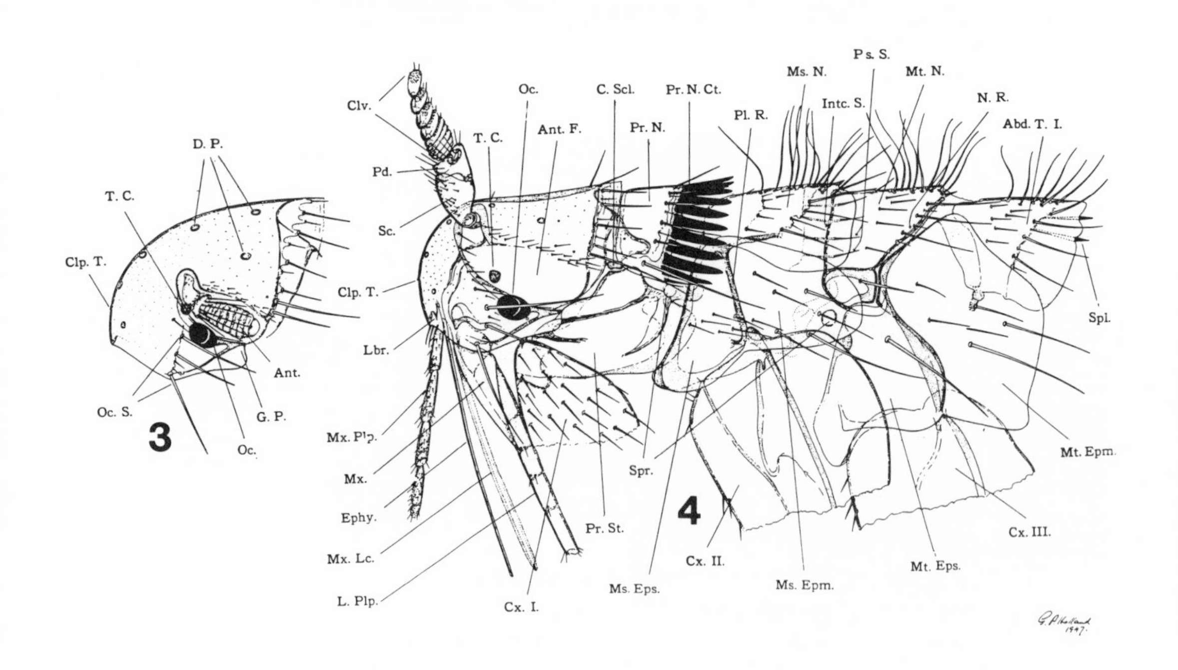

3-4 *Opisodasys (O.) vesperalis* (Jordan). **3** Head capsule of female. **4** Head and thorax of male. After Holland, 1985.

In some genera, the pre- and postantennal portions of the head capsule may be separated dorsally by an interantennal groove, sometimes called the dorsal sulcus or the falx. Earlier students used its presence or absence to divide the order into two suborders. It was later demonstrated that this was an unreliable character and the suborders are no longer recognized.

The portion of the preantennal area projecting caudoventrad below the antennal fossa is called the genal lobe. The genal lobe terminates in an apical spine in some genera, in others the spine is missing. Often the ventral margin of the genal lobe gives rise to a row of stout, heavily pigmented spines, the genal ctenidium. The presence or absence of the genal ctenidium and the configuration and number of spines are of taxonomic significance.

The mouthparts in this group of insects include the maxillary and labial palpi and a fascicle of three stylets. Two of the stylets are derived from the maxillary laciniae and the unpaired stylet is derived from the epipharynx. The maxillary palpi are always 4-segmented, and the maxilla itself is triangular in shape. The labial palpi are usually 5-segmented, although there may be more segments in a few species. The labrum is reduced to a tiny, inconspicuous sclerite at the base of the epipharyngeal stylet.

THE THORAX

Of the three thoracic segments (Fig. 3), the prothorax is more intimately associated with the head than with the remaining two segments. In all segments the tergites are distinct, but the pleurites and sternites are fused and lack sutures that show their original boundaries. In some genera, the caudal margin of the pronotum may bear a row of heavy, pigmented spines, the pronotal ctenidium. This structure is present but much reduced in a few genera, and it is absent in many groups. It is thus possible for a flea to have a pronotal comb with or without a genal comb. However, no fleas possess a genal comb and lack a pronotal comb. The pronotum usually bears a single row of setae, but there may be additional shorter setae cephalad of the main row and, in some genera, there may be two or even three complete rows of setae.

The prothorax of fleas is L-shaped in lateral view. The long axis of the pronotum is perpendicular to the long axis of the body. The fused pleurites and sternite, here called the prosternosome, form a horizontal trough that extends from below the base of the pronotum anteriorly to the base of the mouthparts. The prothoracic legs arise at the apex of the prosternosome. There usually is a sinus in the dorsal margin into which the cervical sclerite fits.

The mesothorax consists of a dorsal mesonotum and a ventral composite sclerite composed of the fused pleurites and sternite. The mesonotum usually bears a single row of long bristles. There may be shorter bristles cephalad of this row, either irregularly scattered or themselves arranged in rows. This sclerite may bear a "mane" in a few species as shown in Figure 3. In most species the pleural sclerites are divided into a mesepisternum and a mesepimeron by a conspicuous internal pleural rod, probably the detached pleural ridge. The chaetotaxy of these sclerites varies among genera but tends to be fairly consistent within a genus. Most fleas possess a variable number of pseudosetae arising under the mesonotal collar.

The metathorax consists of a dorsal metanotum and the fused pleurites and sternite. The metanotum bears an anterior, intercostal sclerotization and a posterior notal ridge, both of which extend ventrally to the apex of the sclerite. The notal ridge in most fleas gives off an anterior branch before reaching the apex of the sclerite, thus separating a small ventral part of the metanotum from the main sclerite. This is variously called the lateral metanotal area, a portion of the metepisternum or the episternum. There is little reason to doubt that it is derived from the metanotum. Lateral metanotal area therefore seems an appropriate appellation. Again there is usually only one main row of bristles present on this sclerite, and there may be a few to many shorter bristles scattered over the segment in front of the main row, or even arranged in irregular rows.

The notal ridge in most fleas terminates at the upper end of the pleural ridge in what, as Holland (1985) put it, appears to be a ball-and-socket joint. This joint contains resilin, an elastic material that aids the flea during the act of jumping. Fleas that have lost the ability to jump have lost this structure and even the pleural ridge may be reduced or missing. When present, the pleural ridge serves to set off the metepisternum from the larger, more conspicuous metepimeron. The long axis of the metathorax is directed obliquely caudad to the extent that much of the dorsal margin of the metepimeron is bordered by the tergite of the first abdominal segment. The chaetotaxy of the metepimeron is sometimes useful in taxonomic discrimination, and it bears a false comb in one genus.

There are two pairs of thoracic spiracles. The first pair is in the pair of link plates connecting the prothorax to the mesothorax. The second pair is situated just below the lateral metanotal area.

THE LEGS

Fleas have three pairs of legs composed of the segments characteristic of the Insecta: coxa, trochanter, femur, tibia and a 5-segmented tarsus. The fore coxae articulate with the apex of the prosternosome and are suspended in an almost transverse plane. Their outer surface usually is provided with oblique rows of short to medium-length setae, and they function much like the blade of a snow plow as the flea moves through the pelage or plumage of its host. The mid and hind coxae are progressively larger and articulate with the meso- and metathorax in a longitudinal plane. Outer and mesal internal ridges are present that serve to stiffen these coxae and provide points of articulation with the meso- and metathorax. In both, the outer rod articulates with an acetabulum at the base of the pleural rod, while the mesal rod articulates with an acetabulum in the meso- and metasterna. The chaetotaxy of these coxae, as well as the presence or absence of spiniform setae on the mesal surface of the hind coxae, is employed in taxonomic discrimination in some cases.

The trochanter is small in all the legs and does not bear structures of taxonomic significance.

The femora are large and the number of setae on the fore femur, combined with the chaetotaxy of the mid and hind coxae, were used by Jordan (1933) in his initial division of the genus *Ceratophyllus s. lat.*

The tibiae are fairly long in most fleas and the number of notches in their dorsal margin, combined with the marginal chaetotaxy, is useful in identification.

Tarsal characters are frequently employed in keys and have proven to be most useful, at least at the generic level. The relative length of the first tarsal segment, for example, separates *Tarsopsylla* from all other ceratophyllid genera. The relative lengths of the apical setae of some tarsal segments are also employed. Perhaps the most useful of the tarsal characters is the number and placement of the lateral plantar bristles on tarsal segment V. This is particularly true for hind tarsal segment V.

PREGENITAL ABDOMINAL SEGMENTS

Following Snodgrass (1946), the first seven abdominal segments of the flea abdomen are termed pregenital. Since it is now generally accepted that the flea abdomen consists of ten segments, the remaining three segments are genital segments.

In general, the pregenital abdominal segments are so unspecialized as to have few taxonomic characters. Except for segment I, which lacks a sternite, each segment consists of a saddle-like tergite bearing one or more rows of bristles. The lower ends of these overlap the upper ends of the corresponding sternite in most fleas, and each segment telescopes into the segment before it. Segment VII bears a variable number of antepygidial setae on its dorsal apical margin. In one genus of bat fleas, these bristles form a false comb and, in another, other terga may have false combs. Frequently the metanotum and at least the anteriormost abdominal tergites will bear short, sharp marginal spinelets, at least dorsally. *Stenoponia* species have a well developed ctenidium on tergum I and subsequent terga bear many apical spinelets.

Behind the antesensilial setae is a saddle-shaped sensory plate called the sensilium or pygidium. This consists of a variable number of hollow, dome-shaped rosettes with holes in their centers from which issue long trichobothria. While their function is not thoroughly understood, their structure is elucidated in Amrine and Lewis (1986), and Amrine and Jerabek (1983) make the argument that the trichobothria function as ultrasonic receptors.

Sternites II-VI are all very similar in size and shape. Sternum II frequently bears characters that are used in identification. These are usually chaetotaxic in nature. However, some fleas have an irregular patch of striation on st. II called a striarium. Its function is not understood, but some workers have postulated that this is to enhance the purchase of the male antennae during copulation. Possibly, but the males in these species have a striarium, too. It also has been said, but never proven, to be an organ of stridulation. The contour of female st. VII is one of the more important taxonomic characters in many genera, and it is frequently the only character available for separating females of related species. Unfortunately, in some species this is a variable character at best, and females in some genera cannot be separated without accompanying males.

Terga I-VII bear laterally placed spiracles. The spiracles of t. I are usually slightly larger and usually arise higher than those on subsequent segments and they may be completely or partly covered by the metepimeron. The spiracles of segment VIII have large stigmatal cavities lined with a pile that is identical in structure to that surrounding the antesensilial cupolas. These lie on either side of the sensilium, and their size and shape are sometimes useful in identification.

GENITAL SEGMENTS OF MALE

It is questionable whether a detailed discussion of the male genitalia (Figs. 5, 6) and intromittent organs would serve a useful purpose in this type of presentation; I will restrict my treatment to those structures of use in identification. For more detailed information, see Traub (1950).

The main male structures employed in identification at the species level are derived from terga and sterna VIII-X. The degree of modification of tergum and sternum VIII varies from family to family and sometimes within the family. As Holland (1985) points out, these structures are similar to those of preceding segments in the Vermipsyllidae. In the Ischnopsyllidae both are about equally expanded posteriorly, enclosing the genitalia. This is also true for the amphipsylline genera. In most of the Hystrichopsyllidae t. VIII is unmodified to reduced and st. VIII is expanded to more or less protect the genitalia. In the Pulicidae t. VIII is very reduced while st. VIII is much expanded to enclose the genitalia. The opposite is true in the Ceratophyllidae where t. VIII is much enlarged and st. VIII is reduced, in some cases to a vestige. In some ceratophyllid genera the inner dorsal surface may bear a conspicuous spiculose area.

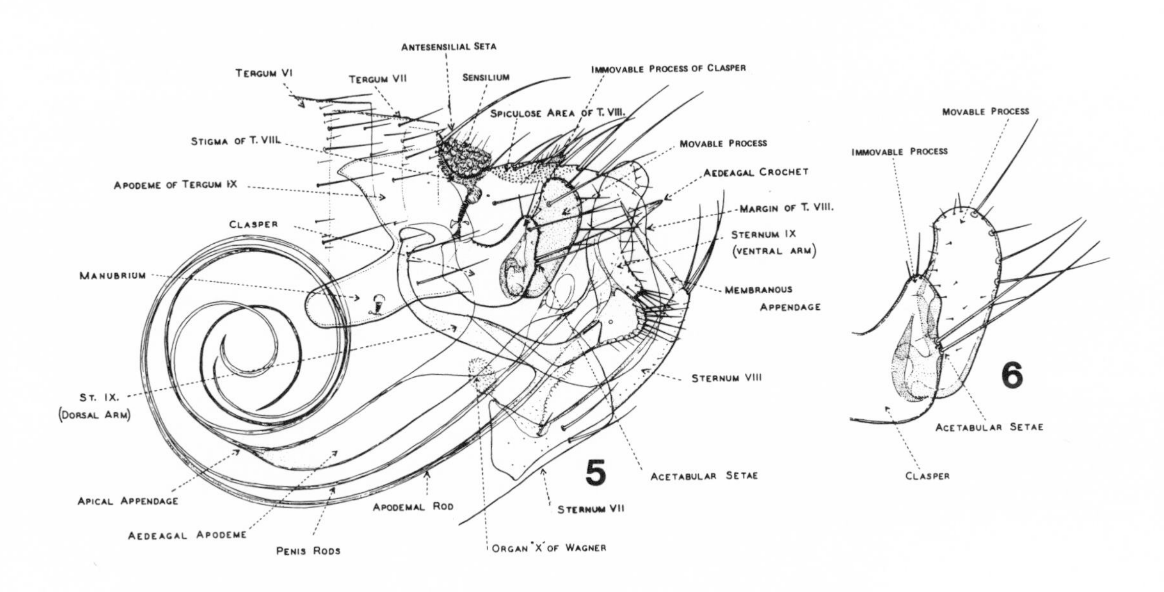

5-6 Details of male genitalia of *Ceratophyllus (C.) niger* C. Fox. After Holland, 1949.

Tergum IX consists of a narrow, strap-like sclerite cephalad of the sensilium, the ends of which extend downward to form the genital claspers. Dorsally this apparatus consists of the apodeme of t. IX that may be variously developed in different families. Ventrally, a pair of arms projects forward to form the manubria. Caudally the arms form the clasper. While the clasper varies from one family to another, it usually consists of one or more fixed processes and a movable process. The apex of the fixed process usually bears characteristic setation, and usually there are one or two acetabular setae arising somewhere on its caudal surface. It also usually is provided with a fovea that accommodates a corresponding incrassation in the anterior face of the movable process. It is this structure that is used to grasp the margin of the female st. VII during copulation. The size, shape, and adornment of these structures are among the most frequently employed in identification at the species and subspecies level.

Sternum IX is branched both dorsally and ventrally. The dorsal arms pass to the outside of the manubria and are internal. The ventral arms are variously shaped and adorned with setae. As with the clasper, the size, shape, and setation of st. VIII and IX are of great importance in identification.

Sternum X and tergum X form small dorsal and ventral plates in the vicinity of the anus.

Studies of the aedeagus, including the penis rods and associated structures, are beginning to indicate that these may be useful in taxonomic discrimination, but to date they have not been used extensively in keys. In view of the complex terminology employed in describing this structure and the fact that its characters are not used in our keys, the aedeagus will not be discussed further.

GENITAL SEGMENTS OF FEMALE

The genital segments of female fleas do not show the high degree of specialization characteristic of the males. As mentioned earlier, st. VII may be variously modified with respect to the contours of its caudal margin. This is one of the best characters for identification of females in many genera, and in some cases subspecies have been based on it. However, although it may be a constant character in some groups, in others it can be highly variable and of little or no value. In addition, it is not uncommon for the segment to be so badly damaged during mating that the identifying characters are obliterated.

Tergum VIII is usually hypertrophied while st. VIII is reduced to a small sclerite. Segment IX is relatively rudimentary. The dorsal and ventral anal flaps are derived from segment X. Except in the Vermipsyllidae, the dorsal anal flap bears a pair of anal stylets.

Female fleas usually have a single spermatheca although in two western genera the normal number is two. This is a small, heavily sclerotized, pigmented internal organ in the abdomen that is the site of sperm storage. Typically the spermatheca consists of an expanded head, or bulga, joined to a more constricted tail, or hilla. In addition to the contours of this organ, characteristics of the junction of the hilla with the bulga and the presence or absence of an apical papilla on the hilla are all useful characters. In some genera the sclerotized duct of the bursa copulatrix also has been employed in combination with other characters.

Fleas and Disease

In reviewing the relationships of fleas to disease it is convenient to break down the various agents of infection into their related groups, as done by Traub (1983). These are listed as bacterial, rickettsial, viral, and other infections. In the latter category are included protozoans (trypanosomes and haemogregarine sporozoans), cestodes *(Dipylidium caninum* and *Hymenolepis diminuta)* and the nematode *Dipetalonema perstans*. The protozoan diseases are not transmitted to humans by fleas, although they are known to be intrazootic. *Dipylidium caninum* is common in various mammals and is occasionally reported from humans, particularly young children. Acquisition is accomplished by ingestion of the infected flea. The obligatory intermediate hosts for *Hymenolepis diminuta* include myriapods and species of the insect orders Lepidoptera and Coleoptera as well as Siphonaptera. As in the preceding species, infection is accomplished by ingestion of the intermediate host. Traub (1983) stated that pulicid fleas serve as the main intermediate hosts for *Dipetalonema [perstans]*. However, human infections are accomplished by bites of infected *Culicoides* in endemic areas.

Allergic dermatitis caused by reaction to flea bites also falls into this category. The degree of response to flea bites varies according to the individual and amount of exposure but after prolonged exposure to their bites, a state of nonreactivity is attained according to Smit (1973).

Finally, a condition known as tungiasis also should be included here. While not a disease as such, the condition is potentially life-threatening if left unattended or improperly treated. In humans, the agent is the female of *Tunga penetrans*. A native of the warmer parts of the Western Hemisphere, the species is now established across tropical Africa as well, presumably having been imported on livestock or pets. Females of this species are small and active. When infesting humans the females usually attach themselves to the soft skin between the toes or under the toenails. Other parts of the body may be attacked in heavy infestations. During the next week to 10 days her abdomen gradually distends to the size of a pea, while at the same time she is being encapsulated by the surrounding tissues of the host. At the completion of this process, only the tip of the abdomen remains in contact with the outside. Contained in these terminal segments are the large spiracles, anus, and the opening to the reproductive system. According to Smit (1973), several thousand eggs are produced during the reproductive life of the female and these are shed in such a way that they are likely to fall to the ground away from the host. The larvae develop in dry, sandy soil and the cycle from egg to adult requires about 3 weeks under optimum conditions. Upon the ultimate death of the female, her presence usually causes a local inflammation culminating in her body being ejected. Frequently the lesion left behind becomes secondarily infected and, if left untreated, may result in the loss of appendages, septicemia, or even tetanus or gangrene.

The relationship between fleas and viruses is somewhat obscure. Traub (1983) indicated that fleas are not known to transmit any viral infection to humans. However, Smit (1973) cited lymphocytic choriomeningitis and Russian spring-summer encephalitis as having been reported from fleas in the U.S.S.R., and Sapegina and Kharitonova (1969) stated that two common bird fleas can transmit the virus for Omsk hemorrhagic fever. Myxomatosis, a viral disease of rabbits and hares, is transmitted by contaminated mouthparts of rabbit fleas, and some workers contend that fleas assist in the perpetuation of viruses in nature far more than has been suspected.

Both Smit (1973) and Traub (1983) indicated that fleas are the most important vectors of *Rickettsia mooseri,* the causative agent of murine typhus. The mode of transmission is thought to be via infective flea feces. Two other rickettsial diseases were cited by Smit (1973) to be associated with fleas, although their normal vectors are ticks. These are Boutonneuse fever in southeastern France and Q-fever reported from Asiatic Russia. Traub (1983) also cited the report of *R. prowazekii,* the causative agent of louse typhus, in Eastern flying-squirrels *(Glaucomys volans),* and the suspicion that squirrel fleas could be the vector. The squirrel flea, *Orchopeas howardi,* has been found naturally infected with rickettsiae, but actual transmission has not been demonstrated.

Several bacterial disease organisms have been associated with fleas in the recent literature including brucellosis, clostridial infections, erysipeloid, glanders, listeriosis, melioidosis, pasteurellosis, pseudotuberculosis, and salmonellosis. None of these is dependent on fleas for its transmission, and in most cases their association with fleas has been little more than that they have been isolated from them in laboratory studies. However, there are two bacterial diseases that are more intimately associated with fleas. These are tularemia, caused by *Francisella tularensis,* and plague, caused by *Yersinia pestis.*

In the case of tularemia, there are conflicting views as to the importance of fleas as vectors of the disease. One theory postulates that this is a disease of leporids and the principal vectors are ticks. This is supported by the indisputable importance of rabbits and hares as a source of the disease in humans and other animals. However, Traub and Jellison (1981) pointed out that if microtine rodents were the reservoir the peculiar geographic distribution of the disease—specifically its absence from Africa south of the Sahara and South America, where leporids and ticks coexist but microtines do not—would be clarified. For the present it seems safe to conclude that fleas, along with a number of other blood-sucking arthropods that feed on small mammals, are important in the epidemiology of this disease.

By far and away the most important disease, bacterial or otherwise, associated with fleas is plague. As pointed out by many authors, the disease has changed the course of history on many occasions and been responsible for the death of tens of millions of people in Eurasia and Africa. In the past fifteen centuries four important pandemics have been recorded. The first began in 542 and before it ended around 600 it had ravaged the whole Roman world. The famed "Black Death" of the 14th century caused an estimated 25 million deaths, one quarter of the entire population of Europe alone. The third extended from the 15th to 17th centuries, culminating in the plague of London in 1664-65. Some scientists believe that we are in the waning days of the fourth pandemic which began in 1894 in western China. Before it began to recede it had spread to India, the Middle East, Africa, parts of Europe, and North, Central, and South America. The history of plague in the United States was thoroughly discussed by Link (1955). The disease is now endemic to North and South America, having been introduced via domestic rodents, to become established in the native rodent populations. Foci of the disease are known from all of the western states, and a variable number of human cases is reported each year, including a few fatalities.

According to records kept by the Plague Branch of the Communicable Disease Center of the U. S. Public Health Service, Fort Collins, Colorado, ten cases have been reported from Oregon. These were: one in 1934, one in 1970, one in 1971, two in 1977, two in 1979, one in 1981, one in 1982 and one in 1983. One fatality each occurred in 1934, 1970, 1981, and 1983. While only one case has been reported from Washington during this period, 38 cases are listed from California, including eleven fatalities.

Plague, also known as bubonic plague, oriental plague, pest, and the black death, is an acute, febrile infection that is highly fatal when untreated. It is characterized by inflammation of the lymphatic system, frequently septicemia, and diffuse hemorrhages into the skin, subcutaneous tissues, and viscera. Usually large buboes develop in the inguinal, axillary, or cervical regions of the body. The disease is caused by *Yersinia pestis,* a Gram negative, pleomorphic bacillus.

Although plague is primarily a disease of rats and other small mammals, especially rodents, it is, when untreated, one of the most virulent infectious diseases in humans. It is known in three forms, separable on the basis of the symptoms. Most frequently encountered is the form known as bubonic plague, deriving its name from its most conspicuous symptom. Primary septicemic plague involves the invasion of the bloodstream by bacilli and constitutes the second form. The third form is primary pneumonic plague and involves the lungs. Plague epidemics are almost always of the bubonic type but always include a few primary septicemic and secondary pneumonic cases. While bubonic and septicemic plague usually are transmitted by fleas, pneumonic plague may be passed directly from person to person via aerosols from infected patients' respiratory tracts.

Plague exists in two forms over much of its range. Epidemic or human plague usually involves murine, synanthropic rodents and their fleas as a part of the reservoir. Rodents of prime importance are *Rattus rattus,* the black rat, and *Rattus norvegicus,* the Norway rat. The flea most commonly associated with these hosts is *Xenopsylla cheopis* which, coincidentally, is the vector *par excellence* of plague. Because of the attraction of these rats to stored products, particularly grain and other foodstuffs, they are cosmopolitan in distribution and have been introduced to all continents through port cities. From here, they have been transported inland by rail and highway traffic, taking their fleas with them.

In contrast, sylvatic plague is the name applied to the disease as it exists in nature in wild rodent populations. In this form it is transmitted from rodent to rodent by fleas without the involvement of humans. Humans acquire the disease accidentally from exposure to infected fleas or contact with diseased rodents. Although many species of rodents are known to function as reservoirs of sylvatic plague, ground squirrels and marmots are the most important in Eurasia and North America, cavies in South America, and various species of murine rodents in southern Africa. The transition from sylvatic plague to murine plague takes place when wild and domestic rodents come in contact with each other and exchange their fleas.

As early as the late 1800s it was suspected that fleas were involved in some way with the transmission of plague. It was not until 1914 that Bacot and Martin described the process in detail. Using *Xenopsylla cheopis* and *Nosopsyllus fasciatus* they were able to demonstrate that fleas feeding on infected rodents, in this case laboratory rats, frequently developed a blockage of the proventriculus due to the growth of plague bacilli. In this condition their ability to transmit the disease was greatly enhanced, for two reasons. First, proventricular blockage prevents blood from entering the midgut where digestion and absorption normally take place. As a result the individual flea becomes ravenously hungry and increases its attempts to feed. Second, due to the cuticular lining of the crop, this organ is capable of only limited expansion. Once filled to capacity, sufficient back-pressure is created to overcome the action of the cibarial and pharyngeal pumps. The result is the injection of contaminated blood back into the host animal.

It is now known that different species of fleas vary in their ability to become blocked. Some species evidently never do so and thus can only transmit the disease by means of contaminated feces or contaminated mouthparts. Blocked fleas usually die of desiccation in a few days, but the best vector species evidently autolyze the blockage before it results in death and thus remain available for reblockage at a later date. Even species that are known to be poor vectors may compensate for this deficiency by building to enormous populations in the nest or lair of the host. It has been speculated that this probably is how the sylvatic phase of plague is maintained in wild rodent populations.

As indicated earlier, sylvatic plague is endemic to the more arid regions of the western United States: east to western Texas, Oklahoma, Kansas, and North Dakota. The first recorded case from Oregon was a sheepherder in Lane County in 1934. Partly as a result of this fatality, the Oregon State Board of Health began plague surveys in August of 1936. The recent history of plague in the Pacific Northwest already has been described.

Literature Review

With the exception of the works by C. A. Hubbard, there have been relatively few publications dealing with the fleas of the area. Hubbard started publishing on the western flea fauna in 1940, his studies culminating in his book *Fleas of Western North America* in 1947. He continued publishing on the North American flea fauna until 1962 when his research interests took him to East Africa.

Hansen (1964) reported on a small collection of fleas taken on Steens Mountain, Harney County, Oregon. He listed 25 species of fleas as well as a number of other ectoparasitic arthropods. In 1968, Egoscue described a new species of *Stenistomera* from near Crane, Oregon, also in Harney County. Lewis, in Maser et al. (1974), reviewed the flea species associated with the sage vole. Ten taxa were discussed, and their possible role in plague transmission was considered. This study was conducted in the Crooked River National Grassland in Jefferson County. Lewis and Maser (1978) described a new species of *Phalacropsylla* collected in Malheur County, Oregon. Egoscue (1980) published the first record of *Anomiopsyllus amphibolus* for the state, having collected this species in Harney County. Hopkins (1980) reported on the fleas associated with the Virginia opossum in the environs of Portland, Oregon (Multnomah County). Lewis and Maser (1981) published an annotated checklist of fleas associated with the mammals indigenous to the H. J. Andrews Experimental Forest in Linn and Lane counties of Oregon. Haas (1982) reported sixteen taxa from Clackamas County from various small mammals. Gresbrink and Hopkins (1982) listed 23 taxa from Crater Lake National Park, Klamath County, Oregon. Thirty-six of these specimens were identified to only five genera. Easton (1983) also reported 23 species of fleas. While this study dealt mainly with collections from Neptune State Park, Lane County, Oregon, and the William L. Finley National Wildlife Refuge in Benton County, Oregon, a few other collecting sites are listed also. Hopkins (1985) listed nineteen taxa collected by him from a study area near Bend, Deschutes County, Oregon. All but 29 of the 954 specimens were identified to species.

Much less work has been done in the states adjacent to Oregon and concentrated studies in these areas would be useful as well as productive. Foremost among the workers in the states adjacent to Oregon was C. A. Hubbard. His 1943 and 1961 papers summarized what was known of the flea fauna of California, though other workers have also dealt with elements of the fauna. Holland (1957) and Holland and Jameson (1950) described new hystrichopsyllids from California. Jameson and Brennan (1957), Mead (1963), Nelson and Smith (1976), Nelson et al. (1979), Stark and Kinney (1969) and a few others have treated various aspects of the fauna, but no systematic survey of the state has been published.

Our knowledge of the flea fauna of Nevada is even less well developed. Again, Hubbard (1949b) is the only complete listing. Jellison and Senger (1979) and Nelson (1972) both deal with faunal aspects of the state, but neither add materially to Hubbard's list. The same is true for Idaho, and only Jellison and Senger (1976) present anything resembling an inventory of the state.

The flea fauna of the state of Washington is certainly best known from the works of Bacon (1953), Bacon and Bacon (1964), O'Farrel (1975) and Poole and Underhill (1953). The only comprehensive listing of the species is that of Svihla (1941), now very much out of date.

In summary, while our knowledge of the species composition and distribution of the flea fauna of Oregon is as good or better than any other of the contiguous states, adjacent areas of the Pacific Northwest are in much need of additional study.

A Note on the Illustrations

In his recent book on the fleas of Canada, Alaska and Greenland, Holland (1985) used a few illustrations from his earlier work (1949) and illustrations from some of his subsequent research papers. The bulk of his plates, however, were new original drawings. A number of taxa exist in the Northwestern United States that have not been reported from Canada, but much of the fauna is shared.

In the interest of saving time on this project and avoiding much duplication of effort with respect to illustrations, Dr. Holland was approached with an offer of co-authorship for permission to use pertinent drawings from his 1985 work. Characteristically, he declined co-authorship but provided both copies of the illustrations and his permission to use them. Permission subsequently has been received from the Entomological Society of Canada.

Illustrations of taxa not known from Canada have been made, when material was available. An effort has been made to simulate Holland's drawing style. These drawings are indicated as being original in their captions.

Glossary of Abbreviations Used in the Keys and Illustrations

Abd. abdomen; abdominal
Ac. acetabulum; acetabular
Aed. aedeagus; aedeagal
An. anus; anal
Ant. antenna; antennal
Ant.F. antennal fossa
Ant.S. antesensilial seta(e)
Ap. apodeme; apodemal
Bu. bulga of spermatheca
B.Cop. bursa copulatrix
Cl. clasper lobe
Clv. clava of antenna
Coll. collar of spermatheca
Crc. crochet of aedeagus ("paramere")
C. Scl. cervical sclerite
Ct. ctenidium
Crib. cribiform area of spermatheca
Cx. coxa
D.O. ductus obturatus
D.P. dermal pits
D.r.s. ductus receptaculi seminis
E.D. ejaculatory duct
Ephy. epipharynx
Epm. epimeron
Eps. episternum
F. movable process(es) of clasper
F.Ct. false ctenidium
Fk. anal fork
Flx. falx (interantennal ridge & suture)
Fm. femur
Fr. frons; frontal
Fr.T. frontal tubercle
Fv. fovea
G. gena; genal
Gl.W. gland of Wagner
G.P. genal process
Hl. hilla of spermatheca
Intc.S. intercostal sulcus
Int.G. interantennal groove
Int.R. interantennal ridge
L. labium; labial
Lc. lacinia
Lb. lobe
Lbr. labrum
M. manubrium of clasper
Mb. vexillum or membranous appendage
Ms. meso- pertaining to thoracic segment II
Mt. meta- pertaining to thoracic segment III
Mx. maxilla; maxillary
N. notum; notal (of thorax)
N.R. notal ridge
Oc. eye; ocular
Oc.S. ocular seta(e)
P. fixed process of clasper
Pap. papilla of spermatheca
Pl. pleuron; pleural
Pl.A. pleural arch
Plp. palpus
Pl.R. pleural ridge
Pl.St. prosternosome
Pn. penis
Pr. pro- pertaining to thoracic segment I
Ps.S. pseudosetae
Pvt. proventriculus
Rd. rod(s)
R.S. spermatheca or receptaculum seminis
S. seta(e)
Sc. scape of antenna
Scl. sclerotized area
Scl.R. sclerotized ridge
Sens. sensilium
Sn. sinus
Spf. spiniforms
Spic. spiculose area of male tergum VIII
Spl. apical spinelets
Spr. spiracle
St. sternum
Stig. stigma cavity of tergum VIII
Stl. anal stylet
Str. striated area of metepimeron or sternum II
Swd. internal sword-like incrassation
Strig. strigulae
T. abdominal tergum; tergal
Tb. tibia
T.C. trabecula centralis
Tent. tentorial arms
Th. thorax; thoracic
Tr. trochanter
Ts. tarsal segments
I, II, III, IV, etc. segmental numbers

Distribution Maps

The inclusion of distribution maps can be quite useful in summarizing, at a glance, the known occurrence of organisms. However, they may also be misleading in at least two ways. First, in areas of high ecological diversity it is obvious to the experienced biologist that species distribution is not continuous in most cases, in spite of what the map may show. Second, such maps only show where collections have been made; or, to put it another way, where collectors have been active. Still, maps do serve a useful purpose, and the maps show the distribution of collections in the Pacific Northwest. Most Oregon collections were made by Maser and associates, but a few are based on literature records. Only a single dot is placed in each county irrespective of how many collections were ultimately reported for that county.

The Pacific Northwest is an area of high physiographic, geological, and ecological diversity. The reader is referred to the work by Franklin and Dyrness (1988) for maps reflecting this diversity.

General Comments Concerning Flea Nomenclature and Collections

Workers in the Siphonaptera are more fortunate than systematic entomologists working on such large orders as the Coleoptera or Lepidoptera. Over 95 percent of the systematic studies in the order are dated since 1895, and these have been performed by a relatively small group of specialists. As a result, nomenclatural problems have been kept to a minimum, and most of the important ones have been submitted to the International Commission on Zoological Nomenclature for adjudication. The late Karl Jordan, and subsequently the late G.H.E. Hopkins, both associated with the Rothschild Collection of fleas in the British Museum (Natural History), were both nomenclators who were well aware of the importance of a stable nomenclature. The tradition was carried forward by F.G.A.M. Smit, who supervised the collection until his retirement in 1980. At present the future of the largest flea collection in the world is very much in question since, at this writing, no professional supervisor has been appointed.

Anticipating the trend toward automation, Smit and Wright (1978a) produced a list of 5-digit code numbers for all known taxa in the order, whether still valid or not. Unlike such ungainly systems as the International Species Inventory System (ISIS) which requires a 19-digit number for each name, the Smit and Wright system is based on the last three digits of the year the description was published, followed by two more digits that are determined by the chronology of description. Thus, the number for *Pulex irritans* Linnaeus, 1758, is 75801, since this was the first flea species listed in *Systema Naturae,* 10th edition. The number for *Pulex penetrans* Linnaeus, 1758, is 75802, since it is listed after the first species. Although automation has not yet been accomplished with fleas, we have included the Smit and Wright numbers after the names in anticipation of the day when it will. With respect to flea collections, most universities with departments of entomology maintain at least small teaching collections, and those employing systematists frequently maintain collections numbering in the millions of specimens. Owing to the relationship of ectoparasitic arthropods to their vertebrate hosts, the presence of a large arthropod collection does not necessarily indicate that fleas, lice, ticks, and mites will be abundantly represented. Major institutional or governmental collections in the United States include those of the United States National Museum of Natural History (USNM), Washington, D.C., the Communicable Disease Center, Fort Collins, Colorado, and the Field Museum of Natural History, Chicago, Illinois. Certainly the most extensive private collection from the Pacific Northwest was that amassed by C. Andreson Hubbard. The bulk of this was sent to the USNM and the British Museum (Natural History), but sets of representative species were prepared by Hubbard and sent to a number of museums and universities in the United States. Although the private collection of the senior author is willed to the Field Museum, voucher specimens of most of the species discussed here have been deposited at Oregon State University in the Systematic Entomology Laboratory. While host and collection data are far too extensive to include here, they are available, upon request, from the senior author.

Systematic Review of the Fleas

The following is a systematic review of the species of fleas known to occur in northern California, Washington, Oregon, and western Nevada. A few species not collected but suspected to occur there also are included. Treatment of the taxa is not exhaustive, but is designed to provide the nonspecialist with the means to identify specimens taken during faunal studies in the area. Species belonging to some genera are extremely similar, particularly in the female sex, and collections of importance in epidemiological studies should be referred to specialists for accurate determination or confirmation of their identification.

Each species account includes the current scientific name, the name of the original describer, a 5-digit computerization number based on Smit and Wright (1978a), reference to figure and map numbers, a brief nomenclatural history, and comments. In cases where a species is known from many collections, information pertaining to host preferences and seasonal occurrence is included. For the most part, specific collection localities and numbers of specimens are not included since this type of information would unduly increase the length of this study. Such information is available upon request from the senior author.

Only six of the fifteen families belonging to the order Siphonaptera are represented in this fauna. The order in which they are listed here is a reflection of their order of treatment in the Hopkins and Rothschild catalogues and does not suggest any particular phylogenetic affinities. The families under consideration may be separated by the following key.

KEY TO THE FAMILIES OF SIPHONAPTERA

(Modified from Hopkins and Rothschild, 1962)

1 Outer internal ridge of mid coxa absent; mesonotum without pseudosetae under the collar; metepimeron extending far upward, its spiracle placed much above the metepisternum; metanotum and abdominal terga without apical spines or spinelets; spiracles circular; abdominal terga II-VII with at most one row of bristles; no bristles above spiracle of t. VIII; sensilium with either eight or fourteen pits each side; hind tibia without an apical tooth on outside .... **Pulicidae**

Outer internal ridge of mid coxa usually present, sometimes short, and absent in a few cases in which both ridges are absent on both the mid and hind coxae; tooth at apex of hind tibia generally pointed, rarely rounded; sensilium sometimes with thirteen or fourteen, but usually with sixteen or more pits each side **2**

2 Of the two dorsal bristles of the fore femur (guarding the femorotibial joint), the outer somewhat shorter than the inner; no fourth link-plate (between metepimeron and basal abdominal sternite); combs, spinelets, antepygidial bristles, spiniform bristles on inner surface of hind coxae, and anal stylet of female all absent .... **Vermipsyllidae**

Not with this combination of characters **3**

3 Metanotum without marginal spinelets, though pseudosetae may be present; the arch of the tentorium visible in frons except in species with a vertical genal comb; in male st. IX without tendon from angle forwards; in female sensilium usually more or less convex and anal stylet usually with one long apical bristle and one or two small or minute subapical ones .... **Hystrichopsyllidae**

Metanotum with spinelets; dorsal side of sensilium straight, in male ending with a transparent collar covering base of anal tergum, in female sensilium not separate from anal tergum and not raised above base of latter; anal stylet of female with one or two rather long lateral bristles in addition to apical one; in male st. IX with a tendon running forward from the junction between the two arms **4**

4 Head fracticipit; genal comb consisting of two (extremely rarely three) broad spines, which may be pointed or extremely obtuse, placed immediately behind the oral angle, parasites exclusively of bats .... **Ischnopsyllidae**

Head fracticipit or integricipit; genal comb (if present) placed behind or below the vestigial eye; of the two dorsal apical bristles of the fore femur, the outer one longer than the inner; spinelets of the metanotum apical, antepygidial bristles well developed in both sexes; female with anal stylet **5**

5 No genal comb; in front of eye (which is well developed except in *Foxella*) no arch of tentorium visible; three bristles in ocular row, the uppermost in front of the eye; no interantennal suture; antennal fossa open; club of male antenna extending on to propleurum; metanotum and some of abdominal terga with spinelets; st. VIII of male narrow, sometimes quite small or vestigial .... **Ceratophyllidae**

With or without genal comb; in front of eye (which may be vesitigial or absent) an arch of tentorium present but sometimes concealed by the genal comb if this is particularly extensive; upper ocular bristle at or near margin of antennal fossa; interantennal suture variable, often well developed; eye often sinuate or even vestigial; st. VIII of male at least moderately large .... **Leptopsyllidae**

FAMILY **PULICIDAE**

Derivation: The family name is derived from the Latin word *pulex* (flea) and the Latin suffix *idae*. The type genus is *Pulex* Linnaeus, 1758.

General Characters: Ocular setal row normally of two setae, the upper usually arising in front of the eye; inner surface of hind coxae with a row or cluster of short, spiniform setae; antesensilial setae one each side; no setae above spiracular fossa of tergum VIII; sensilium with fourteen pits per side; anal stylets present in females; one spermatheca.

World Distribution: As a family, worldwide in distribution with a few cosmopolitan species. There are 26 genera and 190 species.

Number of Northwestern Taxa: 14

General References: Hopkins and Rothschild (1953), Holland (1985).

At least three species in this family are cosmopolitan in distribution as a result of human transport. Two of these, *Pulex irritans* and *Ctenocephalides felis,* are the only fleas that most humans are likely to encounter in most parts of the Northern Hemisphere. The third, *Xenopsylla cheopis,* is the notorious vector of bubonic plague and thus has earned a place in infamy for the entire family. The remaining species tend to be relatively host specific and are seldom encountered by the nonspecialist.

The family is currently divided into two subfamilies; the Tunginae with two tribes and the Pulicinae with four tribes. Both subfamilies are represented in the area under consideration, with representatives in four of the tribes. The indigenous genera may be separated with the following key.

KEY TO THE WESTERN GENERA OF *PULICIDAE*

1 Inner side of hind coxa with spiniform bristles, sensilium bearing fourteen pits per side 2

Inner side of hind coxa without spiniform bristles, sensilium with eight pits per side .... **Hectopsylla**

2 Clavus of the antenna symmetrical, oval or eliptical in outline 3

Clavus of antenna asymmetrical, its anterior segments foliaceous, projecting caudad 6

3 Frontal tubercle absent, ctenidia present or absent 4

Frontal tubercle and ctenidia present .... **Cediopsylla**

4 Pronotal comb present, parasites of rabbits and hares 5

Pronotal comb absent, parasites of sea birds .... **Actenopsylla**

5 Antennal clavus capitate, segments fused anteriorly; prementum with a stout seta extending almost to apex of labial palpus, parasites of ground squirrels .... **Hoplopsyllus**

Antennal clavus clavate, segments distinct anteriorly, prementum lacking long, stout seta, parasites of rabbits and hares .... **Euhoplopsyllus**

6 Mesothoracic pleural rod absent **7**

Mesothoracic pleural rod present **8**

7 Metanotum reduced, much shorter than abdominal tergum I; frons angular .... **Echidnophaga**

Metanotum at least as long as abdominal tergum I; frons smoothly rounded .... **Pulex**

8 Both genal and pronotal comb present .... **Ctenocephalides**

Neither genal nor pronotal comb present .... **Xenopsylla**

GENUS **Hectopsylla** Frauenfeld

Hectopsylla. Frauenfeld, 1860, S. B. Akad. Wiss. Wien 40: 464. (Type species by monotypy: *Hectopsylla psittaci* Frauenfeld.)
Hectoropsylla. Oudemans, 1906, Ent. Ber. 2:103. (*Nomen novum* for *Hectopsylla* Frauenfeld.)

The ten taxa assigned to this genus are mostly associated with rodents and other small mammal hosts in Central and South America. At least three species have been reported from birds, and the species discussed here is typically a bird parasite.

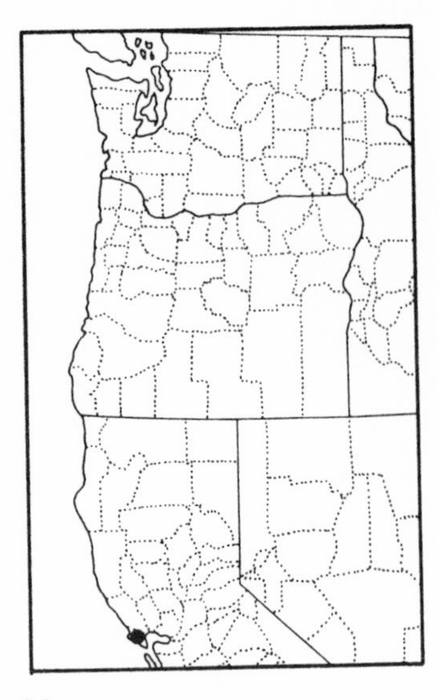

Hectopsylla psittaci Frauenfeld

[86001]

Hectopsylla psittaci. Frauenfeld, 1860, S. B. Akad. Wiss. Wien 40:464 (St. Jago [Santiago de Chile] from Tschoroi [a parrot]).
Pulex testudo. Weyenbergh, 1881, Period. zool. 3: 267 (from *Strix perlata* [= *Tyto alba tuidara*] from Argentina).
Hectopsylla psittaci Frauenfeld. Nelson, Wolf and Sorrie, 1979, J. Med. Ent. 16: 548 (Point Reyes Bird Observatory, Marin Co., California, from *Petrochelidon pyrrhonota*).
Hectopsylla psittaci Frauenfeld. Schwan, Higgins and Nelson, 1983, J. Med. Ent. 20: 690 (6.5 km W. Newark, Alameda Co., California, from nests of *P. pyrrhonota).*

Although this species was reported by Augustson in 1943 from southern California, it was assumed by him and subsequent authors that the record was attributable to the transfer from imported birds. Nelson et al. (1979), in reporting on a collection from California, suggested that these fleas had been brought back by the birds from their wintering grounds in central South America. They based their conclusion on the fact that all specimens were gravid, neosomic, and damaged, suggesting a long association with their hosts. Schwan et al. (1983) proved beyond a reasonable doubt that the species is established in cliff swallow nests as far north as Alameda County, California, in the San Francisco Bay area. They also reported collections from the nests of the black phoebe, and suggested a similar association with English sparrows. Since cliff swallows nest as far north as Alaska, it is quite possible that this flea occurs in the area under consideration under the proper environmental conditions.

GENUS **Echidnophaga** Olliff

Echidnophaga. Olliff, 1886, Proc. Linn. Soc. N. S. W. (2)1:171 (type species by monotypy: *E. ambulans* Olliff).

Argopsylla. Enderlein, 1903, Wiss. Ergebn. 'Valdivia', 1898-9, 3:263 (type species by designation: *Sarcopsyllus gallinaceus* Westwood).

Xestopsylla. Baker, 1904, Proc. U. S. Nat. Mus. 27: 373, 374 (type species by monotypy: *X. gallinacea* (Westwood)).

No species of this genus is native to the Western Hemisphere. Excluding *E. gallinacea,* which is practically cosmopolitan in warmer parts of the world, the remaining 21 species are distributed in the Palaearctic Region from the Mediterranean east to the Orient, Africa south of the Sahara, and the Australian Region, where they are parasites of various rodents, carnivores, and marsupials. Evidently they are of no known economic importance, other than occasional pests of domestic poultry, and none has been associated with disease transmission.

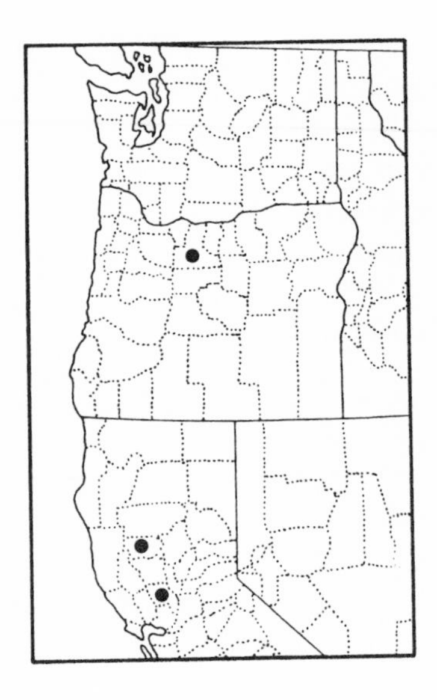

Echidnophaga gallinacea (Westwood)

[87501]

(Figs. 7, 8, 9)

Sarcopsyllus gallinaceus. Westwood, 1875, Ent. mon. Mag. 11: 246 (Ceylon, on domestic fowl).
Echidnophaga gallinacea Westwood, 1875. Hubbard, 1947, Fleas of Western North America, pp. 50-52.

As indicated, this species is widely distributed throughout the warmer parts of the world. As the name suggests, it is frequently associated with gallinaceous birds and occasionally becomes a minor problem with domestic poultry. This association doubtless has played a major role in the broad distribution of the species. Costa Lima and Hathaway (1946) list over seventy species of birds and mammals from which this flea has been taken. Wild carnivores and the larger insectivores seem to be especially preferred hosts.

Echidnophaga gallinacea is frequently a pest of domestic poultry in the southern United States. Reports of its occurrence usually refer to localized infestations, although under these conditions individuals may be plentiful. The species has been reported from Wasco County, Oregon, by Hubbard (1947) from *Spermophilus beecheyi douglasi* and *Peromyscus maniculatus gambeli,* and also from a number of localities in southern California. To date it is not known from north of Oregon.

The life cycle and some behavioral observations for this species were described by Parman (1923). Adults remain inactive for a few days after emergence from the cocoon, usually not attaching to the host until the fifth to eighth day. Females remain attached in the same place for extended periods, while males detach and reattach frequently. Copulation takes place on the host, and subsequently one to four eggs are laid daily. The number of eggs produced by a single female has not been determined.

The incubation period varies from 4 to 14 days, the minimum period being attained at an average temperature of 25°C. According to Parman (1923), the larvae feed exclusively on the excreta of the adults. This is probably an erroneous statement, because the larvae occur in situations relatively remote from the attached adults, and doubtless their diet includes a variety of organic debris, as is true of other species. Larval development requires 14 to 31 days; pupation takes place in a cocoon constructed of silk and debris and lasts 9 to 19 days.

Control measures for this species include dusting infested poultry with various insecticidal powders and spraying breeding areas such as poultry houses and under buildings where the larvae occur.

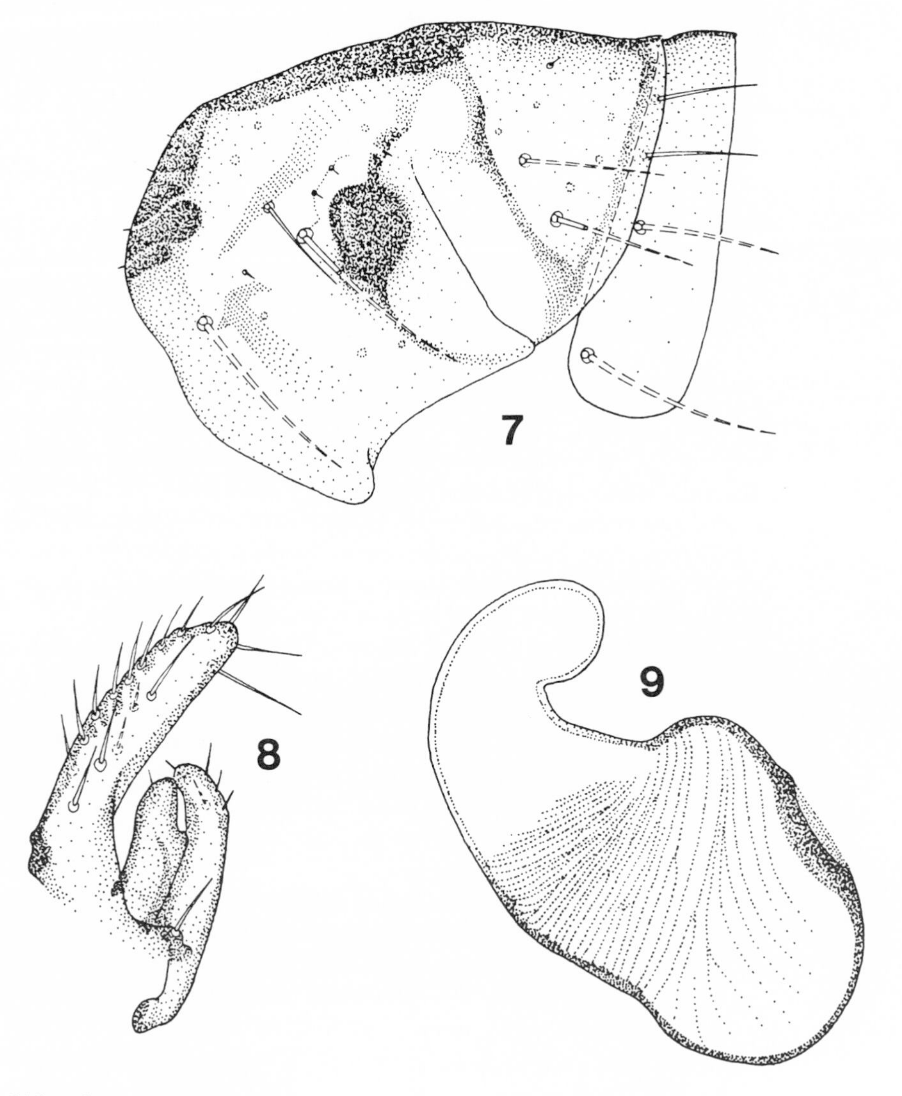

7-9 *Echidnophaga gallinacea* (Westwood). **7** Head of male. **8** Clasper of male. **9** Female spermatheca. Original.

GENUS **Pulex** Linnaeus

Pulex. Linnaeus, 1758, Syst. Nat., 10th ed., I, 614 (type species by subsequent designation: *P. irritans* Linn. by Baker, 1904, p. 371).

This genus at present includes six species, five of which are restricted to North and Central America. The sixth, *P. irritans,* has become established throughout the world, via human transport. *Pulex* is distinguished from related genera by the absence of a mesothoracic pleural rod, an asymmetrical antennal clavus, the absence of a frontal tubercle and the lack of pronotal and genal ctenidia. A vestigial genal comb of short, transparent teeth may be present in some individuals but it is never a conspicuous feature.

The two species reported here are separable only by characters associated with the male aedeagus. Because of this it is likely that many of the published records are based upon misidentifications. A third western species, *P. porcinus,* is a parasite of peccaries and is extralimital to the area under consideration.

KEY TO THE WESTERN SPECIES OF *PULEX*

1 Dorsal aedeagal sclerite broad and blunt, crochet flattened (Fig. 10) .... **simulans**

Dorsal aedeagal sclerite long and narrow, crochet triangular .... **irritans**

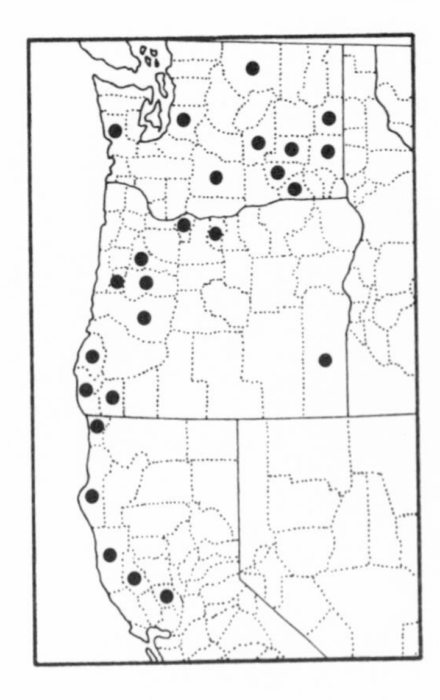

Pulex simulans Baker

[89501]
(Fig. 10)

Pulex simulans. Baker, 1895, Canad. Ent. 27: 65, 67 (U.S.A., from *Didelphys [=Didelphis] virginiana,* [Devil's River, Texas]).

Pulex irritans var. *dugesii.* Baker, 1899, Ent. News 10:37 (Guanajuato, Mexico, from *Spermophilus macrourus* [*Spermophilus variegatus*]).

Pulex irritans Linnaeus. Jordan and Rothschild, 1908, Parasitology I: 7 (*simulans* synonymized).

Pulex simulans Baker (*species revocata*). Smit, 1958, J. Parasitol. 44: 523-526 (*simulans* resurrected).

See discussion of this species under *Pulex irritans.*

Pulex irritans Linnaeus

[75801]

(Figs. 11, 12, 13)

Pulex irritans. Linnaeus, 1758, Syst. Nat., 10th ed., I: 614 (in Europa, America).

Although commonly known as the human flea, this cosmopolitan species is mainly a parasite of large mammals in North America, showing a particular preference for carnivores. However, most of our records of this species from Oregon are from *Odocoileus hemionus columbianus,* the mule deer. Hubbard (1947) also reports collections from this host, although its occurrence is likely accidental, not reflecting true host preference.

The closely related species, *P. simulans,* was reported from Oregon by Smit (1958). These three specimens, listed in Hopkins and Rothschild (1953) under *P. irritans,* were also from deer and were reported collected at Reston, in Douglas County, and Green Mountain. In view of the history of confusion between the two species, *P. simulans* is included as a member of the northwestern fauna on the strength of Smit's records. We have been unable to locate the Green Mountain locality, but Reston is a few miles west-south-west of Roseburg, Oregon. *Pulex irritans* is reported from a number of collections from Washington and northern California.

There is no evidence to indicate that *P. irritans* is important in the transmission of sylvatic plague in wild rodent populations. *Pulex simulans,* on the other hand, seems primarily to parasitize prairie dogs (*Cynomys*) and may play a major role in the transmission of plague in these rodents.

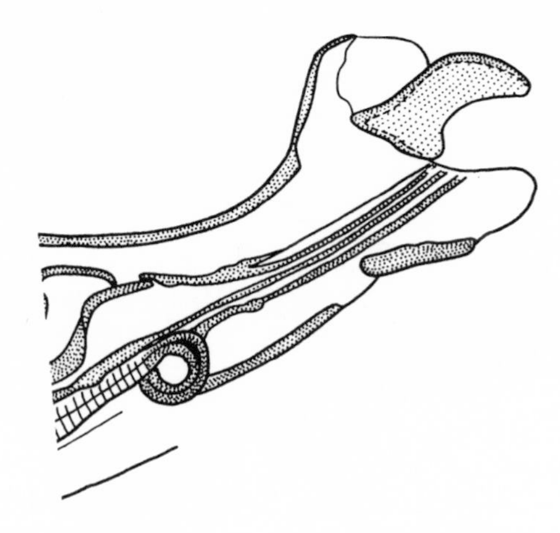

10 *Pulex simulans* Baker. Apex of male aedeagus. After Holland, 1949.

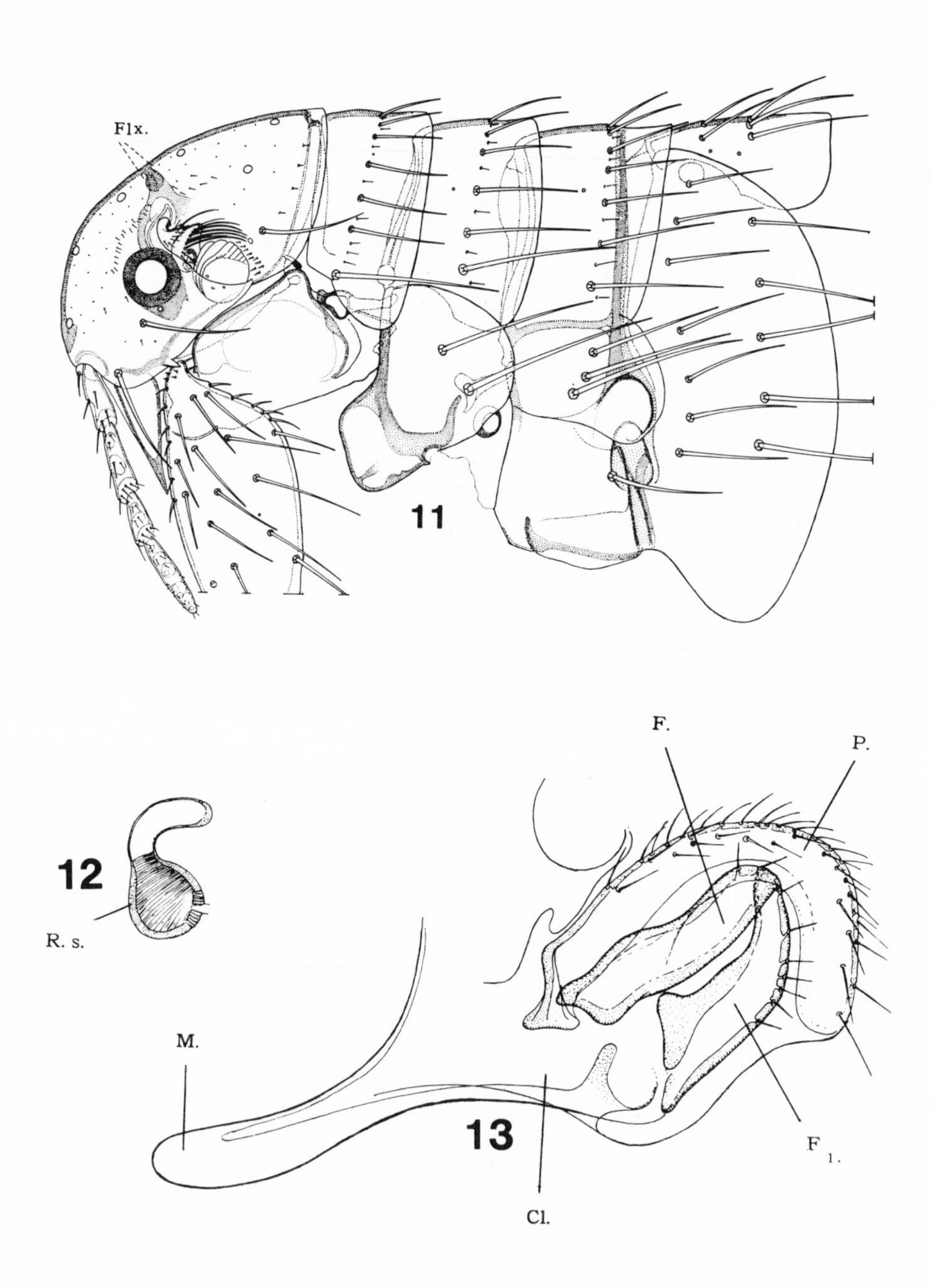

11 Head and thorax of male *Pulex irritans* Linnaeus. After Holland, 1985.

12-13 *Pulex irritans* Linnaeus. **12** Female spermatheca. **13** Male clasper. After Holland, 1949.

Examination of large numbers of specimens of both taxa reveals that the females are essentially impossible to separate without accompanying males. Morphological features useful in separating males of the two species include the shape of the dorsal aedeagal sclerite (d.a.s.) and the crochet (cr.) and are shown in Figure 10 for *P. simulans.* In *P. irritans* the d.a.s. is much narrower and the cr. is triangular.

GENUS **Ctenocephalides** Stiles and Collins

Ctenocephalus. Kolenati, 1857, Wien. ent. Monatsch. I: 65 (type species by subsequent designation: P. canis Curtis, by Baker, 1904, p. 371).

Ctenocephalides. Stiles and Collins, 1930, Publ. Hlth. Repts., Wash. 45: 1308 (type species by designation: *Pulex canis* Curtis).

This genus at present includes fourteen recognized species and subspecies, most of which are restricted to the Ethiopian Region, where they parasitize hyraxes, ground squirrels, and a broad range of carnivores. *Ctenocephalides canis* and the nominate subspecies of *C. felis* are practically worldwide in distribution and are common ectoparasites of domestic cats and dogs. Under conditions of poor sanitation, human habitations may become infested, the adults feeding more or less indiscriminately on humans and animals alike.

Both of the species mentioned above have been reported from the area under consideration and may be separated with the following key.

KEY TO THE WESTERN SPECIES OF *CTENOCEPHALIDES*

1 Frons smoothly and evenly rounded causing the head to appear blunt; space between the apical and postmedian long bristles on dorsal margin of hind tibia with 2 notches, each with a short, stout bristle; metepisternum usually with 3 bristles .... **canis**

Frons oblique, causing the head to appear conical and pointed anteriorly; upper short, stout bristle on dorsal margin of hind tibia absent; metepisternum with one bristle .... **felis felis**

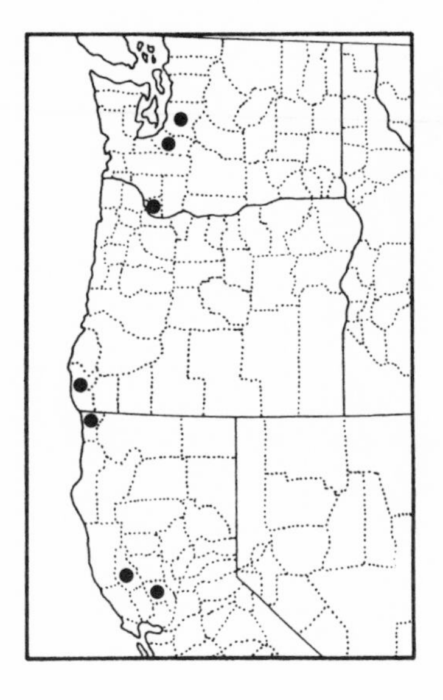

Ctenocephalides felis felis (Bouché)

[83501]

(Figs. 14, 18, 19, 20, 21)

Pulex Felis. Bouché, 1835, Nova Acta Leop.-Carol. 17: 505 (Germany, from house cat).

Ctenocephalides felis Bouché, 1835. Hubbard, 1947, Fleas of Western North America, pp. 60-62.

This ubiquitous species is primarily a parasite of cats and owes its worldwide distribution to human transport on pets. It is an exceedingly adaptable species, and has been reported from a broad range of carnivores, as well as the usual series of accidental hosts. Of the two species of this genus occurring in the northwest *C. felis* is the one most likely to be encountered by humans. Although commonly called the cat flea, this is the species usually encountered on dogs and, occasionally, livestock. Infestation of human habitations is not uncommon and is usually traceable to dirt-floored basements, pet bedding, or dirty rugs that harbor the larvae.

The life cycle and biological requirements of this species were described by Bruce (1948). He determined that optimum temperatures for larval development ranged from 27 to 30°C, with optimum humidities of 65-90 percent. Under these conditions, the life cycle from egg to adult can be completed in 18-21 days.

According to Pollitzer (1954), *C. felis* can be infected with the plague organism, *Yersinia pestis,* when fed on an infected host under laboratory conditions, but its vector capacity under natural circumstances is considered to be negligible. However, this and the following species are vectors of the tapeworm, *Dipylidium caninum,* a parasite of cats and dogs that has also been reported from humans, almost always children. Infection is acquired by ingestion of the infected adult flea.

The other three subspecies of *C. felis* occur in Africa and the Orient. See Hopkins and Rothschild (1953) for a detailed discussion of the genus.

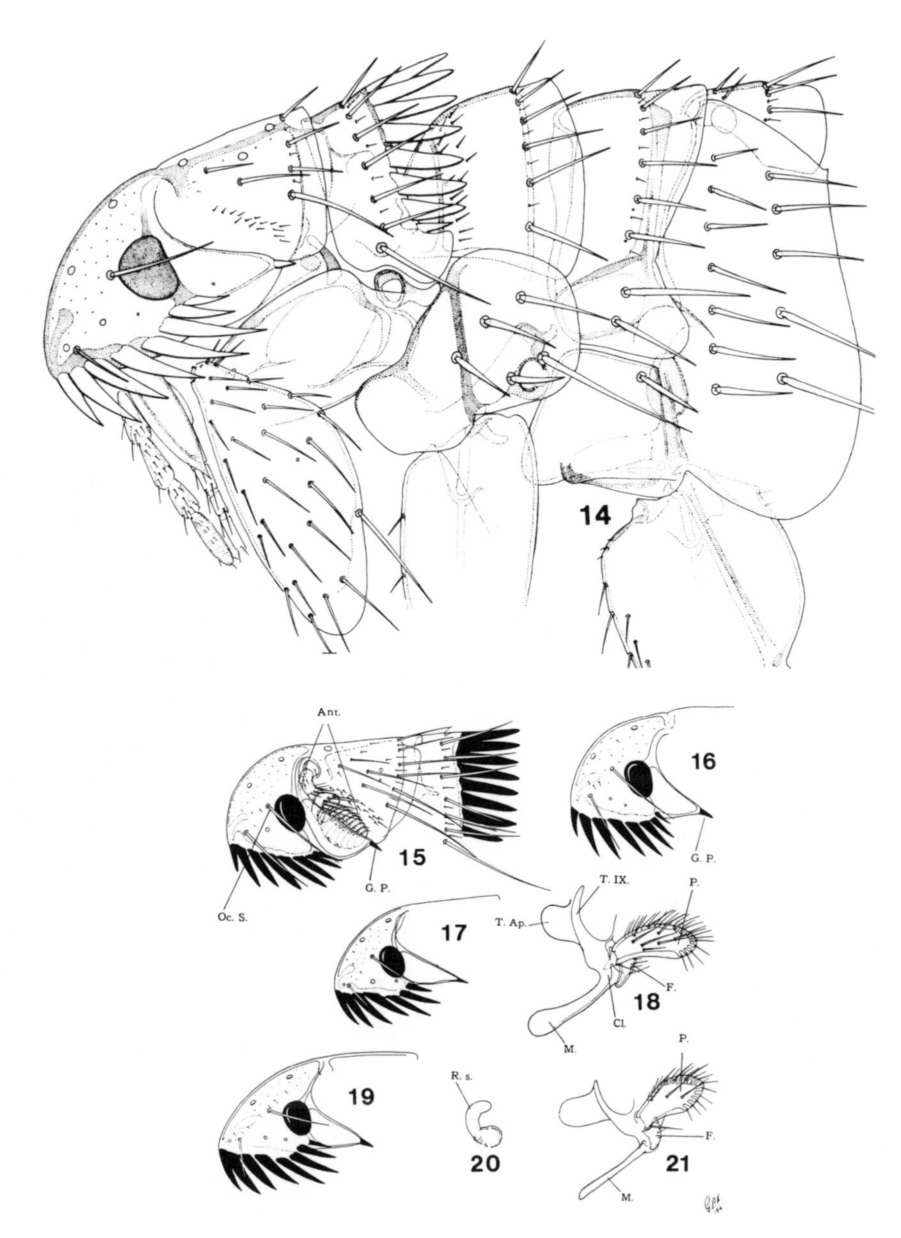

14 Head and thorax of male *Ctenocephalides felis felis* (Bouché). After Holland, 1985.

15-21 *Ctenocephalides* species. **15, 16** Heads of male and female, and **18** male genitalia of *C. canis* (Curtis). **17, 19** Preantennal region of head of male and female. **20** Female spermatheca. **21** Male genitalia of *C. felis felis* (Bouché). After Holland, 1949.

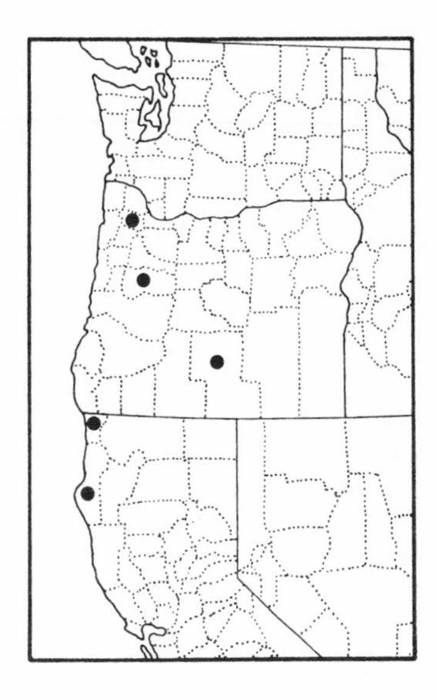

Ctenocephalides canis (Curtis)

[82602]

(Figs. 15, 16, 17)

Pulex canis. Curtis, 1826, British Entomology 3: 114, figs. A-E, 8 (England, from dog).

Ctenocephalides canis Curtis, 1826. Hubbard, 1947, Fleas of Western North America, pp. 62-63.

A close relative of the cat flea, this species shares many biological and distributional characteristics with it. It evidently is not as adaptable, however, and although it has been reported from all the major landmasses of the world, it has failed to become as permanently established as has the cat flea. In addition, it is much more host specific, being mostly restricted to wild members of the Canidae, particularly *Canis* and *Vulpes* species. The species is known from scattered published records from California, Oregon, and Washington, though many of these probably actually refer to cat fleas.

This species seldom assumes the status of a pest to humans, being mainly restricted to wild host animals. It is known to be an intermediate host of the rat tapeworm, *Hymenolepis diminuta,* and a few human cases of parasitism via ingestion of an infected flea have been reported.

GENUS **Cediopsylla** Jordan

Cediopsylla. Jordan, 1925, Novit. Zool. 32: 103 (type species designated as "species identified as *C. simplex* Baker, 1895").

Acediopsylla. Ewing, 1940, Proc. biol. Soc. Wash. 53: 37 (type species by designation: *Ctenocephalus inaequalis* Baker, [1904]).

This is a small genus of only five recognized taxa which parasitizes lagomorphs in the Americas. A single species, known from two subspecies, occurs commonly in the northwest. The fact that these subspecies are sympatric, frequently being found on the same host individual and showing considerable intergradation in their diagnostic characters, suggests that subspecific status may not be justified.

At least one species of this genus, *C. simplex* (Baker, 1895), has been shown to be dependent upon sex hormones in the blood of the host for the completion of its own reproductive cycle. This dependence on the host for hormonal as well as nutritional support was originally elucidated by Rothschild and Ford (1973) using *Spilopsyllus cuniculi* (Dale, 1878). An excellent summary of these studies appears in Rothschild (1975).

KEY TO THE WESTERN SPECIES OF *CEDIOPSYLLA*

1 Long bristles on caudal margin of male clasper forming a continuous row; both sexes with scattered bristles between dorsolateral and subventral rows on hind tibia .... **inaequalis inaequalis**

Long bristles on caudal margin of male clasper divided into a dorsal group of four to six and a ventral group of two to three; both sexes lacking bristles between dorsolateral and subventral rows on hind tibia .... **inaequalis interrupta**

Cediopsylla inaequalis inaequalis (Baker)

[89510]
(Fig. 22)

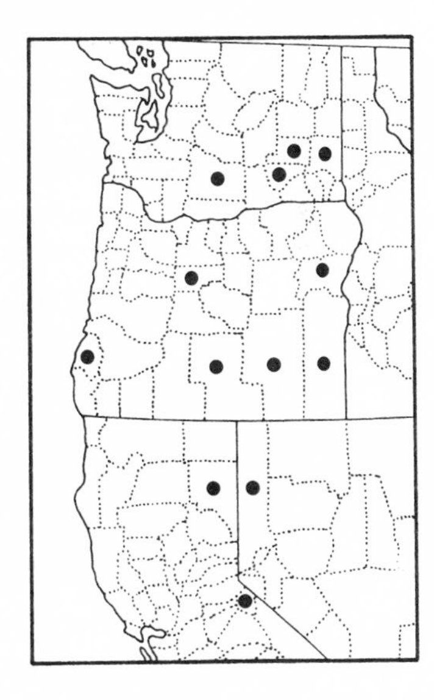

Pulex inaequalis. Baker, 1895, Canad. Ent. 27: 163, 164 (near Grand Canyon in Arizona from cottontail and jack rabbits).
Ctenocephalus inaequalis Baker. Baker, 1904, Proc. U. S. Nat. Mus. 27: 385.
Cediopsylla inaequalis inaequalis Baker, 1895. Hubbard, 1947, Fleas of Western North America, pp. 74-76.

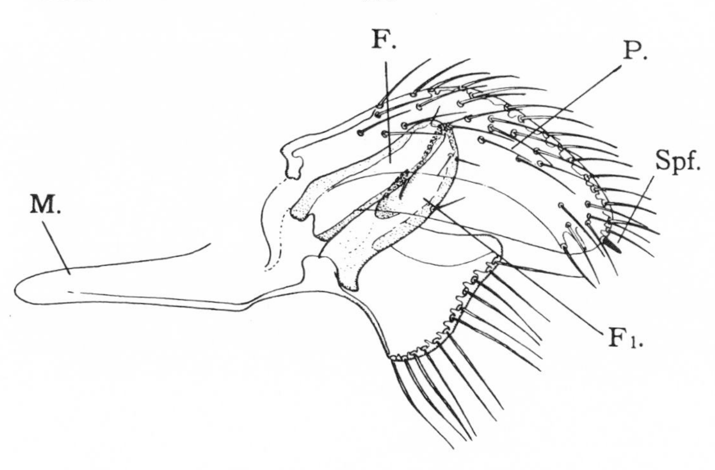

22 Clasper of male *Cediopsylla inaequalis inaequalis* (Baker). After Holland, 1985.

and

Cediopsylla inaequalis interrupta Jordan

[92517]
(Figs. 23, 24)

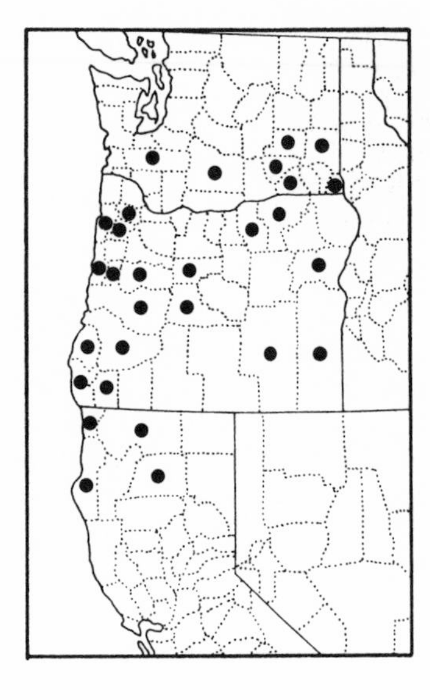

Cediopsylla inaequalis interrupta. Jordan, 1925, Novit. Zool. 32: 103 (Palo Alto, California, from fox).
Cediopsylla inaequalis interrupta Jordan, 1925. Hubbard, 1947, Fleas of Western North America, pp. 76-78.

This species ranges from Montana, western Nebraska, and New Mexico, west to the Pacific Ocean. The nominate subspecies occupies the bulk of this area, with *C. i. interrupta* being mainly confined to the area west of the Cascade Mountains of Washington and Oregon, south to San Francisco, California. Hubbard (1947) reports a few collections of this subspecies from east of the Cascades in the Great Basin area. The preferred hosts of both subspecies are rabbits and hares, particularly the nest-building species of the genus *Sylvilagus.* Predator records are common, especially from coyotes, foxes, and other large carnivores. Both subspecies are known from a number of published records from Washington, Oregon, California, Nevada and Idaho. These fleas are not known to be capable of transmitting either plague or tularemia, although both diseases have been reported from rabbits and hares.

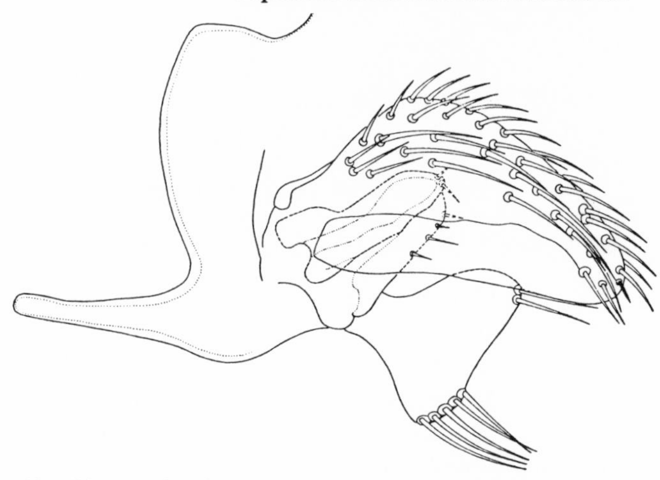

23 Clasper of male *Cediopsylla inaequalis interrupta* Jordan. After Holland, 1985.

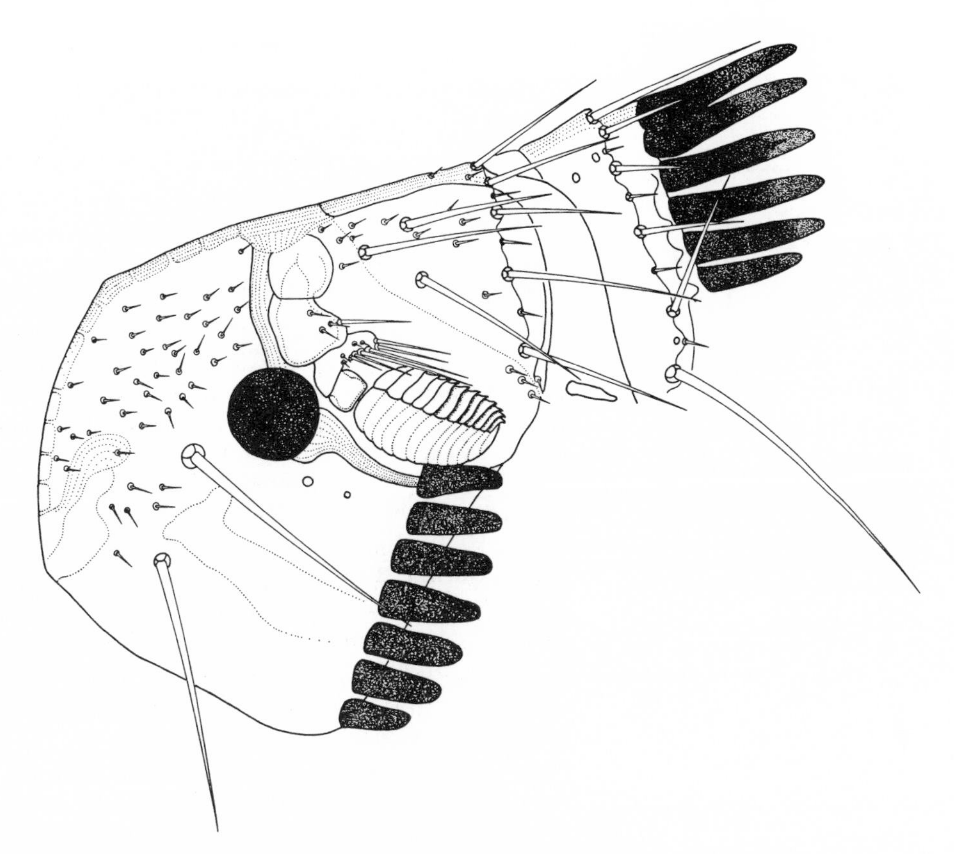

24 Head of male *Cediopsylla inaequalis interrupta* Jordan. Original.

GENUS **Hoplopsyllus** Baker

Hoplopsyllus. Baker, 1905, Proc. U. S. Nat. Mus. 29: 128, 144 (type species: *Pulex anomalus* Baker, 1904).

Euhoplopsyllus. Ewing, 1940, Proc. biol. Soc. Wash. 53: 37 (subgenus of *Hoplopsyllus*. Type species: *Hoplopsyllus affinis* Baker).

Hoplopsyllus Baker. Hopkins and Rothschild, 1953, An Illustrated Catalogue... Vol. I, pp. 182-183.

Euhoplopsyllus Ewing. Smit, 1967, Mitt. zool. Mus. Berlin 43: 86 (elevated to full generic status).

There are two species assigned to this genus. One was described from the Federal District of Mexico from the Volcano rabbit, *Romerolagus diazi*. The other species is known from a number of collections from Oregon and Utah, south to northern Mexico, where its preferred hosts are species of *Spermophilus*.

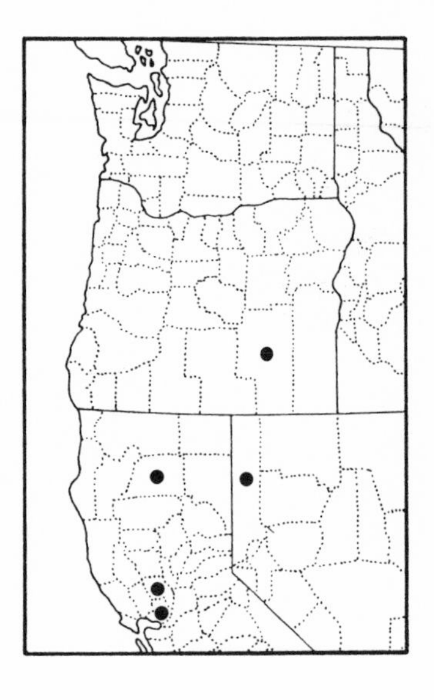

Hoplopsyllus anomalus (Baker)

[90409]
(Fig. 25)

Pulex anomalus. Baker, 1904, Proc. U. S. Nat. Mus. 27: 378, 381 (Arboles, southern Colorado, from "large gray-brown spermophile").
Hoplopsyllus anomalus Baker. Baker, 1905, Proc. U. S. Nat. Mus. 29: 130, 144.
Hoplopsyllus anomalus Baker, 1904. Hubbard, 1947, Fleas of Western North America, p. 72.
Hoplopsyllus (Hoplopsyllus) anomalus (Baker), 1904. Hopkins and Rothschild, 1953, An Illustrated Catalogue... Vol. I, pp. 184-185.

The range of this species includes an area from Utah and Colorado, west to the Pacific coast, and south to northern Mexico. Although it was not taken during any of the surveys reported here, Hansen (1964) collected fifteen specimens during his study in the Steens Mountain area of Harney County, Oregon, and there are literature records from California and Nevada. The only host reported for the species by Hansen was *Spermophilus leucurus* but it is also known from *S. beecheyi* and *S. variegatus,* as well as accidental records from predators and other small mammals.

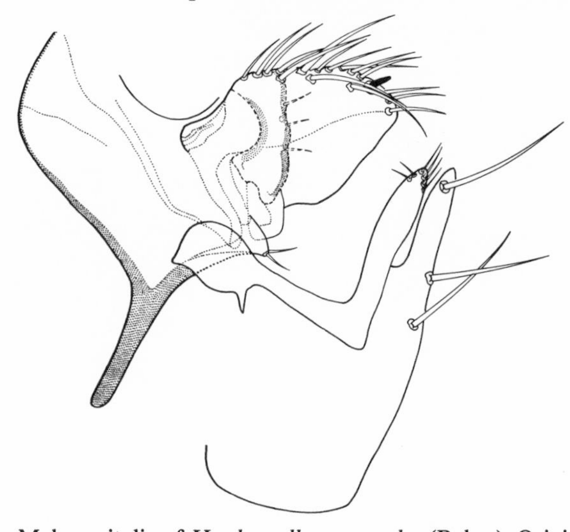

25 Male genitalia of *Hoplopsyllus anomalus* (Baker). Original.

GENUS **Euhoplopsyllus** Ewing

Euhoplopsyllus. Ewing, 1940, Proc. biol. Soc. Wash. 53: 37, as a subgenus of *Hoplopsyllus* (type species by designation: *Hoplopsyllus affinis* Baker).

Euhoplopsyllus Ewing. Smit, 1967, Mitt. zool. Mus. Berlin 43: 86 (elevated to generic rank).

This genus includes four recognized species, one of which contains five subspecies. All but one, *E. glacialis profugus,* occur in the Nearctic or Neotropical Regions where they are parasites of rabbits and hares. Predator records are also common.

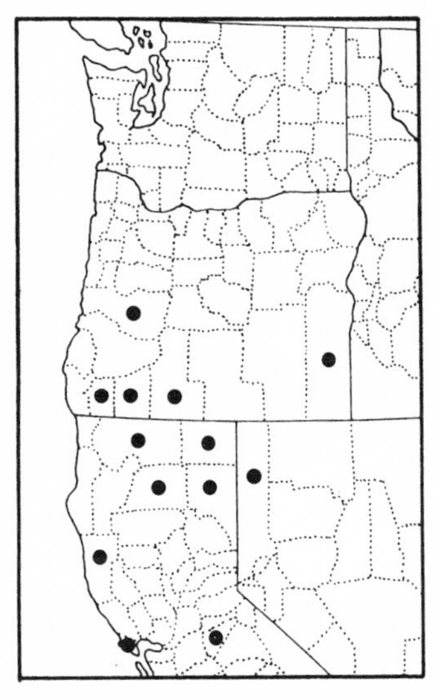

Euhoplopsyllus glacialis foxi (Ewing)

[92401]
(Fig. 26)

Hoplopsyllus foxi. Ewing, 1924, Parasitology 16: 350, pl. 14, fig. 4 (San Francisco, California, from *Lepus bachmani* [= *Sylvilagus bachmani*]).

Hoplopsyllus powersi. C. Fox, 1926, Pan-Pacific Ent. 2: 184 (Los Angeles, California, from *Sylvilagus bachmani*).

Hoplopsyllus minutus. C. Fox, 1926, *op. cit.,* p. 185 (San Francisco, California, from *Sylvilagus bachmani*).

Hoplopsyllus foxi Ewing. Kohls, 1939, Publ. Hlth. Rep., Wash. 54: 2019 (sinks *powersi* and *minutus*).

Hoplopsyllus glacialis foxi Ewing, 1924. Hubbard, 1947, Fleas of Western North America, pp. 68-69.

Hoplopsyllus glacialis tenuidigitus Stewart, 1940. Hubbard, 1947, *op. cit.,* pp. 69-70.

Hoplopsyllus glacialis tenuidigitus Stewart. Hopkins and Rothschild, 1953, An Illustrated Catalogue ...Vol. I., pp. 189-190.

Euhoplopsyllus glacialis glacialis (Taschenberg). Smit, 1967, Mitt. zool. Mus. Berlin 43: 86 (elevates *Euhoplopsyllus* to full genus).

and

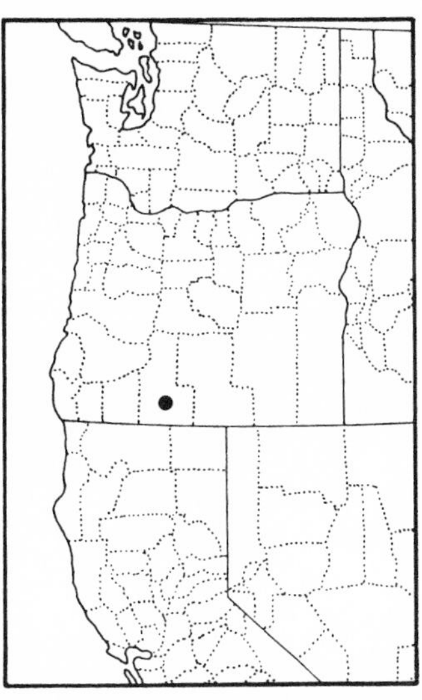

Euhoplopsyllus glacialis lynx (Baker)

[90411]

Pulex lynx. Baker, 1904, Proc. U. S. Nat. Mus. 27: 373, 383 (Moscow, Idaho, from *Lynx canadensis*).

Hoplopsyllus lynx Baker. Baker, 1905, Proc. U. S. Nat. Mus. 29: 130, 144.

and

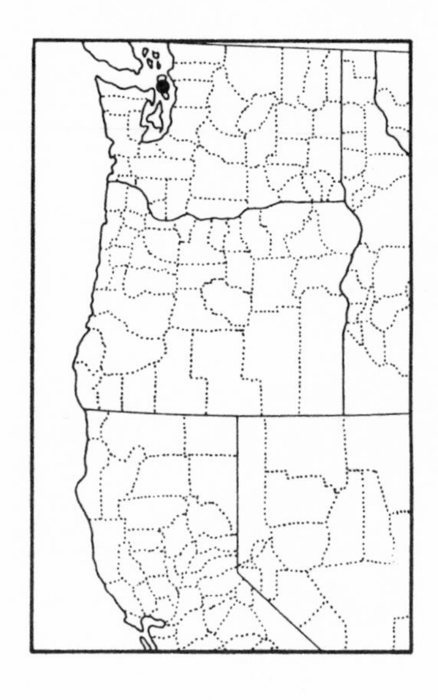

Euhoplopsyllus glacialis affinis (Baker)

[90410]

Pulex glacialis Taschenberg. Baker, 1895 (*nec* Taschenberg, 1880), Canad. Ent. 27: 108 (near Grand Canyon, from cottontail rabbit).

Pulex affinis. Baker, 1904, Proc. U. S. Nat. Mus. 27: 378, 382.

Hoplopsyllus affinis Baker. Baker, 1905, Proc. U. S. Nat. Mus. 29: 130, 144.

Although the subspecies of *E. glacialis* are not especially difficult to distinguish, the species is not a particularly common member of the northwestern fauna, unlike the other rabbit fleas, *Cediopsylla inaequalis* and *Odontopsyllus dentatus.* Hubbard (1947) reports *E. g. foxi* from southern Oregon. Additional records in the Lewis Collection are from Klamath, Lane, and Malheur counties, also in southern Oregon. The bulk of the published records for this taxon are from California. The subspecies *E. g. lynx* was reported from Oregon by Kohls (1940) and *E. g. affinis* from Whitby Island, Washington, by Svihla (1941).

There is still no evidence that this species plays a role in the transmission of either plague or tularemia.

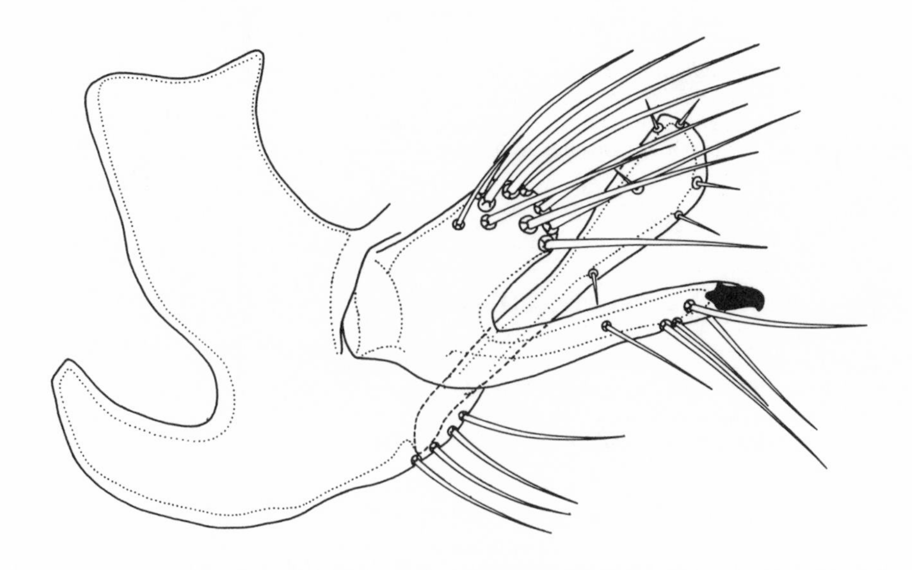

26 Male clasper of *Euhoplopsyllus glacialis foxi* (Ewing). Original.

GENUS **Actenopsylla** Jordan and Rothschild

Actenopsylla. Jordan and Rothschild, 1923, Ectoparasites I: 309 (type species by designation: *Actenopsylla suavis* Jordan and Rothschild).
Actenopsylla Jordan and Rothschild. Hopkins and Rothschild, 1953, An Illustrated Catalogue... Vol. I, p. 198.

Actenopsylla differs from other genera of the Spilopsyllinae in lacking both genal and pronotal ctenidia and a frontal tubercle. Only a single species is known.

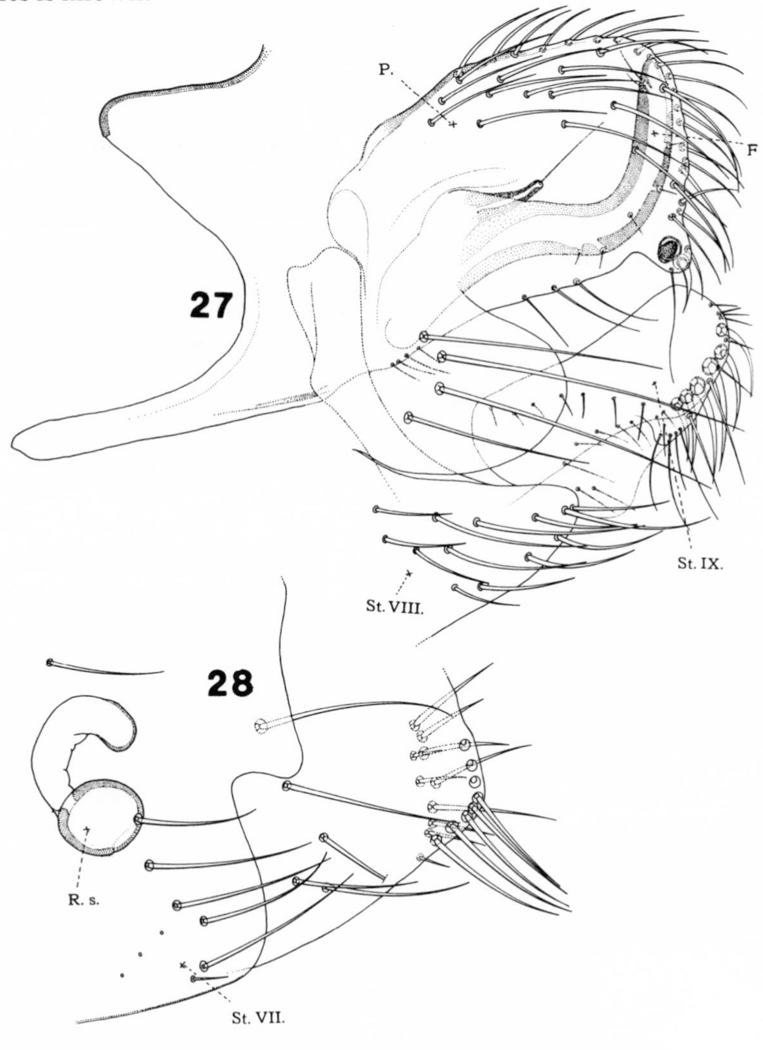

27 Genitalic segments of male *Actenopsylla suavis* J. & R. After Holland, 1985.
28 Genitalic segments of female *Actenopsylla suavis* J. & R. After Holland, 1985.

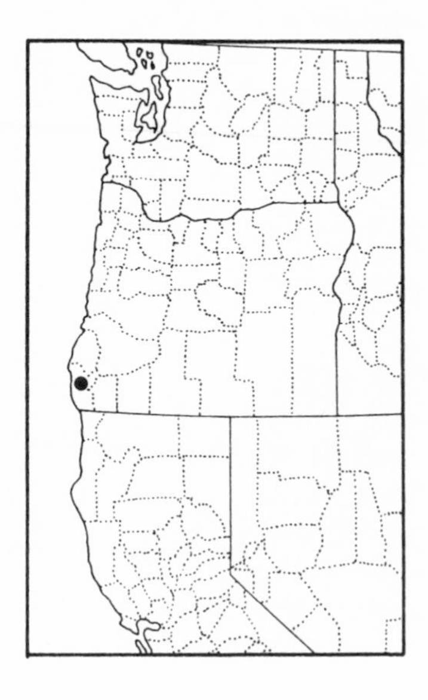

Actenopsylla suavis Jordan and Rothschild

[92311]

(Figs. 27, 28, 29)

Actenopsylla suavis. Jordan and Rothschild, 1923, Ectoparasites I: 309 (Coronado Island [probably Los Coronados Islands], Mexico, in nest burrow of *Ptychorhamphus aleuticus)*.

Actenopsylla suavis Jordan and Rothschild. Hopkins and Rothschild, 1953, An Illustrated Catalogue... Vol. I, p. 199-200.

Actenopsylla suavis Jordan and Rothschild. Marshall and Nelson, 1967, J. Med. Ent. 4: 335-338.

Actenopsylla suavis Jordan and Rothschild. Clifford et al., 1970, J. Med. Ent. 7: 438-445.

This is an ectoparasite of burrow-nesting sea birds along the Pacific coast of North America, described from material thought to have come from Los Coronados Islands, off the coast of Tijuana, Mexico. Marshall and Nelson (1967) reported the species from South Farallon Island, California, off the coast of San Francisco. It has since been reported from Goat and Hunter's Islands, Curry County, Oregon, by Clifford et al. (1970) from a number of different species of sea birds. As Holland (1985) opined, since three of the host species occur on the British Columbian and Alaskan coasts, there is every reason to believe that *Actenopsylla suavis* occurs there, too.

29 Head and thorax of male *Actenopsylla suavis* J. & R. After Holland, 1985.

GENUS **Xenopsylla** Glinkiewicz

Xenopsylla. Glinkiewicz, 1907, S. B. Akad. Wiss. Wien, Abt. 1, 116: 385 (type species by monotypy: *X. pachyuromydis* Glink. [= *cheopis* Rothschild, 1903]).
Loemopsylla. Jordan and Rothschild, 1908, Parasitology I: 15 (type species by designation and monotypy: *cheopis* Roths.).
Rooseveltiella. C. Fox, 1914, Bull. U. S. Hyg. Lab. No. 97: 7 (type species by monotypy: *R. georychi* C. Fox).
Alaopsylla. Jordan, 1933, Novit. Zool. 39: 55 (type species by monotypy: *A. papuensis* Jordan).

Although this is one of the largest genera in the order, containing 83 currently recognized species and subspecies, not one is native to continents of the Western Hemisphere or the European portion of the Palaearctic Region. One or two species are cosmopolitan in distribution due to human transport, and the species reported here has become well established, at least in local populations, throughout most of the United States. Most members of the genus are parasites of rodents, although many feed readily on other hosts. A few species are parasites of birds.

Xenopsylla cheopis (Rothschild)

[90312]

(Figs. 30, 31, 32, 33)

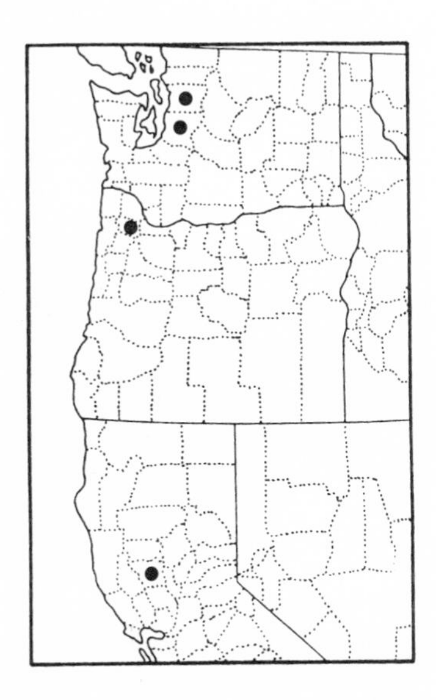

Pulex cheopis. Rothschild, 1903, Ent. mon. Mag. 39: 85, pl. I, figs. 3, 9; pl. 2, figs. 12, 19 (near Shendi, Sudan, from *Acomys witherbyi* [= *A. cineraceus*]).
Xenopsylla cheopis Rothschild, 1903. Hubbard, 1947, Fleas of Western North America, pp. 65-67.

The Oriental rat flea is thought to have had its origins in the Nile Valley, probably as a parasite of *Arvicanthus niloticus* and other rodents. It subsequently has been transported throughout the world on *Rattus norvegicus,* the Norway rat, and *Rattus rattus* subspecies, the brown or roof rat, both species that infest marine shipping. Once established in port cities it extended its range inland, again by human transport of rats, and today may be found in at least local populations throughout the world.

It is this species that originally was associated with the transmission of Bubonic Plague by the British Indian Plague Commission. The mechanics of transmission were ultimately described by Bacot and Martin in 1914. This disease and its

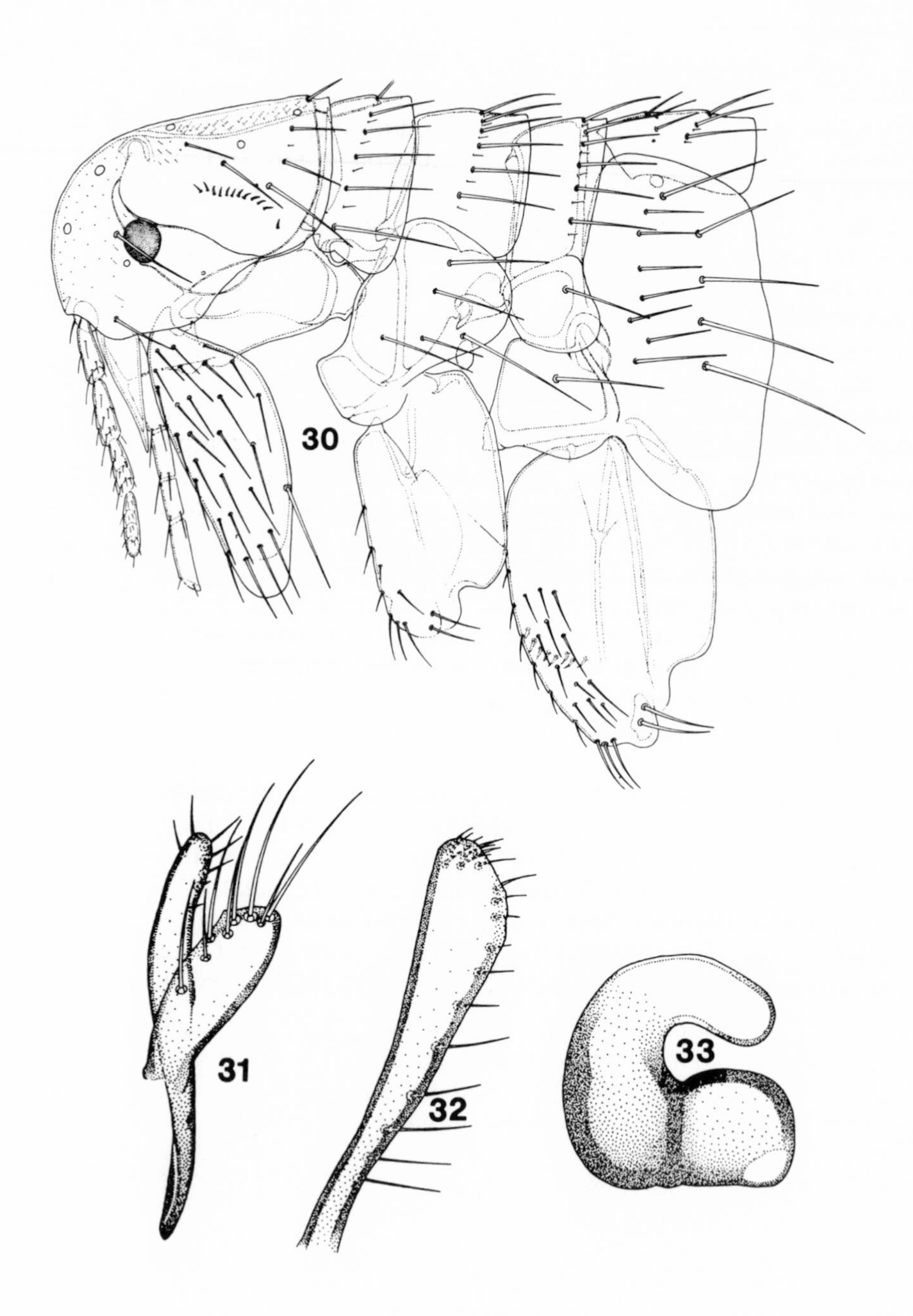

30 Head and thorax of male *Xenopsylla cheopis* (Rothschild). After Holland, 1985.

31-33 *Xenopsylla cheopis* (Rothschild). **31** Male clasper. **32** Male sternum IX. **33** Female spermatheca. Original.

epidemiology are discussed in detail elsewhere (see pages 15-17) but it is accepted that it was *X. cheopis* which was responsible for both the establishment of plague in the domestic rat population and the introduction of the disease into wild rodent populations in various parts of the world. The result is that sylvatic plague in the western United States and Central Asia is a serious concern to public health officials in these areas.

Xenopsylla cheopis is considered to be the best vector of plague of any species of flea. This is due to a unique combination of anatomical, physiological, and behavioral characteristics. First, a high percentage of individuals become "blocked" shortly after having fed on a diseased host. Frequently this blockage is only temporary, disappearing in a few days, before its effects are fatal to the flea. Such individuals may thus become blocked again at a later date and again transmit the disease to other healthy hosts. Second, unlike most fleas that are quite host specific, *X. cheopis* has catholic tastes and will feed on a wide variety of hosts if its preferred hosts are not available. As a result, what is essentially a disease of rodents may be transmitted to a broad range of host animals, including humans.

The life cycle and biological requirements of this species are well known, the flea having been kept in culture for many years for vector and insecticide studies. Under optimum conditions of 25°C and 80 percent relative humidity, the whole cycle from egg to adult can be completed in about 21 days, but doubtless requires longer under natural conditions. A number of investigators have noticed that males in culture require 2-3 days longer to mature than do females.

There is a vast literature dealing with this species and its role in disease transmission. For current information the reader is referred to the various publications of the United States Public Health Service.

FAMILY **VERMIPSYLLIDAE**

Derivation: This family name is derived from the Latin word *vermis* (a worm), the Greek *psylla* (a flea) and the Latin suffix *idae,* and refers to the worm-like appearance of the neosomic, gravid females in the genus *Vermipsylla* Schimkewitsch, 1885, the type genus.

General Characters: Tergal spinelets and ctenidia absent; eye and anterior tentorial branch well developed; antennal fossae not fused internally to form a trabecula centralis; occipital collar reduced or practically absent; pseudosetae under the mesonotal collar; hind coxae lacking spiniform setae on inner surface; hind tibia lacking an apical tooth on outer side; four pairs of lateral plantar bristles; several setae above spiracular fossa VIII; antesensilial setae absent; sensilium transverse; females lack an anal stylet and have a single spermatheca.

World Distribution: Holarctic, with three genera and about 37 species and subspecies. Principally found on Carnivora, Artiodactyla, and Perissodactyla. All North American species are associated with carnivores.

Number of Northwestern Taxa: 1

General References: Hopkins and Rothschild (1956), Holland (1985).

Two of the three known genera are restricted to Central Asia where they parasitize ungulates, usually at high elevations. The third genus, *Chaetopsylla,* is mainly restricted to Eurasia, where its members, with one exception, parasitize small to medium-sized carnivores. However, four species belonging to this genus are known from North America, and one has recently been taken in the northwest.

GENUS **Chaetopsylla** Kohaut

Chaetopsylla. Kohaut, 1903, Alatt. Kozlem. 2: 37 (type species by subsequent selection by de Cunha, 1914, Contribuição para oestudo dos sifonapteros do Brasil, p. 105: *C. rothschildi* Koh.).

Oncopsylla. Wahlgren, 1903, Ark. Zool. 1: 186 (type species: *Pulex globiceps* Taschenberg).

Subgenus *Chaetopsylla* Kohaut. Wagner, 1906, Horae soc. ent. Ross. 37: 453 (*Oncopsylla* sunk).

Trichopsylla Kolenati. Jordan and Rothschild, 1920, Ectoparasites 1: 63 (*Chaetopsylla* Kohaut and sunk).

Trichopsylla Kolenati. Hopkins, 1952, Bull. zool. Nomencl. 6: 349-52 (request for suppression of *Trichopsylla* in favor of *Chaetopsylla* Kohaut).

Chaetopsylla Kohaut. Opin. int. Comm. Zool. Nomencl. 12(9): 253-264 (Opinion 388) (*Trichopsylla* Kolenati suppressed under Plenary Powers).

Of the 25 currently recognized species and subspecies belonging to this genus, four are known to occur in North America. All are parasites of larger carnivores. *Chaetopsylla tuberculaticeps* is circumpolar as a parasite of bears. *Chaetopsylla setosa* is also mainly associated with bears, but seems to be restricted to British Columbia. Holland (1985) lists a series from a porcupine. *C. lotoris* is a monoxenous parasite of raccoons in the eastern United States, west to at least Iowa. Only *C. floridensis* has been taken in the Pacific Northwest.

Chaetopsylla floridensis (I. Fox)

[93906]

(Figs. 34, 35)

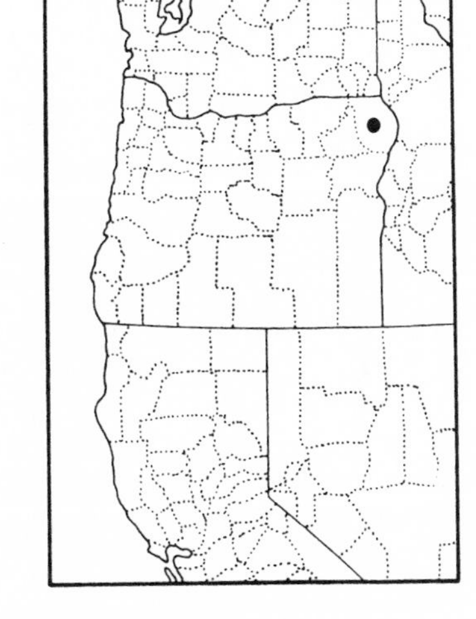

Trichopsylla floridensis. I. Fox, 1939, Proc. ent. Soc. Wash. 41: 45, pl. 6, fig. 6 (Gainesville, Florida, from "garden truck leaf mold").

Chaetopsylla (Chaetopsylla) floridensis (I. Fox), 1939. Hopkins and Rothschild, 1956, An Illustrated Catalogue...Vol. II, pp. 93-96, figs. 171-173.

Chaetopsylla floridensis I. Fox, 1939. Hopla, 1965, The Siphonaptera of Alaska, AAL-TR-64-12 Vol. I: 293-296, pl. 33, figs. A-I, map 14.

Since its description, this species has been taken in localities and from hosts that suggest that the information about the type series is in error. An alternative hypothesis offered by Dr. Allen Benton is that the "garden truck leaf mold" consisted largely of peat moss imported from western Canada. At this point it is too late to prove which explanation is correct. Hopla (1965) and Holland (1985) both list it from southeastern Alaska, and there is material in our collection from Colorado, as well as two females from Wallowa County, Oregon. A few specimens have been reported from species of *Mustela,* but most collections are from the pine marten, *Martes americana.*

Nothing is known of the importance of this species in the transmission of disease, but other members of the genus do not seem to play a role elsewhere.

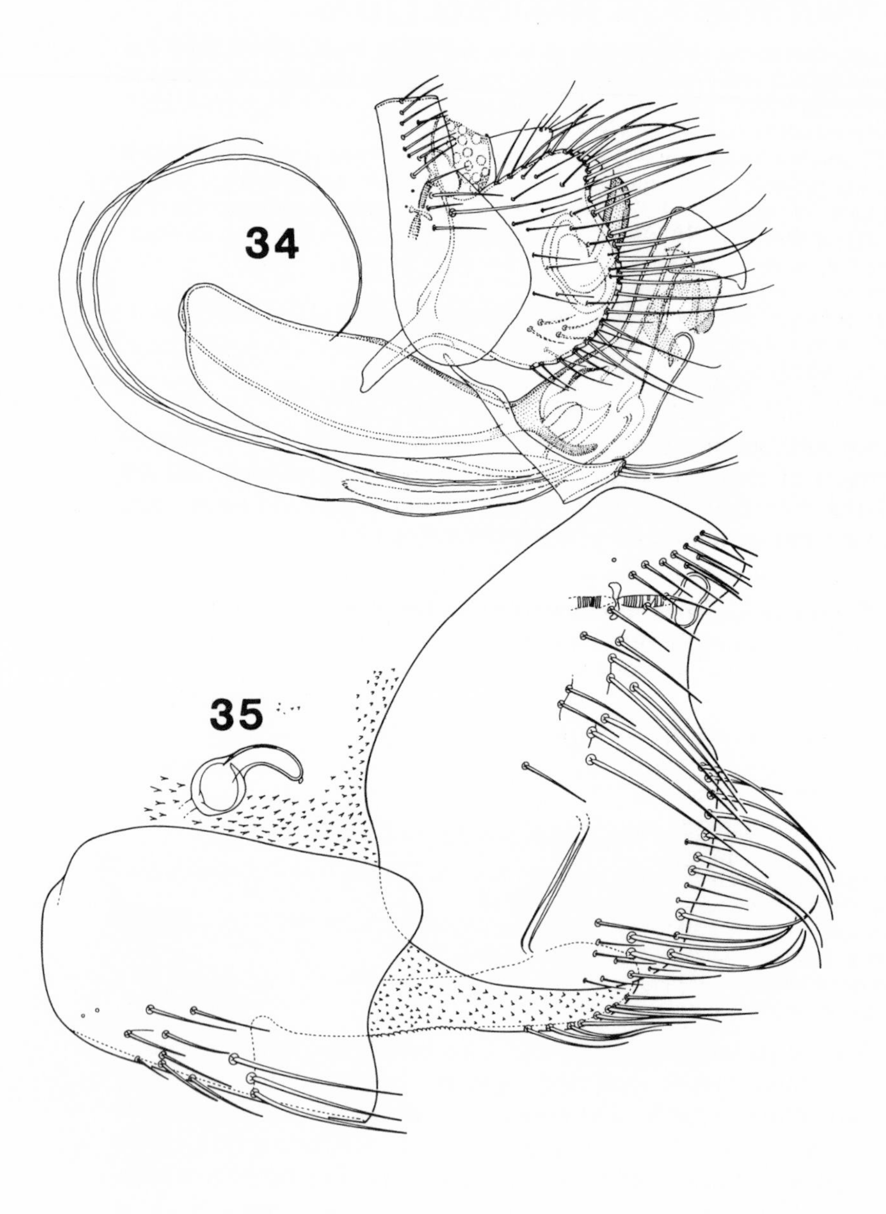

34-35 *Chaetopsylla floridensis* (I. Fox). **34** Modified abdominal segments of male. **35** Modified abdominal segments of female. After Holland, 1985.

FAMILY **ISCHNOPSYLLIDAE**

Derivation: The family name is derived from the Greek words *ischno* (thin) and *psylla* (a flea) and the Latin suffix *idae* and alludes to the long, thin, graceful bodies and legs of these fleas. The type genus is *Ischnopsyllus* Westwood, 1883, and is extralimital to the area under consideration here.

General Considerations: Preoral ctenidium present, of two spines; trabecula centralis usually present; frontal tubercle small, permanent or deciduous; eye usually vestigial; pronotal ctenidium always present; one or more abdominal terga with marginal spinelets, ctenidia or pseudoctenidia; lateral metanotal area short; legs long and slender; normally with five pairs of lateral plantar setae; metepisternum with a squamulum.

World Distribution: Worldwide, with nineteen genera and about 115 species. Ectoparasites of bats, particularly insectivorous species.

Number of Northwestern Taxa: 4

General References: Hopkins and Rothschild (1956), Holland (1985).

Only three genera belonging to this family occur in North America and species of two of these are known from the Northwest. All are associated with bats belonging to the family Vespertilionidae, and the genera may be separated with the following key.

KEY TO THE NORTHWESTERN GENERA OF *ISCHNOPSYLLIDAE*

1 Antepygidial bristles modified to form a false comb on abdominal tergum VII; genal ctenidium of two narrow, sharp-pointed spines .... **Nycteridopsylla**

Antepygidial bristles not so modified; genal ctenidium of two blunt spines .... **Myodopsylla**

GENUS **Nycteridopsylla** Oudemans

Nycteridopsylla. Oudemans, 1906, Tijdschr. Ent. 49: 68 (type species by designation: *Ceratopsyllus pentactenus* Kolenati, 1856).

Eptescopsylla. I. Fox, 1940, Fleas of Eastern United States, p. 107 (type species by designation: *Nycteridopsylla chapini* Jordan, 1929).

Eptescopsylla I. Fox. Hopkins, 1952, J. Wash. Acad. Sci. 42: 364 (placed as a synonym of *Nycteridopsylla*).

This is a Holarctic genus of bat fleas containing eighteen named taxa. The three North American species were recently treated by Lewis and Wilson (1982). The genus is unique in having the multiple antepygidial bristles shortened and spiniform, forming a false comb on the caudal margin of abdominal tergum VII. The eastern species, *Nycteridopsylla chapini* Jordan, 1929, is known only from collections east of the 100th parallel, mainly as a winter parasite of *Eptesicus fuscus* (big brown bat). *Nycteridopsylla intermedia* Lewis and Wilson, 1982, is currently known from Texas, Colorado, and Nevada. It appears to prefer *Plecotus townsendi* as a host and, although it has not been collected in the Northwest, it might be expected since this bat occurs throughout the area. The third species, *N. vancouverensis* Wagner, 1936, ranges throughout the western tier of states from central California north to British Columbia and is known from a number of collections from Oregon.

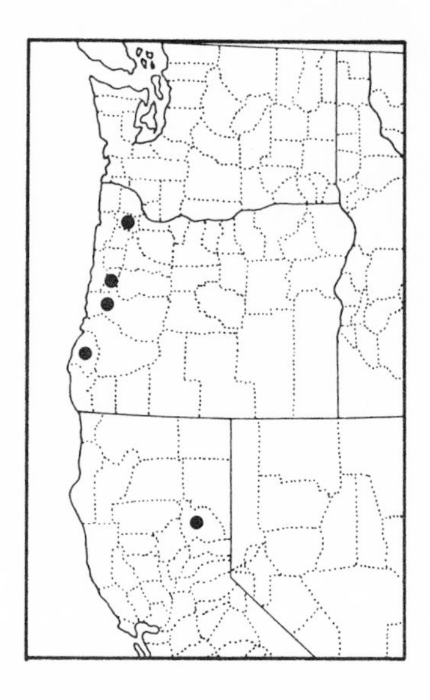

Nycteridopsylla vancouverensis Wagner

[93638]

(Figs. 36, 37, 38, 39)

Nycteridopsylla vancouverensis. Wagner, 1936, Z. Parasitenk. 8: 658, figs. 7, 8 (Vancouver, B. C. from *Lasionycteris noctivagans).*

Eptescopsylla vancouverensis (Wagner). Ewing and Fox, 1943, Misc. Publ. U. S. Dept. Agric. No. 500, p. 99.

Nycteridopsylla vancouverensis Wagner. Hopkins and Rothschild, 1956, An Illustrated Catalogue... Vol. II, pp. 233-235.

Lewis and Wilson (1982) listed all of the published collection records for this species. It has been collected from Coos, Lane, Benton, and Washington counties in Oregon, and Plumas County in California. Its preferred host is evidently *Myotis californicus,* although it has been taken from *Plecotus townsendi* and *Lasionycteris noctivagans,* the type series having been collected from the latter host species. All collection records date from November to May and suggest that this is a winter flea. Since its preferred host is known to hibernate, it is likely that it has a life cycle similar to that of *N. chapini* in the eastern United States.

GENUS **Myodopsylla** Jordan and Rothschild

Myodopsylla. Jordan and Rothschild, 1911, Novit. Zool. 18: 88 (type species by designation: *Ceratopsylla insignis* Rothschild, 1903).

Myodopsylloides. Augustson, 1941, Bull. S. Calif. Acad. Sci. 40: 104 (type species by designation: *M. piercei* Augustson [= *M. palposa* Rothschild, 1904]).

Myodopsylloides Augustson. Hopkins, 1952, J. Wash. Acad. Sci. 42: 364 (placed as a synonym of *Myodopsylla*).

There are currently twelve recognized species belonging to this genus. One of these is known from two subspecies. Except for *M. trisellis* Jordan, 1929, known only from China, members of this genus are restricted to the New World where they are associated with

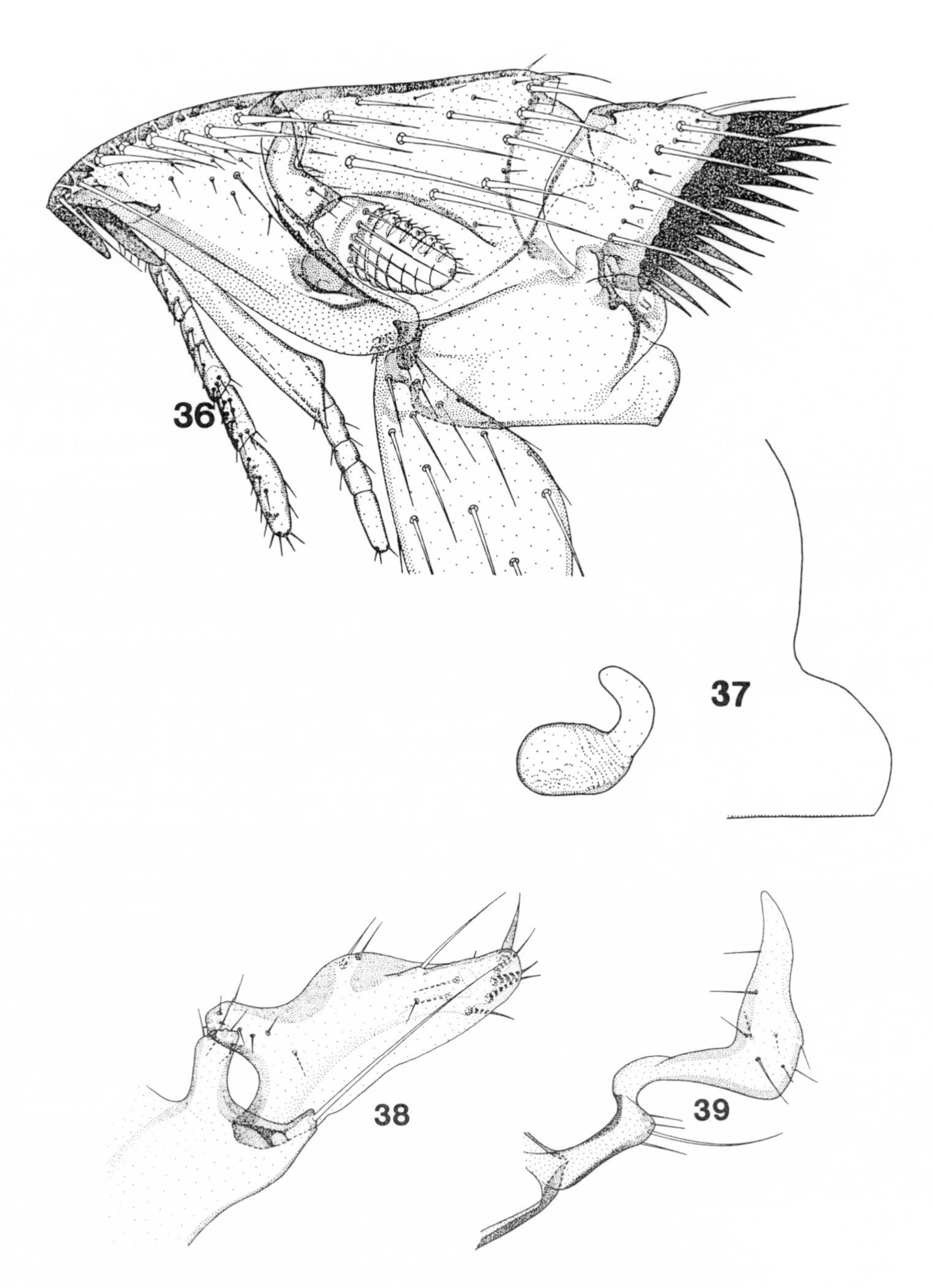

36-39 *Nycteridopsylla vancouverensis* Wagner. **36** Head of male. **37** Modified abdominal segments of female. **38** Clasper of male. **39** Sternum IX of male. Original.

vespertilionid bats, especially species of *Myotis*. Six species are known from North America north of Mexico, but further collections in the southwest will almost certainly add *M. globata* Holland, 1971, to our fauna.

KEY TO THE WESTERN SPECIES OF *MYODOPSYLLA*

1 False combs of thickened bristles present on abdominal tergites I through VI; st. VIII of male with dense tufts of modified bristles on inner surface .... **gentilis**

False combs of modified bristles absent from abdominal tergites; bristles of st. VIII of male not modified **2**

2 Sternum VIII of male with continuous row of long bristles along apical margin to base of segment, lacking a dense basal cluster of setae on inner surface; st. VII of female with distinct caudoventral concavity; apex of hilla of spermatheca obliquely flattened .... **borealis**

Marginal row of long bristles on st. VIII of male interrupted caudoventrally, with a dense cluster of setae on inner surface basally; st. VII of female lacking distinct concavity; apex of hilla of spermatheca smoothly rounded .... **palposa**

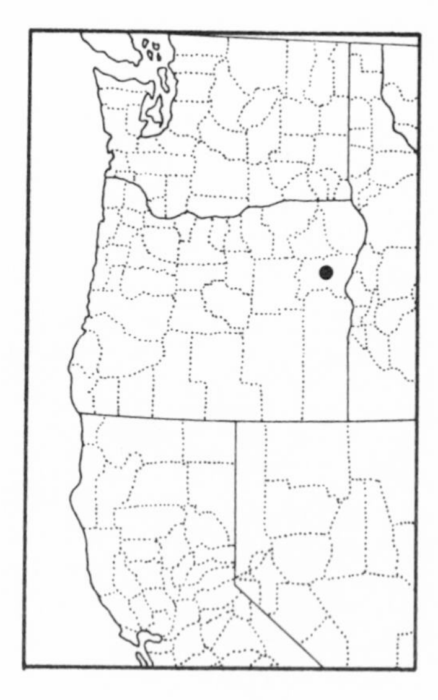

Myodopsylla borealis Lewis

[97826]

(Figs. 40, 41, 42, 43)

Myodopsylla borealis. Lewis, 1978, J. Parasitol. 64: 524-527, figs. 1-4 (7 mi. N, 10 mi. W Camp Crook, 3800 ft, Carter Co., Montana, from *Eptesicus fuscus)*.

Myodopsylla borealis Lewis. Benton, 1980, An Atlas of the Fleas of the Eastern United States, pp. 145-146.

This species recently was described from collections from Montana and Minnesota. Benton (1980) indicates a collection from southern Illinois and one male and three females were taken in Baker County, Oregon, on 11 July 1977. To date the only recorded host has been *Eptesicus fuscus* (big brown bat). As this is one of the commonest bats throughout most of the United States, it seems unusual that this flea should be so rare in collections.

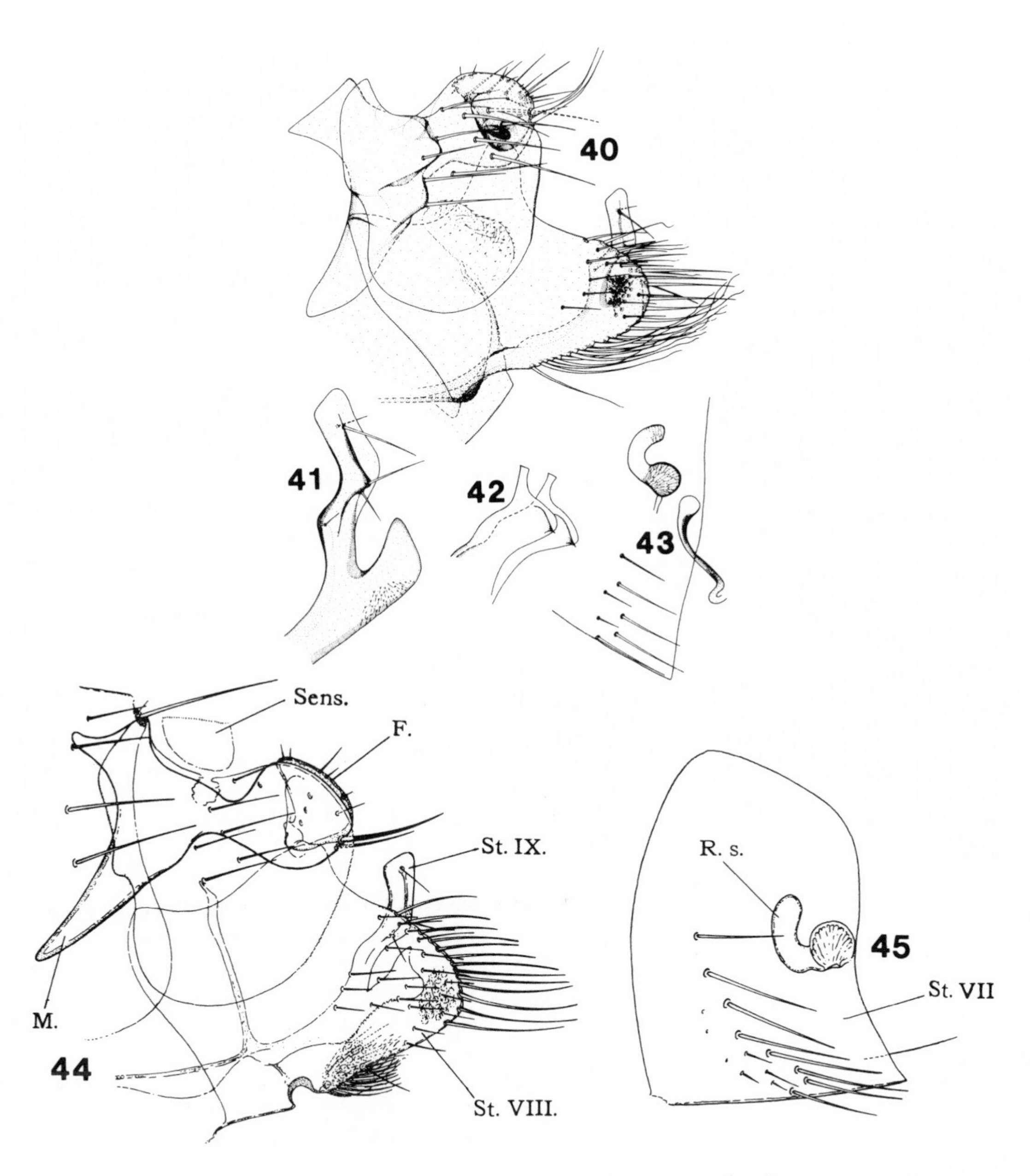

40-43 *Myodopsylla borealis* Lewis. **40** Modified abdominal segments of male. **41** Apex of sternum IX of male. **42** Crochets of male. **43** Modified abdominal segments of female. Original.

44-45 *Myodopsylla palposa* (Rothschild). **44** Modified abdominal segments of male. **45** Modified abdominal segments of female. After Holland, 1985.

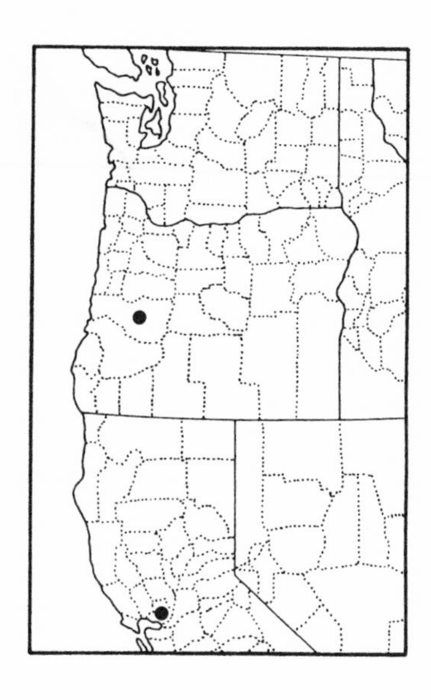

Myodopsylla palposa (Rothschild)

[90478]

(Figs. 44, 45)

Ceratopsylla palposus. Rothschild, 1904, Novit. Zool. 11: 652 (Cowichan, near Duncan, Vancouver Island, from *Eptesicus fuscus*).

Rhinolophopsylla palposus Rothschild. Wagner, 1936, Canad. Ent. 68: 206.

Myodopsylloides piercei. Augustson, 1941, Bull. S. Calif. Acad. Sci. 40: 104, pl. 7, figs. 1-5 (Santa Cruz Island, California, from *Antrozous pallidus pacificus)*.

Myodopsylloides palposa Rothschild. Hubbard, 1947, Fleas of Western North America, p. 376, fig. 233 (*piercei* synonymized).

Myodopsylla palposus Rothschild. Jordan, 1947, Tijdsch. Ent. 88: 91, fig. 9A.

Myodopsylloides palposus (Rothschild). Holland, 1949, The Siphonaptera of Canada, p. 182, pl. 42, figs. 345-348, map 43.

Myodopsylla palposa (Rothschild). Hopkins and Rothschild, 1956, An Illustrated Catalogue... Vol. II, pp. 238-240.

The range of this rare flea extends south from British Columbia to southern California and it appears to be restricted to the coastal portion of the continent. One specimen each was collected on *Eptesicus fuscus* and *Antrozous pallidus* from Lane County, Oregon, 13 and 14 May 1970. All other known records are those cited in Hubbard (1947), Holland (1949, 1985) and Hopkins and Rothschild (1956).

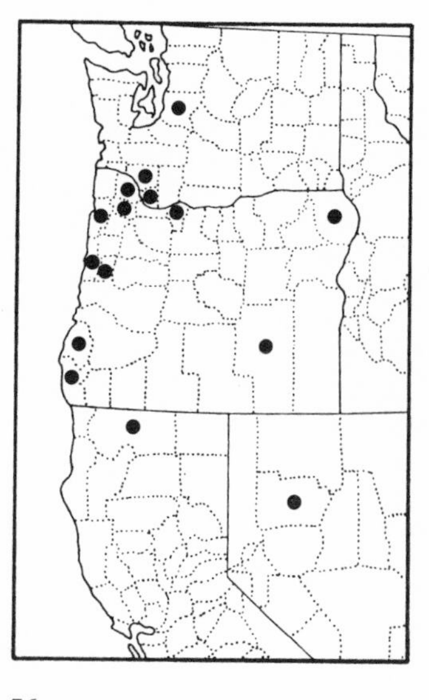

Myodopsylla gentilis Jordan and Rothschild

[92113]

(Figs. 46, 47)

Myodopsylla gentilis. Jordan and Rothschild, 1921, Ectoparasites I: 152, fig. 131 (Okanagan Landing, British Columbia, from bat).

Myodopsylla gentilis Jordan and Rothschild. Hopkins and Rothschild, 1956, An Illustrated Catalogue...Vol. II, pp. 249-251.

Of the species of bat fleas native to the area this is by far the commonest. Its preferred hosts seem to be *Myotis yumanensis* and *M. lucifugus*. Both of these host species occupy hot attics and lofts during the warmer months and are similar in other biological and behavioral characteristics. Although the winter haunts of *M. yumanensis* are unknown due to migration, *M. lucifugus* retires to caves and mines to hibernate. It therefore seems likely that this species of flea is associated with the sites occupied by the maternity colonies of the hosts. Since both species of bat return to these sites in April and early May, according to Barbour and Davis (1969), the high infestation rate in April may be a reflection of the population levels at these roosting sites.

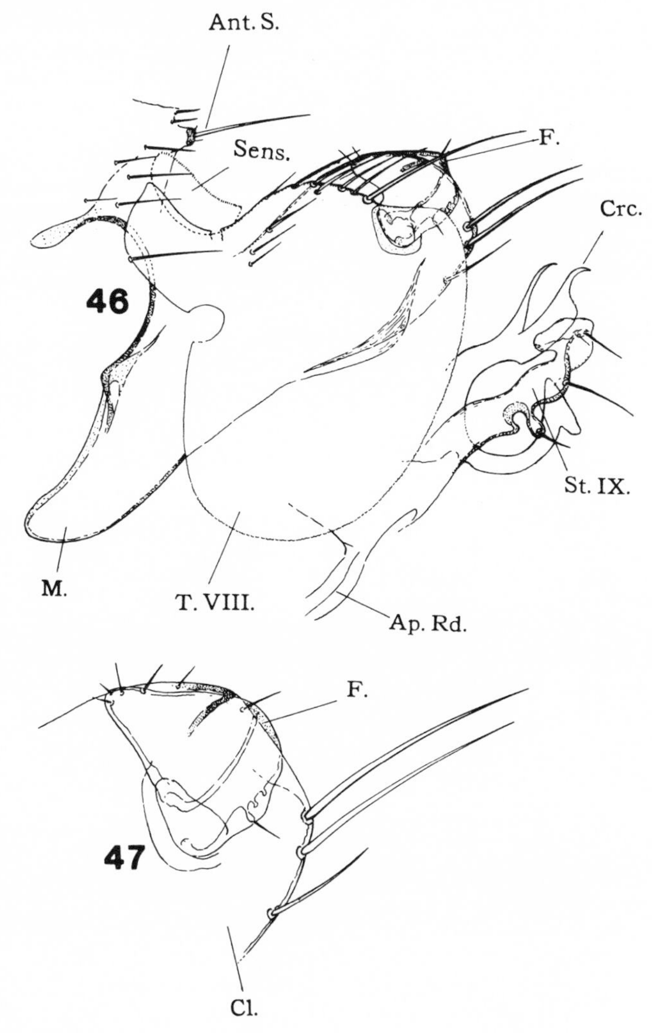

46-47 *Myodopsylla gentilis* J. & R. **46** Modified abdominal segments of male. **47** Enlarged clasper of male. After Holland, 1985.

Host and seasonal distribution for *Myodopsylla gentilis* in Oregon

Host Species	Males	Females
Myotis yumanensis	52	106
Myotis lucifugus	36	64
Myotis evotis	1	3
Myotis volans	1	3
Eptesicus fuscus	0	2
Lasionycteris noctivagans	1	0
*Neotoma fuscipes**	0	1

	April	May	June	July	August	September	October
Male	44	6	8	8	25	0	0
Female	94	10	12	12	49	1	1

* A woodrat. An accidental host.

Whitaker and Easterla (1975) reported this species from a number of bats collected at Big Bend National Park, Texas, but there also *Myotis yumanensis* was obviously the preferred host.

FAMILY **HYSTRICHOPSYLLIDAE**

Derivation: The family name is derived from the Greek words *hystri* (a porcupine) and *psylla* (a flea) and the Latin suffiix *idae,* and alludes to the bristly appearance of many members of this family. The type genus is *Hystrichopsylla* Taschenberg, 1880.

General Characters: Pronotal ctenidium present (except in *Anomiopsyllus* and *Jordanopsylla*); arch of tentorium visible in genal region; antennal fossae usually open; metanotum without marginal spinelets; spiniform setae on inner surface of hind coxae present or absent; hind tibia with apical tooth on outside; male sensilium strongly convex; females with one or two spermathecae.

World Distribution: As a family, worldwide, mainly as ectoparasites of rodents. There are about 44 genera and approximately 764 species and subspecies.

Number of Northwestern Taxa: 58

General References: Hopkins and Rothschild (1962, 1966), Holland (1985).

This is a large family. Of the 44 currently recognized genera, eighteen of these, representing five subfamilies, are found in the area under consideration. In general, members of this family are parasites of small rodents, but many are associated with insectivores. With rare exceptions, all are considered to be nest fleas, and individuals of many of the species are extremely rare in collections. As a result, many new species are being discovered and described, even from relatively well known localities.

Two subfamilies are restricted to the Ethiopian Region, one to the Australian Region, and the remainder are mainly Holarctic in distribution, with members of one subfamily spilling into the Oriental Region and another being represented on all continents except Australia. By far the best key to the subfamilies is that of Hopkins and Rothschild (1966) and the following is a modification of it.

KEY TO THE SUBFAMILIES OF THE *HYSTRICHOPSYLLIDAE* OF NORTH AMERICA

1 Pleural ridge of metathorax short, interrupted or missing; antennal clavus with some segments partly or completely fused, so that it appears to consist of only seven or eight segments .... **Rhadinopsyllinae**

Pleural ridge of metathorax complete; antennal clavus of nine distinct segments 2

2 Genal comb absent; pronotum and t. II-VII each usually with only one row of bristles, the anterior row vestigial if present at all. **Anomiopsyllinae**

Either genal comb well developed or t. II-VII each with at least two well developed rows of bristles, or both 3

3 Genal comb of two overlapping spines or absent; in the latter case a striarium present on basal abdominal sternite .... **Neopsyllinae**

Genal comb of some other type 4

4 Segment V of all tarsi with five pairs of lateral bristles, no bristles of similar size on plantar surface of this segment; female with two spermathecae .... **Hystrichopsyllinae**

Segment V of all tarsi rarely with more than four pairs of lateral bristles, female with one spermatheca 5

5 Genal comb of four well developed, uniformly spaced spines, none arising behind the eye; sensilium with thirteen pits each side ..... **Doratopsyllinae**

Genal comb of three spines, one of which is partly or completely concealed beneath another; sensilium with more than thirteen pits each side .... **Ctenophthalminae**

SUBFAMILY **Hystrichopsyllinae**

Of the four recognized genera belonging to this subfamily, one is restricted to the Neotropical Region, another to the western Palaearctic Region and the remaining two are Holarctic. The principal hosts for both genera are small rodents.

KEY TO THE NORTHWESTERN GENERA OF *HYSTRICHOPSYLLINAE*

1 Large fleas with a well developed genal comb of at least five spines .... **Hystrichopsylla**

Somewhat smaller fleas which lack a genal comb .... **Atyphloceras**

GENUS **Hystrichopsylla** Taschenberg

Hystrichopsylla. Taschenberg, 1880, Die Flöhe, pp. 63, 83 (type by designation: *P. obtusiceps* Ritsema).

This genus contains thirty presently recognized species and subspecies. It is Holarctic in distribution, the various species being loosely associated with small rodents and insectivores and their predators.

KEY TO THE NORTHWESTERN SPECIES OF *HYSTRICHOPSYLLA*

1 Both sexes exceedingly large (ca. 8 mm) and with four antepygidial bristles per side; parasites of *Aplodontia rufa* .... **schefferi**

Both sexes much smaller and with three antepygidial bristles per side; not found on *Aplodontia rufa* 2

2 Genal comb of seven or more spines; caudal margin of female st. VII angular, incised ventrally .... **occidentalis** ssp.

Genal comb of six spines; caudal margin of female st. VII undulate, rounded ventrally .... **dippiei** ssp.

Hystrichopsylla dippiei dippiei Rothschild

[90203]

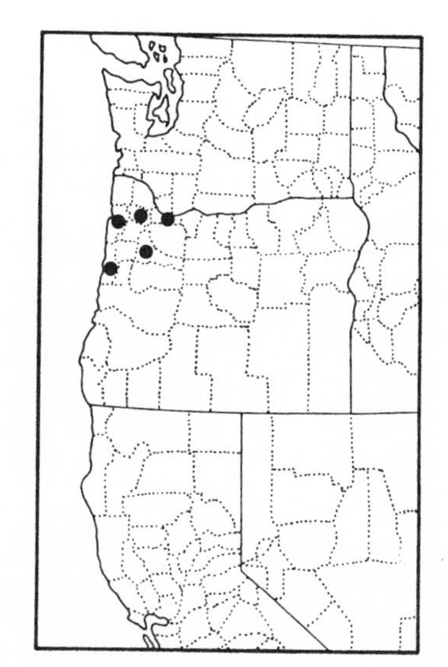

Hystrichopsylla dippiei. Rothschild, 1902, Ent. Rec. 14: 63, pl. 2, fig. 2 (Alberta, Canada, from *Putorius longicaudatus*) [= *Mustela frenata* ssp.].

Hystrichopsylla dippiei Rothschild. Jordan, 1929, Novit. Zool. 35: 174 (placed as a subspecies of *Pulex gigas* Kirby, 1837).

Hystrichopsylla gigas dippiei Rothschild, 1902. Hubbard, 1947, Fleas of Western North America, p. 357, fig. 219.

Hystrichopsylla dippiei Rothschild. Holland, 1949, The Siphonaptera of Canada, p. 75, pl. 9, figs. 42-44 (*partim*).

Hystrichopsylla dippiei dippiei Rothschild. Holland, 1957, Canad. Ent. 89: 315, figs. 2, 10, 12, 22 (lectotype and restricted locality designated).

References to the occurrence of this taxon are those of Hubbard (1947, 1949) and Svihla (1941) who list it as a subspecies of *H. gigas,* now considered a *nomen dubium.* Of Hubbard's illustrations of the female st. VII, one resembles *H. dippiei,* the other *H. occidentalis.* His illustrations of the male are not diagnostic. There are published records of *H. dippiei* from the Northwest, but collection records suggest that the nominate taxon is extralimital to the area.

Hystrichopsylla dippiei spinata Holland

[94935]

(Fig. 48)

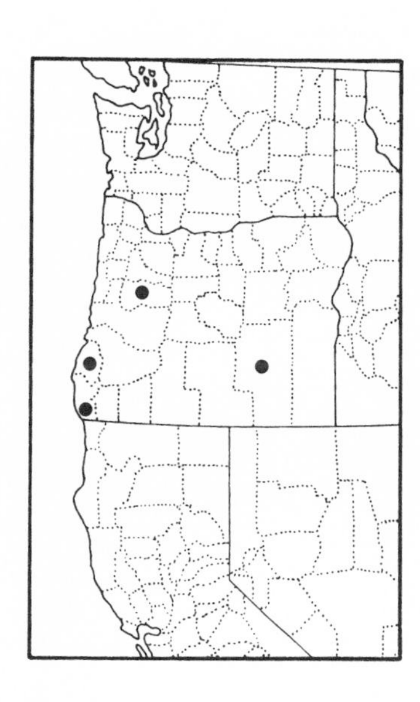

Hystrichopsylla spinata. Holland, 1949, The Siphonaptera of Canada, p. 77, pl. 10, figs. 50-52 (Vancouver, British Columbia, from *Spilogale gracilis olympica*).

Hystrichopsylla dippiei spinata Holland. Holland, 1957, Canad. Ent. 89: 316, figs. 9, 15, 16, 24.

Evidently this is not a dominant element in the fauna, since only seventeen specimens have been collected in the last 12 years. This is the dominant subspecies in the area under consideration at elevations below 1000 m. Most collections reported here came from *Spilogale putorius,* with only three males collected on *Eutamias townsendi.* All were taken during the cooler months.

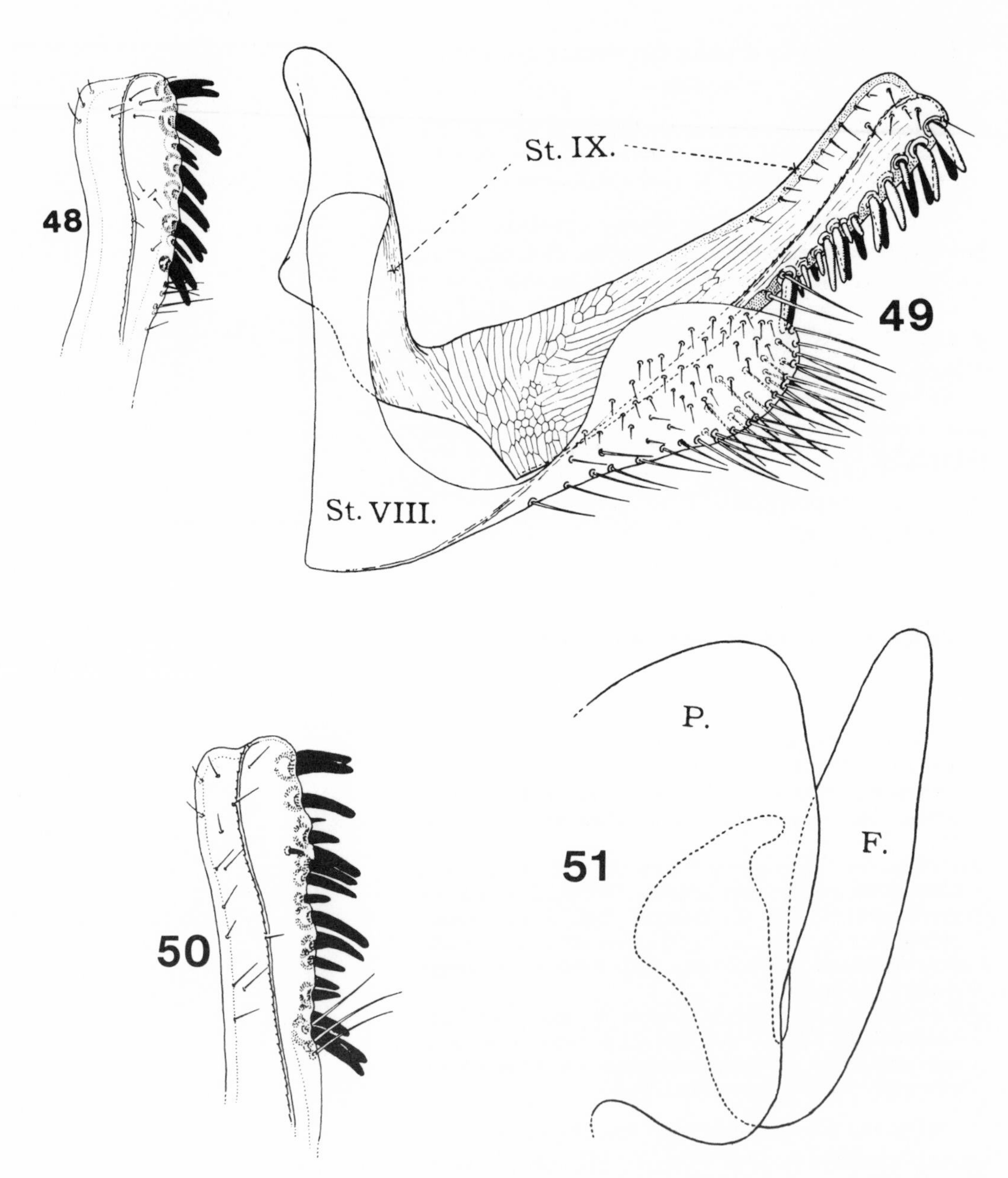

48 Apex of male sternum IX of *Hystrichopsylla dippiei spinata* Holland. After Holland, 1985.

49-51 *Hystrichopsylla occidentalis occidentalis* Holland. **49** Male sternites VIII and IX. **50** Apex of male sternite IX. **51** Outline of male clasper. After Holland, 1985.

Hystrichopsylla dippiei neotomae Holland

[95746]

Hystrichopsylla dippiei neotomae. Holland, 1957, Canad. Ent. 89: 316, figs. 4, 11, 13, 25, 28 (Strawberry Canyon, Berkeley, Alameda Co., California, from nest of *Neotoma* sp.).

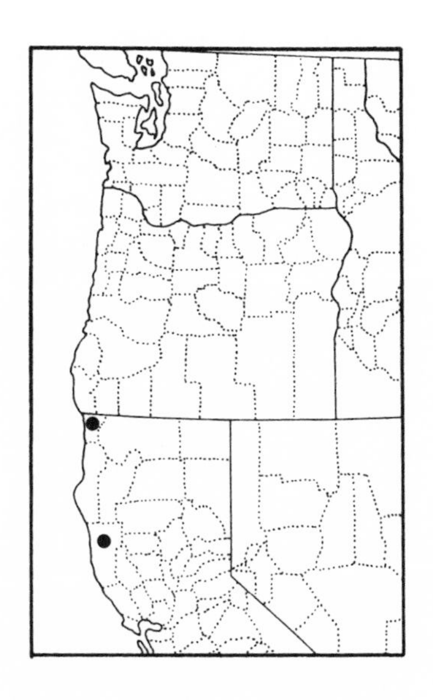

This taxon is known from collections from Del Norte and Mendocino counties in California, and the latter specimens may belong to this subspecies or to *H. d. spinata* according to Hopkins and Rothschild (1962). They are separable from this subspecies in having the apex of the distal arm of st. IX with an oblique margin and an apical dilation. Females are inseparable from those of other subspecies.

Hystrichopsylla occidentalis occidentalis Holland

[94934]

(Figs. 49, 50, 51)

Hystrichopsylla "dippiei". Rothschild, 1902 (*partim*), Ent. Rec. 14: 63 (female described as that of *dippiei*).

Hystrichopsylla "dippiei" Rothschild. Ewing, 1929 (*nec* Rothschild, 1902), Manual of External Parasites, p. 158, fig. 89.

Hystrichopsylla "gigas dippiei" Rothschild. Hubbard, 1947, Fleas of Western North America, p. 357, fig. 219 (*partim*).

Hystrichopsylla occidentalis. Holland, 1949, The Siphonaptera of Canada, p. 76, pl. 10, figs. 47-49 (Mount Seymour, near Northlands, Burrard Inlet, B. C., from *Clethrionomys gapperi caurina*).

Hystrichopsylla occidentalis occidentalis Holland, 1949. Campos and Stark, 1979, J. Med. Ent. 15: 431-444 (*H. occidentalis* redescribed, *H. linsdalei* Holland reduced to subspecific status, *H. o. sylvaticus* described).

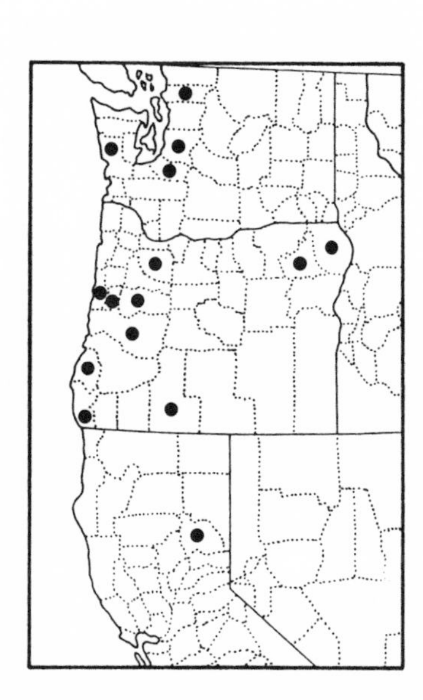

With the exception of two females collected in the northeastern part of Oregon, all collections of this species are from the western tier of counties west of the Cascades. The majority of collections come from microtine rodents, and other host records probably reflect accidental host associations.

Some records of *H. dippiei* subspecies published prior to 1949 certainly refer to this species.

This is the northernmost subspecies of the complex and it ranges north to southern Alaska and east to central Alberta. While not a common species, it is not rare in collections; it has been taken every month of the year, although it is most frequently encountered from August to April.

There are no published records of this species being associated with the transmission of disease.

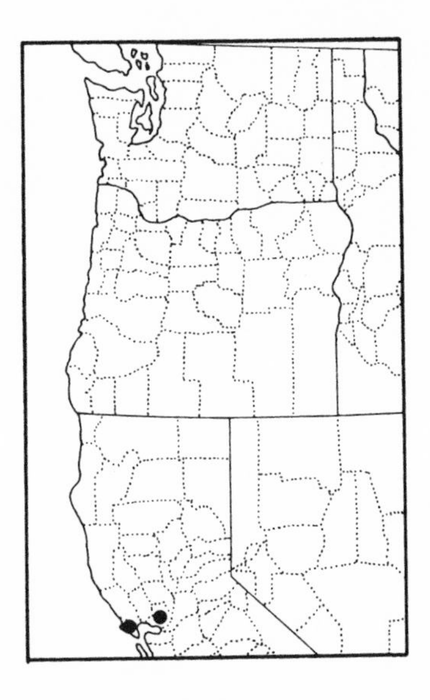

Hystrichopsylla occidentalis linsdalei Holland

[95748]

Hystrichopsylla linsdalei. Holland, 1957, Canad. Ent. 89: 322 (Berkeley Hills, Alameda Co., California, from nest of *Microtus* sp.).

Hystrichopsylla occidentalis linsdalei Holland. Campos and Stark, 1979, J. Med. Ent. 15: 434 (demoted to subspecific status).

This taxon is only known from the northern two thirds of California. According to Campos and Stark (1979), it intergrades with the nominate subspecies in the interior of California. The subspecies are separable in the males by the following characters. The nominate subspecies has all of the setae on the apical lobe of st. VIII directed posteriorly, while the anterior setae on this segment are directed ventrally. Also, the apex of st. IX is much broader in *H. o. linsdalei* than in the nominate taxon.

Hystrichopsylla schefferi Chapin

[91901]
(Fig. 52)

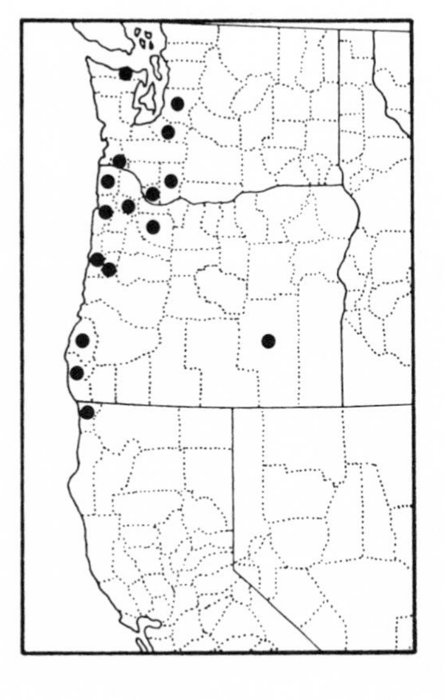

Hystrichopsylla schefferi. Chapin, 1919, Bull. Brooklyn ent. Soc. 14: 50 (Puyallup, Washington, from nest of *Aplodontia rufa*).

Hystrichopsylla mammoth. Chapin, 1921, Proc. ent. Soc. Wash. 23: 25 (Mammoth, Mono Co., California, from *Aplodontia [rufa] californica*).

Hystrichopsylla schefferi schefferi Chapin, 1919. Hubbard, 1947, Fleas of Western North America, p. 359, fig. 220.

Hystrichopsylla schefferi mammoth Chapin, 1921. Hubbard, 1947, *op. cit.,* p. 361, fig. 221.

Hystrichopsylla schefferi Chapin. Holland, 1949, Proc. ent. Soc. B. C. 45: 8 (*mammoth* synonymized).

Hystrichopsylla hubbardi. Augustson, 1953, Bull. S. Calif. Acad. Sci. 52: 119, pl. 22, figs. 1, 2 (Coos Co., Oregon, from *Aplodontia rufa pacifica*).

Hystrichopsylla schefferi Chapin. Holland, 1957, Canad. Ent. 89: 314, 319, figs. 6, 17 (*mammoth* and *hubbardi* synonymized).

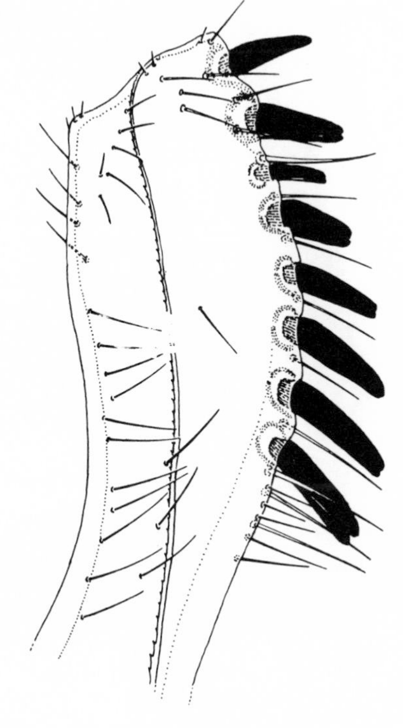

52 Apex of male sternite IX of *Hystrichopsylla schefferi* Chapin. After Holland, 1985.

This enormous flea (6-8 mm) is a specific parasite of the mountain beaver, and thus restricted by the distribution of its host. It is known from the Pacific Northwest west of the Cascade Mountains, usually at lower elevations. A few specimens also have been taken from *Spilogale putorius,* the spotted skunk, probably acquired from the burrows of the preferred host. This is a rare species, and Hubbard (1947) reported the average of only one flea per four host animals. The 28 specimens reported here came from fourteen hosts, and August and December were the months of greatest abundance. The species is of no known medical importance. Its range extends from northern Califonia to southern British Columbia.

GENUS **Atyphloceras** Jordan and Rothschild

Atyphloceras. Jordan and Rothschild, 1915, Ectoparasites I: 59 (type species by designation: *Ceratophyllus multidentatus* C. Fox, 1909).

Saphiopsylla. Jordan, 1931, Novit. Zool. 36: 227 (type species by designation: *S. nupera* Jordan, 1931).

Atyphloceras Jordan and Rothschild. Hopkins, 1952, J. Wash. Acad. Sci. 42: 365 (*Saphiopsylla* synonymized).

This is a small Holarctic genus consisting of eight described species. Two of these are restricted to the Palaearctic Region, the remaining six are found in the Nearctic. The genus is distinct from others in the subfamily in lacking a genal ctenidium. All species are typically ectoparasites of small rodents.

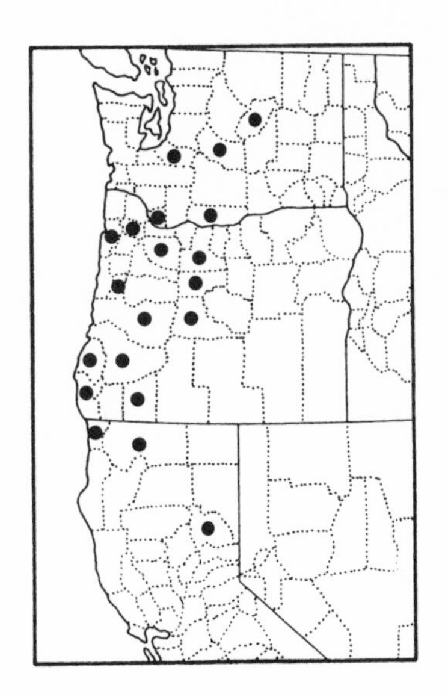

Atyphloceras multidentatus multidentatus (C. Fox)

[90905]
(Figs. 53, 54)

Ceratophyllus multidentatus. C. Fox, 1909, Ent. News 20: 107 (San Francisco, California, from *Microtus californicus* and nest of *Neotoma*).

Atyphloceras multidentatus Fox. Jordan and Rothschild, 1915, Ectoparasites I: 59.

Atyphloceras felix. Jordan, 1933, Novit. Zool. 39: 69, fig. 18 (Cuddy Valley, Ventura Co., California, from *Peromyscus truei*).

Atyphloceras artius. Jordan, 1933, *op. cit.,* 39: 69, fig. 19 (Kelowna, British Columbia, from *Peromyscus*).

Atyphloceras multidentatus (C. Fox). Hopkins, 1952, J. Wash. Acad. Sci. 42: 365 (*artius* and *felix* synonymized).

This is one of the common fleas of small rodents in the area. It occurs most commonly on species of *Microtus* and *Peromyscus,* but predator records are not uncommon and there are a few collections from shrews, tree squirrels, and other small mammals.

As pointed out by Lewis (1974), the role of this flea in disease transmission is not certain. It has been experimentally infected in laboratory tests, and has been reported to be naturally infected with *Yersinia pestis* in nature, but natural transmission of plague still remains to be demonstrated.

Collection records indicate that adults of this species occur during the cooler months of the year, particularly from October through March. Not a single specimen was taken in Oregon during June, July, or August but adults occur throughout the area under consideration during the cooler months.

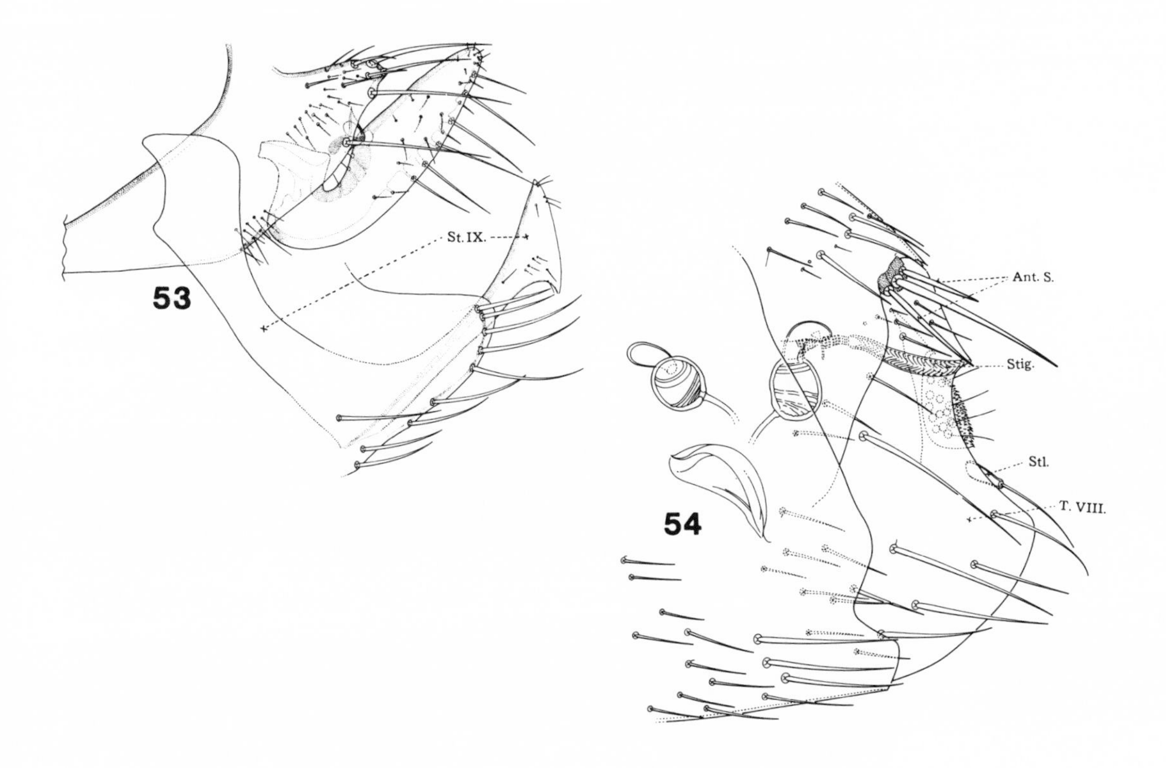

53-54 *Atyphloceras multidentatus multidentatus* (C. Fox). **53** Modified abdominal segments of male. **54** Modified abdominal segments of female. After Holland, 1985.

SUBFAMILY **Neopsyllinae**

This subfamily is well represented in North America by six genera, five of which are unique to the continent. They may be separated by the following modification of the key taken from Hopkins and Rothschild (1962).

KEY TO THE GENERA OF *NEOPSYLLINAE*

1 Lacking a genal comb **2**

Genal comb of two crossed spines present **3**

2 Frons with two rows of bristles; labial palpi not extending to apex of fore coxa .... **Catallagia**

Frons with three rows of bristles; labial palpi extending to apex of fore coxa .... **Delotelis**

3 Frontal tubercle and striarium both lacking **4**

Frontal tubercle and striarium both present **5**

4 Inner surface of hind coxae with an elongate patch of small bristles, some of which are spiniform; anterior abdominal terga with marginal spinelets .... **Phalacropsylla**

Inner surface of hind coxae with a single row of stout spiniforms; abdominal terga without marginal spinelets. .... **Meringis**

5 Fourth link plate present; tarsal segment V of fore and mid tarsi with five pairs of lateral plantar bristles, hind tarsi with four; three antepygidial bristles per side in both sexes .... **Neopsylla**

Fourth link plate absent; four pairs of lateral plantar bristles on all fifth tarsal segments; three antepygidial bristles per side in females, two to three in males .... **Epitedia**

GENUS **Neopsylla** Wagner

Neopsylla. Wagner, 1903, Horae. Soc. ent. ross. 36: 138, 140 (type species by subsequent designation: *bidentatiformis* by Baker, 1905, p. 129).

This is a large genus currently consisting of more than sixty described taxa. It is mainly Palaearctic in distribution, with a few species occurring in the Oriental Region, and one native to western North America, where it parasitizes ground squirrels.

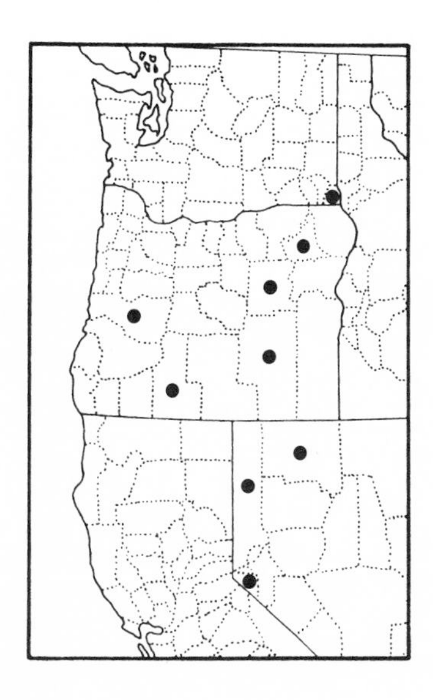

Neopsylla inopina Rothschild

[91523]
(Figs. 55, 56)

Neopsylla inopina. Rothschild, 1915, Ectoparasites I: 30, 39, figs. 32, 33 (Calgary, Alberta, Canada, from *Spermophilus richardsoni*).
Neopsylla texanus. Stewart, 1930, Canad. Ent. 62: 179 (Houston, Texas, from *Rattus norvegicus*).
Tamiophila (?) texana (Stewart). Ewing and I. Fox, 1943, Miscl. Publ. U. S. Dept. Agric. No. 500, p. 76.
Epitedia inopina (Rothschild). Ewing and I. Fox, 1943, *op. cit.*, p. 78.
Neopsylla inopina Rothschild. Hubbard, 1947, Fleas of Western North America, p. 314, fig. 186.
Neopsylla texana Stewart. Hubbard, 1947, *op. cit.*, p. 316, fig. 187.
Neopsylla inopina Rothschild, 1915. Hopkins and Rothschild, 1962, An Illustrated Catalogue... Vol. III, p. 193 (*texana* synonymized).

This is typically an ectoparasite of ground squirrels in northwestern North America, east as far as North Dakota. It is relatively rare in collections and seldom occurs in large numbers on individual hosts. All collections known to us have been made during the months when the host species has been active and out of hibernation, and probably show a distorted picture of the prevalence of this species both spatially and temporally. There is some indication that species of this genus are more common in the nest of the host than on the host itself. Most of the known species are distributed in the colder parts of the Northern Hemisphere, and it is tempting to speculate that peak populations of adults occur during periods when the host is in hibernation and thus inactive.

The species has been infected with plague micro-organisms in the laboratory, but its role in the maintenance of the disease in nature is still unresolved.

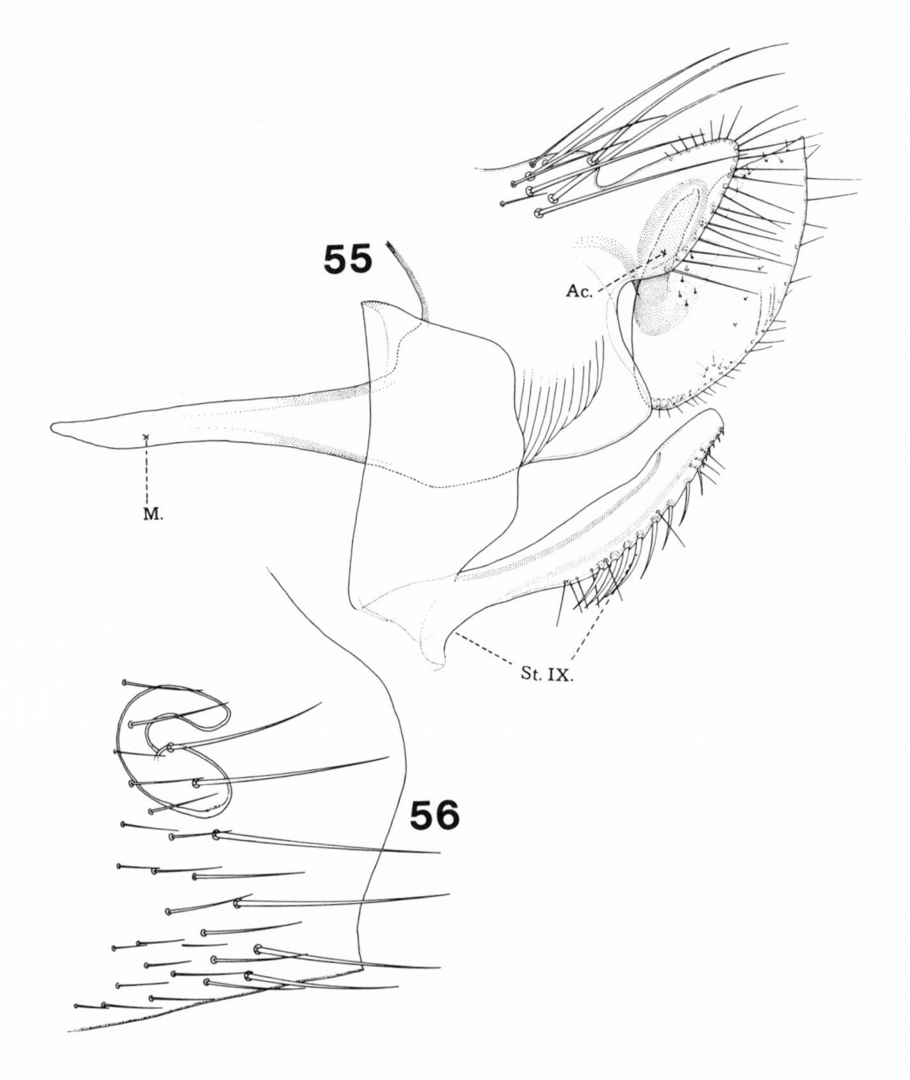

55-56 *Neopsylla inopina* Rothschild. **55** Modified abdominal segments of male. **56** Modified abdominal segments of female. After Holland, 1985.

GENUS **Epitedia** Jordan

Epitedia. Jordan, 1938, Novit. Zool. 41: 124 (type species by designation: *Ctenophthalmus wenmanni* Roths., 1915).

This small genus contains seven species, all of which are restricted to North America. Four of these occur in the area under consideration as parasites of small rodents and insectivores.

KEY TO THE WESTERN SPECIES OF *EPITEDIA*

1 Apical margin of t. VII immediately below the antepygidial setae forming a small, wedge-shaped lobe in both sexes.......................... **stanfordi**

Apical margin of t. VII lacking lobe.. **2**

2 Fixed process of the male clasper divided into two lobes by a deep sinus; hilla of spermatheca of female projecting into bulga............. **wenmanni**

Fixed process of male clasper not divided into two lobes by a deep sinus; hilla of spermatheca of female not projecting into bulga.................... **3**

3 Apex of movable process of male wide, its anterior margin bearing an angled convexity; apical margin of female st. VII lacking a prominent lobe... **scapani**

Apex of movable process of male narrower, its anterior margin only slightly convex; apical margin of female st. VII with a prominent, angular lobe.. **stewarti**

Epitedia scapani (Wagner)

[93636]

(Figs. 57, 58)

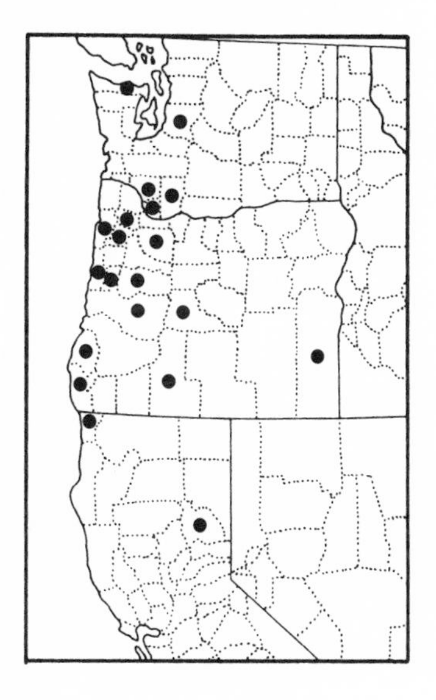

Neopsylla scapani. Wagner, 1936, Z. Parasitenk. 8: 657, fig. 5 (Vancouver, B. C., from *Scapanus orarius*).

Epitedia (Neopsylla) scapani Wagner. Wagner, 1940, Z. Parasitenk. 11: 465, fig. 4.

Epitedia jordani. Hubbard, April, 1940, Pacific Univ. Bull. 37(2): 10 (Newberg, Oregon, from *Sorex trowbridgei*).

Epitedia scapani (Wagner). Holland, 1942, Canad. Ent. 74: 157 (*jordani* placed as a synonym).

Epitedia jordani Hubbard. Hubbard, 1947 (*partim*), Fleas of Western North America, pp. 312-314 (*stewarti* erroneously synonymized).

As the name implies, this is mainly a parasite of insectivores, most commonly perhaps shrews, but it is also frequently taken on small rodents, particularly *Peromyscus* and *Microtus* species. Stray specimens are also common on predators.

The 128 specimens reported here suggest that adults of this species occur more commonly during the warmer months and the bulk of our collections were taken from April to October. The species is distributed throughout the area under consideration.

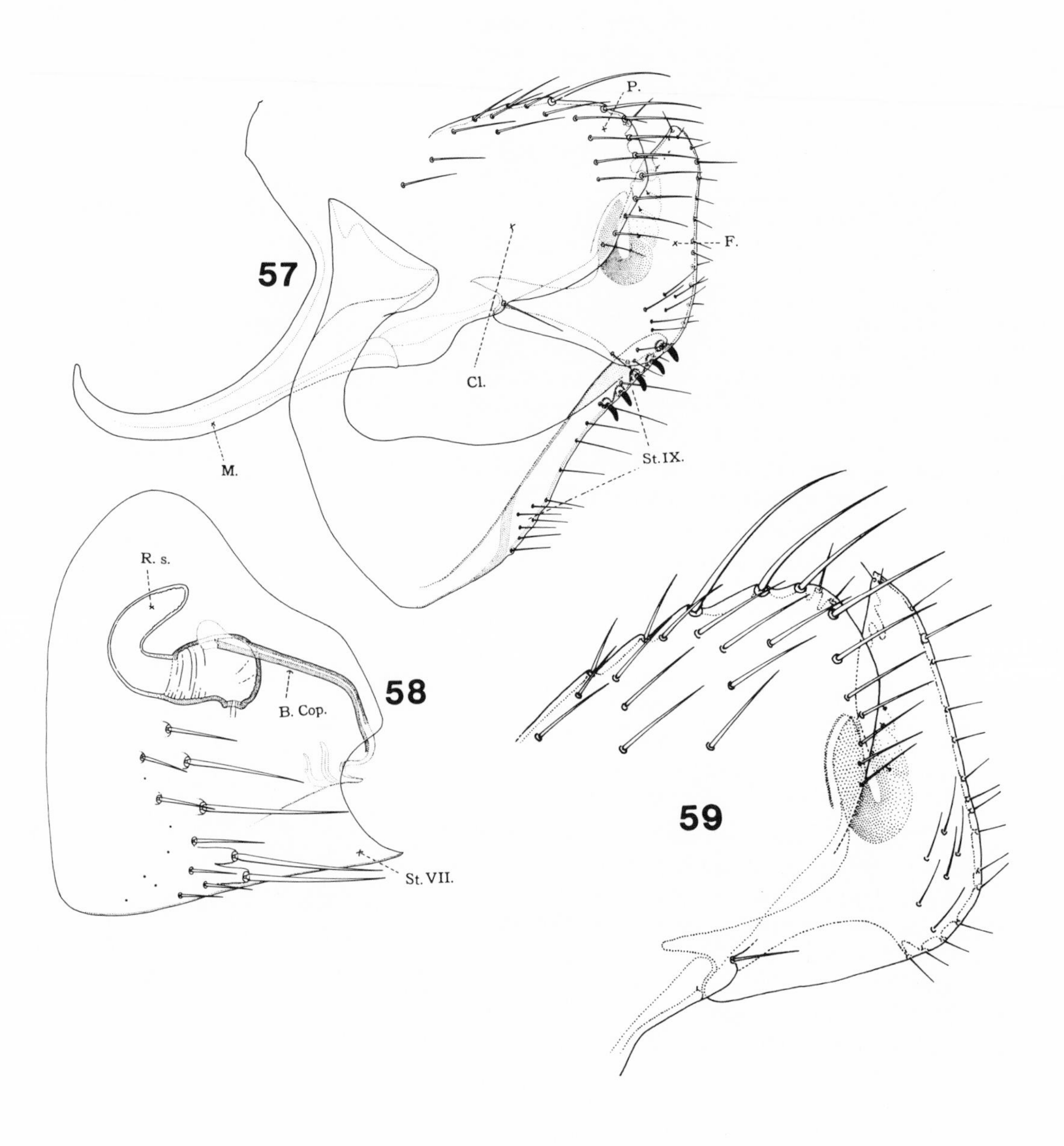

57-58 *Epitedia scapani* (Wagner). **57** Modified abdominal segments of male. **58** Modified abdominal segments of female. After Holland, 1985.

59 Male clasper of *Epitedia stewarti* Hubbard. After Holland, 1985.

Epitedia stewarti Hubbard

[94027]
(Fig. 59)

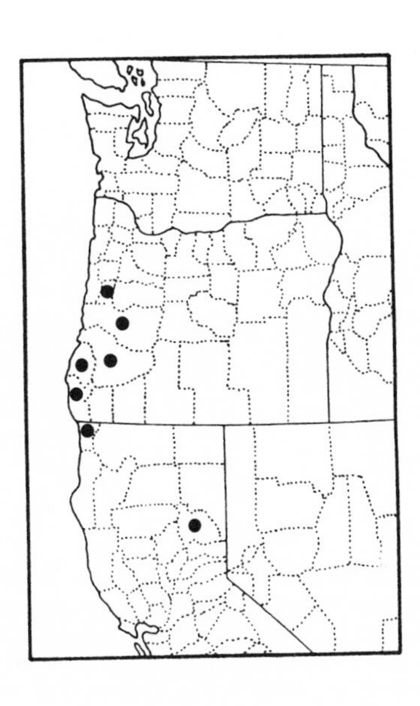

Epitedia stewarti. Hubbard, April, 1940, Pacific Univ. Bull. 37(2): 11 (Smith River, California, from *Sorex pacificus*).
Epitedia stewarti Hubbard. Hubbard, 1947, Fleas of Western North America, p. 312 (synonymized with *jordani*).
Epitedia stewarti Hubbard, 1940. Hopkins and Rothschild, 1962, An Illustrated Catalogue ... Vol. III, pp. 238-240.

This taxon is very similar to the preceding species in both appearance and distribution. However, it is much less frequently collected and shows a much stronger preference for rodent hosts (particularly species of *Microtus*) over insectivores. Our limited collections were taken throughout the year and show no obvious seasonal peak in the adult population. The species is known from Oregon south into northern California.

Epitedia stanfordi Traub

[94412]
(Figs. 60, 61)

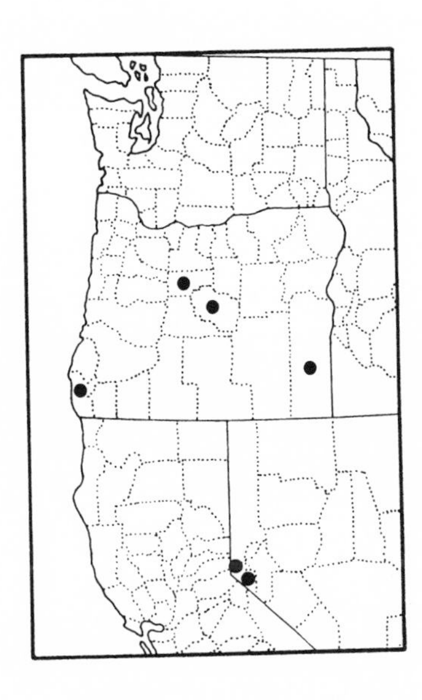

Epitedia stanfordi. Traub, 1944, Zool. Ser. Field Mus. 29: 214 (Fillmore, Millard Co., Utah, from *Peromyscus truei*).
Epitedia stanfordi Traub. Hopkins and Rothschild, 1962, An Illustrated Catalogue ... Vol. III, pp. 244-246.

This species seems to be confined to rodent hosts and was the least commonly collected member of the genus. While Hopkins and Rothschild (1962) suggest that it is specific to *Peromyscus* species, our limited records show an equal affinity for microtines.

This is a western flea, as are the preceding two, but *E. stanfordi* occurs inland to Utah and New Mexico. Our records are too few to indicate seasonal prevalence since the nine specimens reported here were taken in December and March.

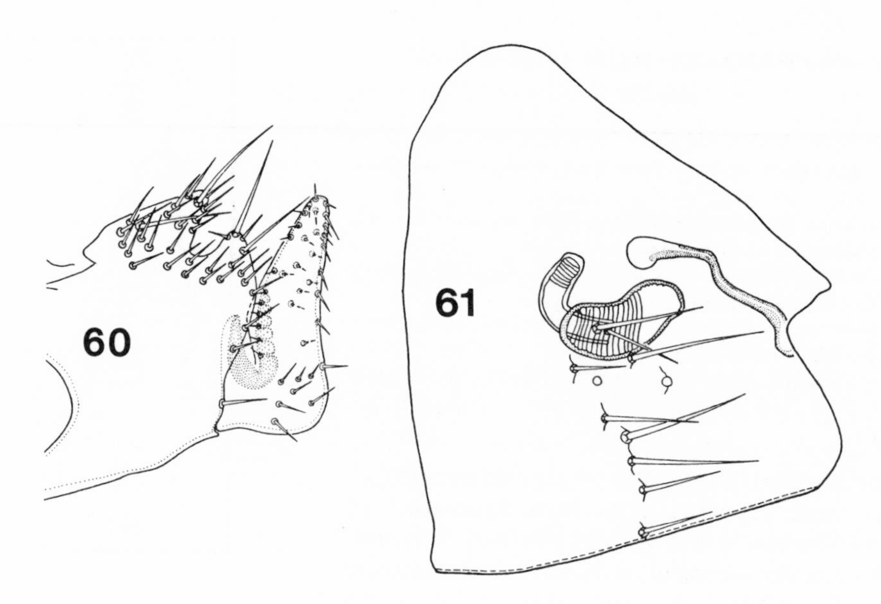

60-61 *Epitedia stanfordi* Traub. **60** Male clasper. **61** Female sternum VII and spermatheca. Original.

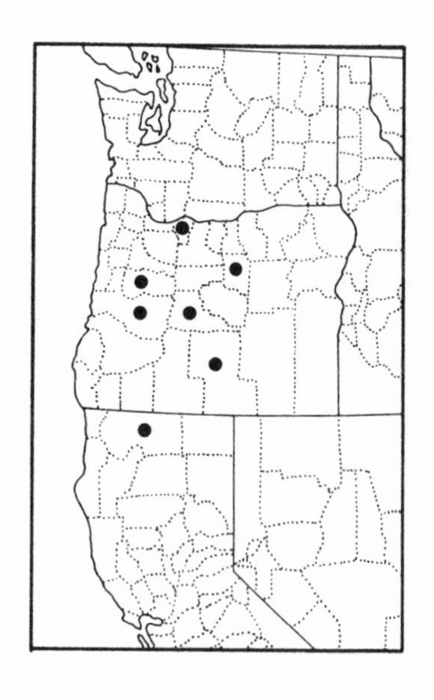

Epitedia wenmanni wenmanni (Rothschild)

[90470]

(Figs. 62, 63)

Ctenophthalmus wenmanni. Rothschild, 1904, Novit. Zool. 11: 642, pl. 14, figs. 75, 77, 79 (British Columbia from *Peromyscus leucopus* and *Neotoma cinerea*).

Neopsylla wenmanni Rothschild. Rothschild, 1915, Ectoparasites I: 30, 32, 39.

Neopsylla similis. Chapin, 1919, Bull. Brooklyn ent. Soc. 14: 50, 58 (Lake Burford, N. M., from *Peromyscus* sp.).

Neopsylla wenmanni Rothschild. Jordan, 1929, Novit. Zool. 35: 172, 175, 177 (*similis* synonymized).

Ctenophthalmus wenmanni Rothschild. Jordan, 1938, Novit. Zool. 41: 124 (made the type species of *Epitedia*).

Epitedia wenmanni wenmanni (Rothschild). Hopkins, 1954, Entomologist 87: 217, fig. 1.

Epitedia wenmanni wenmanni (Rothschild). Hopkins and Rothschild, 1962, An Illustrated Catalogue ... Vol. III, pp. 246-249.

and

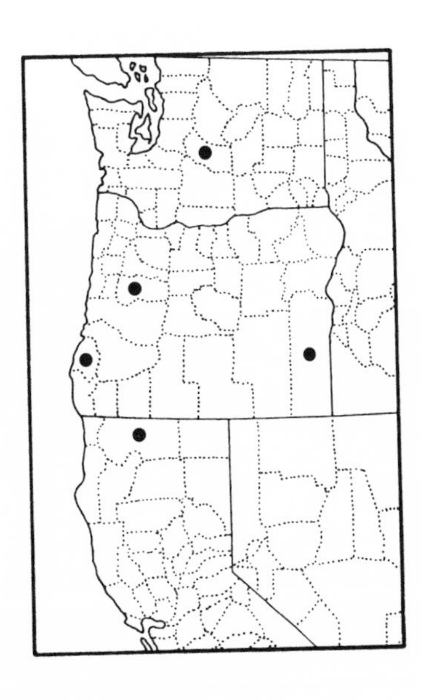

Epitedia wenmanni testor (Rothschild)

[91525]

Neopsylla testor. Rothschild, 1915, Ectoparasites I: 34, fig. 36 (Lansingburgh, near Troy, New York, from nest in hollow pine tree).

Neopsylla testor Rothschild. Jordan, 1938, Novit. Zool. 41: 124 (transferred to *Epitedia*).

Epitedia testor (Rothschild). I. Fox, 1940, Fleas of Eastern United States, p. 99, pl. 27, fig. 140.

Epitedia wenmanni testor (Rothschild). Hopkins, 1954, Entomologist 87: 217, fig 2.

Epitedia wenmanni testor (Rothschild). Hopkins and Rothschild, 1962, An Illustrated Catalogue ... Vol. III, pp. 249-251.

This species, under one or the other subspecies, has been reported from New Brunswick to British Columbia, south to Virginia and Arizona. While it is never plentiful, it is not an uncommon parasite of small rodents, particularly *Peromyscus* species and their ecological associates.

There is some difference of opinion concerning the validity of the subspecies *testor*. Benton (1955) makes a case for its retention, based on a study of 544 specimens collected throughout North America, and notes at least a rough similarity between its range and the Transition Zone boundary of Merriam (1894). An equally valid hypothesis is that the species shows clinal variation from north to south. Both subspecies are retained here since both seem to occur in the area. However, the nominate taxon is represented by a single male (the diagnostic sex) in our collections and the validity of its distinctness is conjectural without more material. Traub (*in lit.*) indicates that individuals referable to both subspecies have been taken from the same host individual.

This species has been reported from a multitude of hosts, including a few birds, but there is little doubt that it is primarily a parasite of *Peromyscus* species throughout its range.

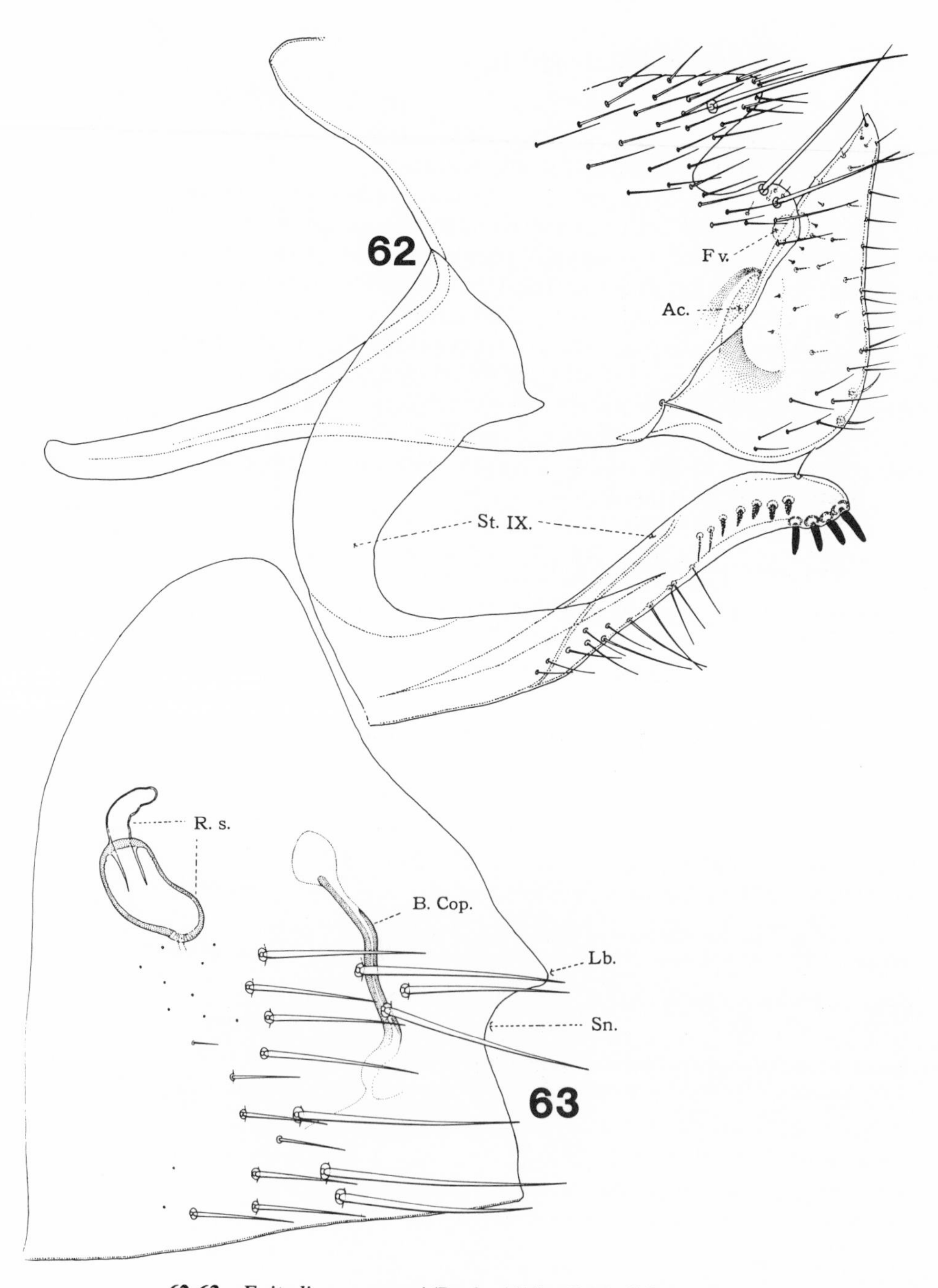

62-63 *Epitedia wenmanni* (Rothschild). **62** Modified abdominal segments of male. **63** Female sternum VII and spermatheca. After Holland, 1985.

GENUS **Catallagia** Rothschild

Catallagia. Rothschild, 1915, Ectoparasites I: 41 (type species: *C. charlottensis* Baker, 1898).

There are seventeen named taxa in this genus, most of which occur in North America. Of these, four species are known from the Pacific Northwest. Adults are primarily ectoparasites of small rodents, especially microtines, but occasionally they are taken on carnivores and insectivores as accidental associates. Females of the North American species are so similar to each other that a satisfactory key has not been devised. This problem is further compounded by the fact that the females of a given population tend to be rather variable in the characters usually used in taxonomic discrimination. As a result, the following key, adapted from Hopkins and Rothschild (1962), relies on characters found in the males. Females without accompanying males cannot be identified using it.

KEY TO THE WESTERN SPECIES OF *CATALLAGIA*

1 Distal arm of male st. IX with subparallel margins and three spiniforms near apex, the second being strongly bent ventrad .... **mathesoni**

Distal arm of male st. IX expanded subapically with three or four spiniforms near apex 2

2 Ventral margin of distal arm of male st. IX lacking a patch of bristles .... **charlottensis**

Ventral margin of distal arm of male st. IX with a patch of subapical bristles 3

3 With three spiniforms near apex of male st. IX, subtended by three or four setae, the dorsalmost of which is bent ventrad .... **decipiens**

With four spiniforms near apex of male st. IX, ventral margin of this segment with a distinct bulge bearing three to eight marginal setae .... **sculleni—4**

4 Bulge in ventral margin of male st. IX not pronounced, bearing only three to four bristles .... **sculleni rutherfordi**

Bulge in ventral margin of male st. IX more pronounced, bearing more than four bristles some of which are spiniform 5

5 Ventral bulge in margin of male st. IX well developed, several of its distal marginal bristles distinctly spiniform .... **sculleni chamberlini**

Ventral bulge in margin of male st. IX less pronounced, no distal bristles spiniform, though one or two may be subspiniform ...... **sculleni sculleni**

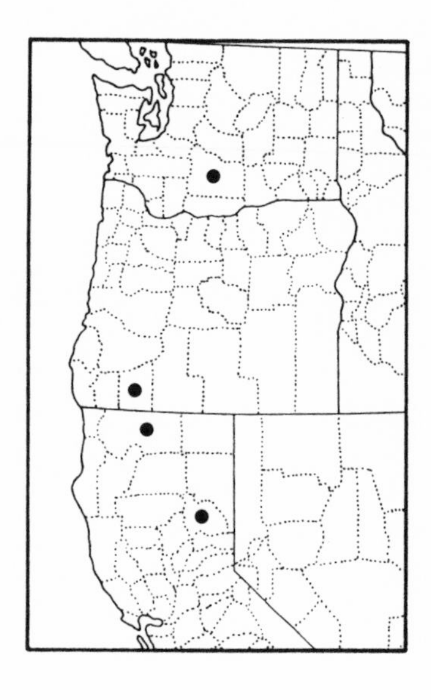

Catallagia mathesoni Jameson

[95072]

(Figs. 64, 65, 76)

Catallagia mathesoni. Jameson, 1950, J. Kansas ent. Soc. 23: 94, fig. I A-F (Quincy, Plumas Co., California, from *Peromyscus boylei*).

This species was described from specimens collected from *Peromyscus boylei* but the author did not designate a type host since additional material had been taken on other small rodents and insectivores. All known records from Oregon are from *P. truei* and it is evident that *Peromyscus* species are the preferred hosts for this flea. Only eight specimens are reported here, six males and two females, all taken from Jackson County, collected on 19-20 February 1972. However, there are literature records from California and Washington from a number of other hosts.

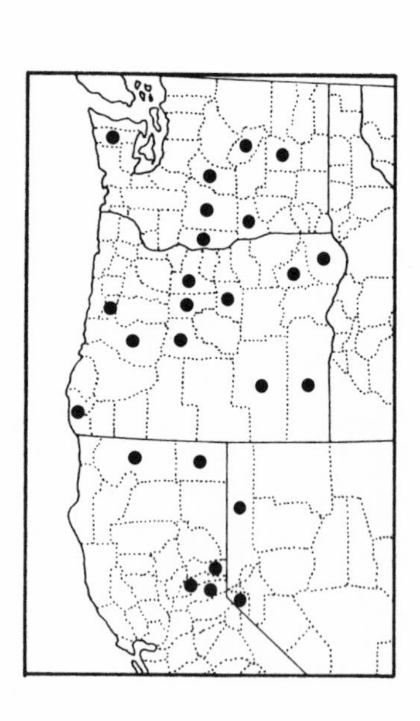

Catallagia decipiens Rothschild

[91530]

(Figs. 66, 67, 75)

Ceratophyllus charlottensis 'Baker'. Rothschild, 1905, (*nec* Baker, 1898), Novit. Zool. 12: 174, pl. 9, fig. 33.

Catallagia decipiens. Rothschild, 1915, Ectoparasites I: 43, figs. 45, 47 (Horse Creek, Upper Columbia Valley, British Columbia, from *Peromyscus*).

Catallagia moneris. Jordan, 1937, Novit. Zool. 40: 267, fig. 50 (Ravalli Co., Montana, from *Marmota flaviventris*).

Catallagia moneris Jordan. Hubbard, 1947, Fleas of Western North America, pp. 284, 285, 291 (said to be a synonym of *decipiens* but not formally synonymized).

Catallagia decipiens Rothschild. Hopkins and Rothschild, 1962, An Illustrated Catalogue ... Vol. III, pp. 271-273.

Catallagia decipiens. Hansen, 1964, Great Basin Naturalist 24: 80.

This is a widely distributed species, ranging from British Columbia and Alberta, south to Arizona, New Mexico, and California. Although it has been reported from a broad range of host animals, its preferred hosts are evidently species of *Peromyscus* and *Microtus*. A high percentage of the specimens taken in Oregon were from *Lagurus curtatus* but this may reflect the fact that nests of this host were also examined. As is commonly the case with many flea species, more specimens may be obtained from the nest of the host than from the host itself. The seasonal distribution of Oregon collections is shown below. Totals: 31 males, 48 females, total 79.

	J	F	M	A	M	J	J	A	S	O	N	D
Males	0	9	2	2	0	0	3	3	1	5	0	6
Females	0	17	2	0	2	4	2	7	4	4	1	5

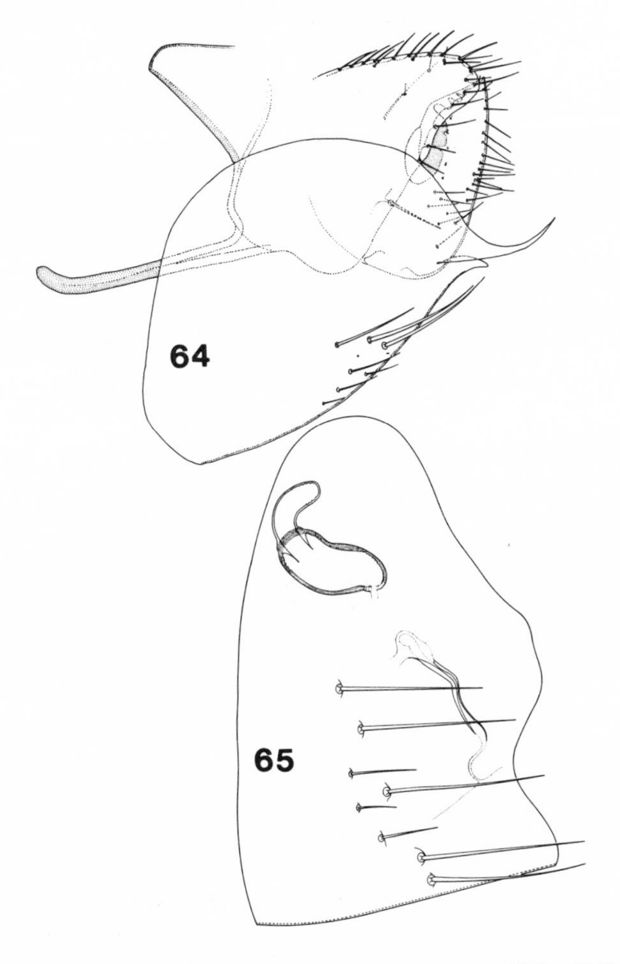

64-65 *Catallagia mathesoni* Jameson. **64** Male clasper and sternum VIII. **65** Female sternum VII and spermatheca. After Holland, 1985.

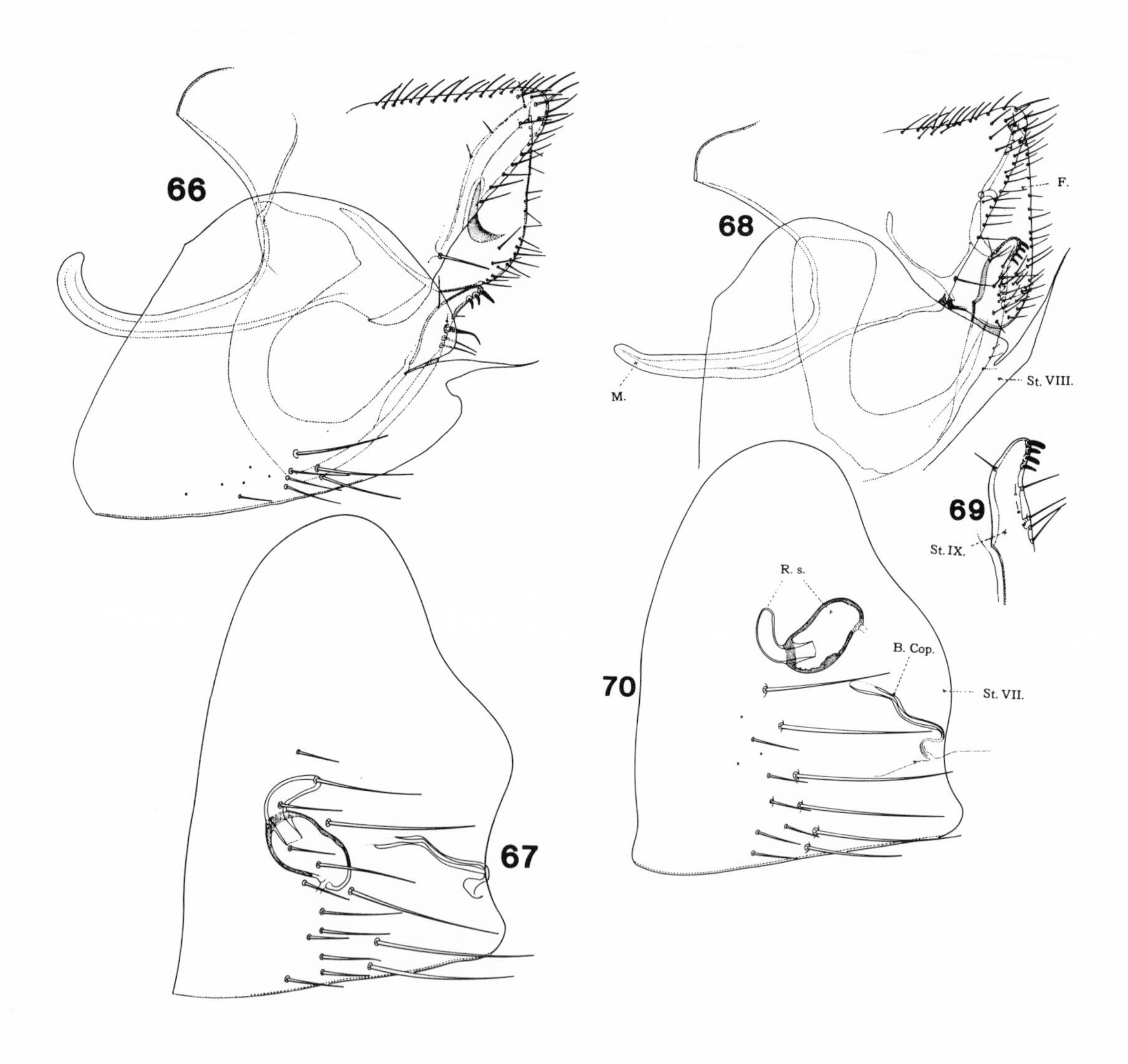

66-67 *Catallagia decipiens* Rothschild. **66** Modified abdominal segments of male. **67** Female sternum VII and spermatheca. After Holland, 1985.

68-70 *Catallagia charlottensis* (Baker). **68** Modified abdominal segments of male. **69** Detail of apex of male sternum IX. **70** Female sternum VII and spermatheca. After Holland, 1985.

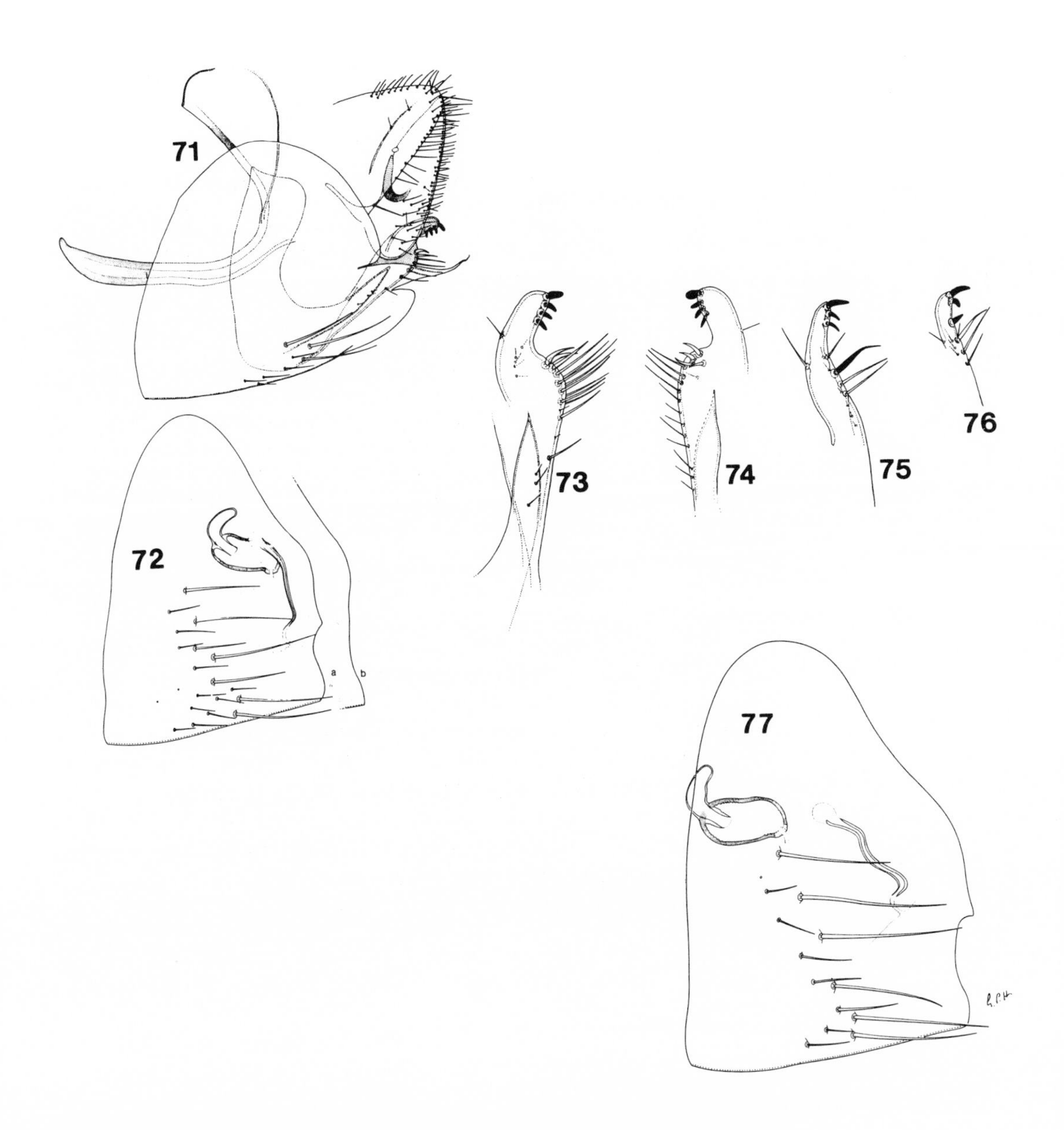

71-72 *Catallagia sculleni sculleni* Hubbard. **71** Modified abdominal segments of male. **72** Female sternum VII and spermatheca. After Holland, 1985.

73-77 *Catallagia* species. **73-76** Posterior and apex of male sternum IX. **73** *C. sculleni sculleni* Hubbard. **74** *C. sculleni chamberlini* Hubbard. **75** *C. decipiens* Rothschild. **76** *C. mathesoni* Jameson. **77** Female sternum VII and spermatheca of *C. sculleni chamberlini* Hubbard. After Holland, 1985.

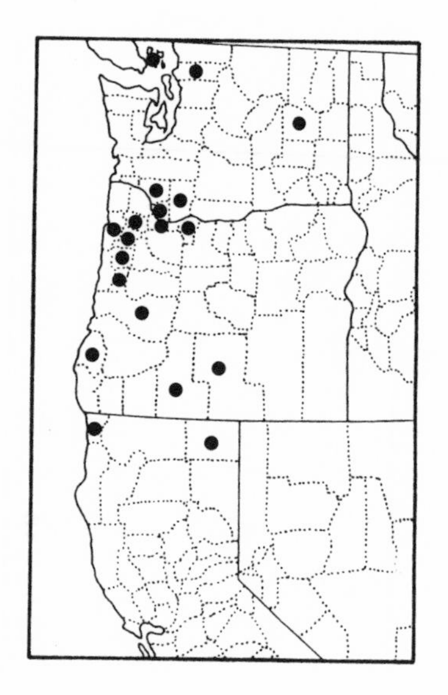

Catallagia charlottensis (Baker)

[89824]

(Figs. 68, 69, 70)

Typhlopsylla charlottensis. Baker, 1898, J. N. Y. ent. Soc. 6: 56 (Massett, Queen Charlotte Islands, from a mouse nest).

Ceratophyllus charlottensis Baker. Baker, 1904, Proc. U. S. Nat. Mus. 27: 386, 390, 441, pl. 12, figs. 6-10.

Odontopsyllus charlottensis Baker. Baker, 1905, Proc. U. S. Nat. Mus. 29: 131, 145.

Catallagia charlottensis Baker. Rothschild, 1915, Ectoparasites I: 43, figs. 44, 46.

Catallagia motei. Hubbard, 1940, Pacific Univ. Bull. 37: 4 (Banks, Oregon, from nest of *Microtus*).

Catallagia motei Hubbard. Hubbard, 1947, Fleas of Western North America, pp. 284, 290, fig. 164 (said to be a synonym of *charlottensis* but not formally synonymized).

Catallagia charlottensis Baker. Hopkins and Rothschild, 1962, An Illustrated Catalogue ... Vol. III, pp. 275-277.

Catallagia charlottensis is certainly not a conspicuous element in the flea fauna of the area according to our collection records. Hubbard (1947) states that it is an extremely abundant species during February and March in the nests of its preferred hosts, species of *Microtus*. It has a limited range from British Columbia to northern California, but is evidently restricted to localities west of the Cascade and Sierra Mountains. Hansen (1964) does not report this species. In addition to collections cited by Hubbard (1947), we report five males and five females from Benton and Lane counties of Oregon from January through March and August through October. Preferred hosts were *Microtus oregoni* and *M. canicaudus* with strays from *Scapanus townsendi* and *Eutamias townsendi.*

Catallagia sculleni sculleni Hubbard
[94030]
(Figs. 71, 72, 73)

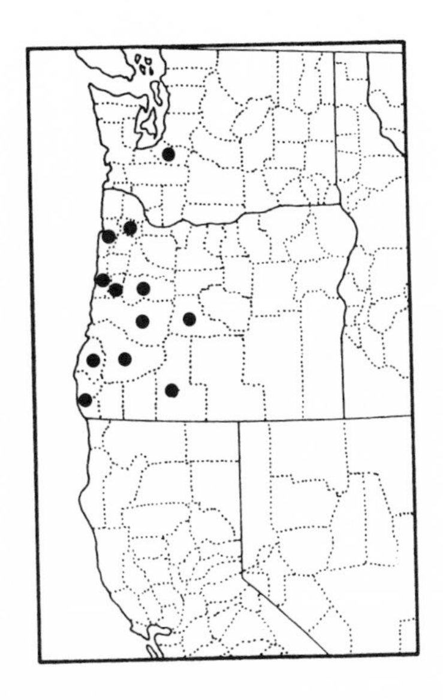

Catallagia sculleni. Hubbard, 1940, Pacific Univ. Bull. 37: 3 (Forest Grove, Oregon, from deer mouse).
Catallagia vonbloekeri. Augustson, 1941, Bull. S. Calif. Acad. Sci. 40: 103, pl. 6, figs. 1, 2 (Santa Rosa Island, Santa Barbara Co., California, from *Peromyscus maniculatus sanctarosae*).
Catallagia sculleni Hubbard. Hubbard, 1947, Fleas of Western North America, pp. 284, 297, fig. 161.
Catallagia vonbloekeri Augustson. Hubbard, 1947, *op, cit.*, pp. 284, 285, 292, fig. 168 (said to be a synonym of *sculleni* but not formally synonymized).
Catallagia sculleni sculleni Hubbard. Hopkins and Rothschild, 1962, An Illustrated Catalogue ... Vol. III, pp. 277-279.
Catallagia sculleni Hubbard. Gresbrink and Hopkins, 1982, Northwest Sci. 56: 178.

and

Catallagia sculleni chamberlini Hubbard
[94031]
(Figs. 74, 77)

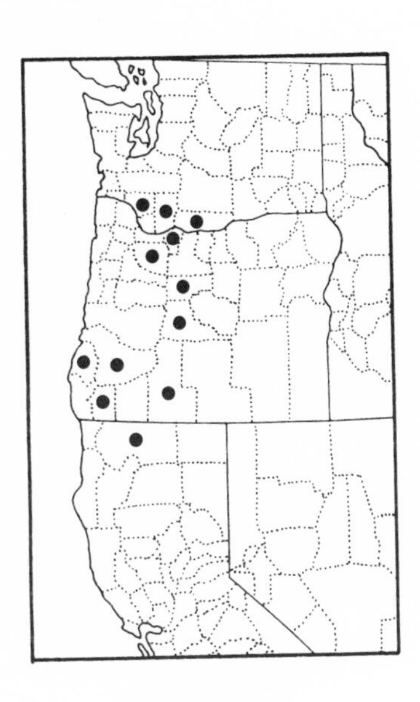

Catallagia chamberlini. Hubbard, 1940, Pacific Univ. Bull. 37: 4 (Rocky Point, north of Klamath Falls, Oregon, from deer mouse).
Catallagia chamberlini Hubbard. Hubbard, 1947, Fleas of Western North America, pp. 284, 285, 289, fig. 163.
Catallagia sculleni chamberlini Hubbard. Hopkins and Rothschild, 1962, An Illustrated Catalogue ... Vol. III, pp. 279-280.
Catallagia chamberlini Hubbard. Gresbrink and Hopkins, 1982, Northwest Sci. 56: 178.

and

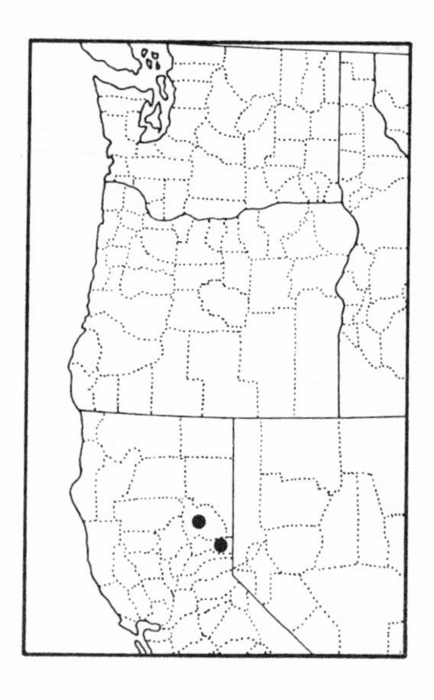

Catallagia sculleni rutherfordi Augustson

[94203]

Catallagia rutherfordi. Augustson, 1942, Bull. S. Calif. Acad. Sci. 40: 142, pl. 16, figs. 5, 6 (Tully's Hole, Fresno Co., California, from *Microtus montanus dutcheri*).

Catallagia rutherfordi Augustson. Hubbard, 1947, Fleas of Western North America, pp. 284, 285, 292 (synonymized with *sculleni*).

Catallagia sculleni rutherfordi Augustson. Hopkins and Rothschild, 1962, An Illustrated Catalogue ... Vol. III, p. 280.

The three taxa currently treated as subspecies of *C. sculleni* have a restricted range from the Sierra Nevada Mountains of California to the coastal belt and Cascade and Siskiyou Mountains of Oregon. The two Oregon taxa, *C. s. sculleni* and *C. s. chamberlini,* are very similar in both sexes and seem to show a clinal variation from the coastal lowlands in the nominate form to the mountains in *C. s. chamberlini.* The following table illustrates host associations for the two taxa.

	C. s. sculleni			*C. s. chamberlini*		
Host	M	F	%	M	F	%
Peromyscus	33	50	42.2	1	11	57.1
Microtus	17	16	17.0	0	6	28.5
Clethrionomys	20	12	16.4	0	1	4.7
Eutamias	4	7	5.6	0	0	0
Miscellaneous	14	21	18.0	0	2	9.5
Subtotals	88	106		1	20	

Hosts grouped in the miscellaneous category include one genus and species of marsupial, three genera and seven species of insectivores, two genera and species of carnivores and six genera and seven species of rodents. Percentages are rounded to the nearest decimal. With the two subspecies combined, 44.18 percent of collections were taken from *Peromyscus* compared to 55.82 percent taken from all other hosts combined.

With respect to seasonal occurrence, the following table shows collection data for these two subspecies. Since the subspecies are inseparable in the female, those of *C. s. chamberlini* were separated on the basis of collection locality.

Seasonal distribution of the subspecies of *C. sculleni*

Subspecies		J	F	M	A	M	J	J	A	S	O	N	D	Totals
sculleni	M	3	3	3	8	0	0	9	24	20	3	7	8	88
	F	3	4	7	11	2	2	7	20	21	6	7	16	106
chamberlini	M	0	0	0	0	1	0	0	0	0	0	0	0	1
	F	2	0	1	0	1	0	5	1	0	0	1	9	20
Totals		8	7	11	19	4	2	21	45	41	9	15	33	215

Catallagia s. rutherfordi might be expected to intergrade with *C. s. chamberlini* in northern California or southern Oregon. It is currently known from the Sierra Nevada Mountains of California where it is evidently a parasite of *Microtus montanus.* This vole also occurs throughout the eastern three quarters of Oregon in the appropriate habitat and a few of the *C. s. sculleni* records listed here were collected from it in Jefferson and Benton counties, Oregon. Since the males are distinct from the other two subspecies, there should be little chance of confusing them. We feel confident that *Catallagia sculleni rutherfordi* has not yet been collected north of the California/ Oregon state line. It is quite possible that this taxon should be elevated to the rank of a full species, but more collections are needed before this can be decided.

GENUS **Delotelis** Jordan

Delotelis. Jordan, 1937, Novit. Zool. 40: 267 (type species by designation: *Ceratophyllus telegoni* Roths., 1905).

This is another small genus of rodent fleas with a restricted distribution in the extreme northwestern United States and southwestern Canada. Their preferred hosts seem to be microtine rodents, but they are also occasionally taken on insectivores and nonmicrotine rodents. Their rarity suggests that they are probably nest fleas that may be more common as adults during the winter months. There are only two known species and both have been collected in the area under consideration.

KEY TO THE SPECIES OF *DELOTELIS*

1 Movable process of male bearing a dense clump of small bristles apically on inner surface, the two spiniform bristles on the dorsal portion of the distal arm of st. IX strongly bent and tapering gently to apex; female st. VII weakly convex apically with a shallow, subventral depression .... **telegoni**

Movable process of male lacking the dense clump of small bristles apically on the inner surface, the two spiniform bristles on the dorsal portion of the distal arm of st. IX not strongly bent, tapering abruptly near apex; female st. VII strongly convex near middle with well developed concavity ventrally .... **hollandi**

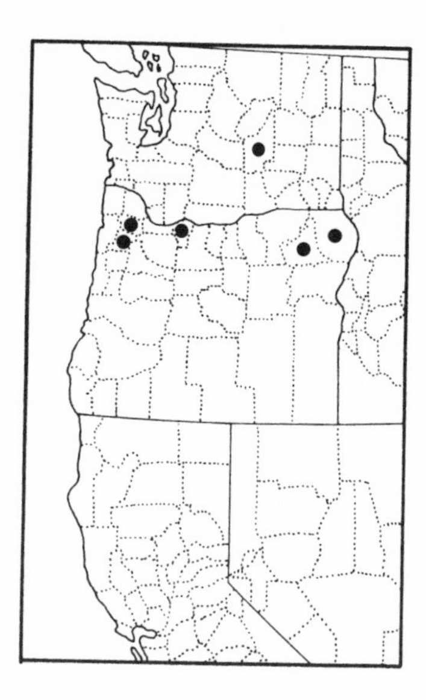

Delotelis telegoni (Rothschild)

[90516]

(Figs. 78, 80)

Ceratophyllus telegoni. Rothschild, 1905, Novit. Zool. 12: 172, pl. 9, figs. 27, 30 (Horse Creek, Upper Columbia Valley, British Columbia, from *Microtus drummondi*).
Odontopsyllus telegoni (Rothschild). Baker, 1905, Proc. U. S. Nat. Mus. 29: 131, 146.
Catallagia telegoni Roths. Rothschild, 1915, Ectoparasites I: 42-43.
Ceratophyllus telegoni Rothschild. Jordan, 1937, Novit. Zool. 40: 267 (designated type species of *Delotelis* Jordan).
Delotelis telegoni (Rothschild). Hubbard, 1947, Fleas of Western North America, p. 294 (*partim,* figures are of *D. hollandi*).
Delotelis telegoni (Rothschild). Holland, 1949, The Siphonaptera of Canada, p. 83 (*partim,* figures are of *D. hollandi*).
Delotelis telegoni (Rothschild). Smit, 1952, Proc. ent. Soc. Wash. 54: 269, figs. 2, 4, 6.

To date this species has only been taken from two counties in the extreme northeastern part of Oregon. Other records exist, but the failure to recognize two species in the genus until 1952 make earlier citings suspect. Both Hubbard (1947) and Holland (1949) figure *D. hollandi* while discussing *D. telegoni.* Here cited are four males and twelve females from the same locality in Union County, Oregon, from *Clethrionomys gapperi,* collected during October 1976. Two additional males collected in Wallowa County, Oregon, during March of the same year were taken from *Glaucomys sabrinus* and may represent an accidental association or an error in data transcription. An additional male and four females in the senior author's collection were taken from *C. gapperi* and *Microtus longicaudus* from Missoula County, Montana, on various dates by J. M. Kinsella, and Bacon (1953) cites a record from near Goose Lake, Grant County, Washington. These and many following records are doubtful. A number of collections of this species are also known from British Columbia. While this is the more eastern of the two species, Hopkins and Rothschild (1962) suggest the possibility that they may be sympatric. The Augustson record reported by Hubbard (1947) probably belongs to this taxon rather than to the following one.

Delotelis hollandi Smit

[95228]

(Figs. 79, 81)

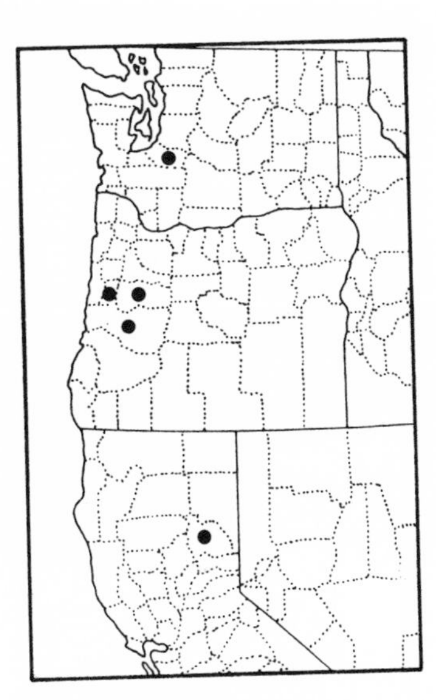

Delotelis 'telegoni (Rothschild).' Hubbard, 1947 (*nec* Rothschild, 1905), Fleas of Western North America, p. 294, fig. 170 (*partim*).

Delotelis 'telegoni (Rothschild).' Holland, 1949, The Siphonaptera of Canada, p. 83, pl. 12, figs. 68-70, map 8 (*partim*).

Delotelis hollandi. Smit, 1952, Proc. ent. Soc. Wash. 54: 269, figs. 1, 3, 5 (Gaston, Oregon, from *Microtus townsendi*).

Delotelis hollandi Smit, 1952. Hopkins and Rothschild, 1962, An Illustrated Catalogue ... Vol. III, pp. 283-284.

Oregon records for this species reported here are from Benton, Lane, and Linn counties, in the west-central part of the state. Of the fifteen specimens (six males, nine females) included, all but one female were collected in the H. J. Andrews Experimental Forest, as reported by Lewis and Maser (1981). This species is seldom collected for probably the same reasons as cited for the preceding taxon. Hopkins and Rothschild (1962) report additional specimens from Oregon, as well as collections from California and British Columbia. The species is also certain to occur in Washington. Holland (1985) lists numerous records from British Columbia, as well as from Alberta and Alaska. Hopla (1965) also notes Alaskan records. Evidently this and the preceding species are sympatric, at least in the north.

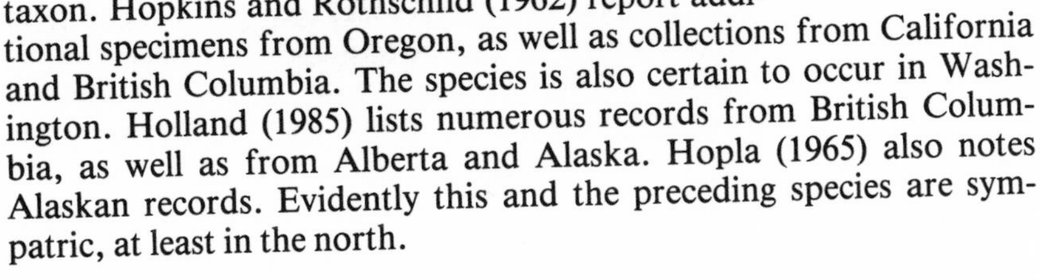

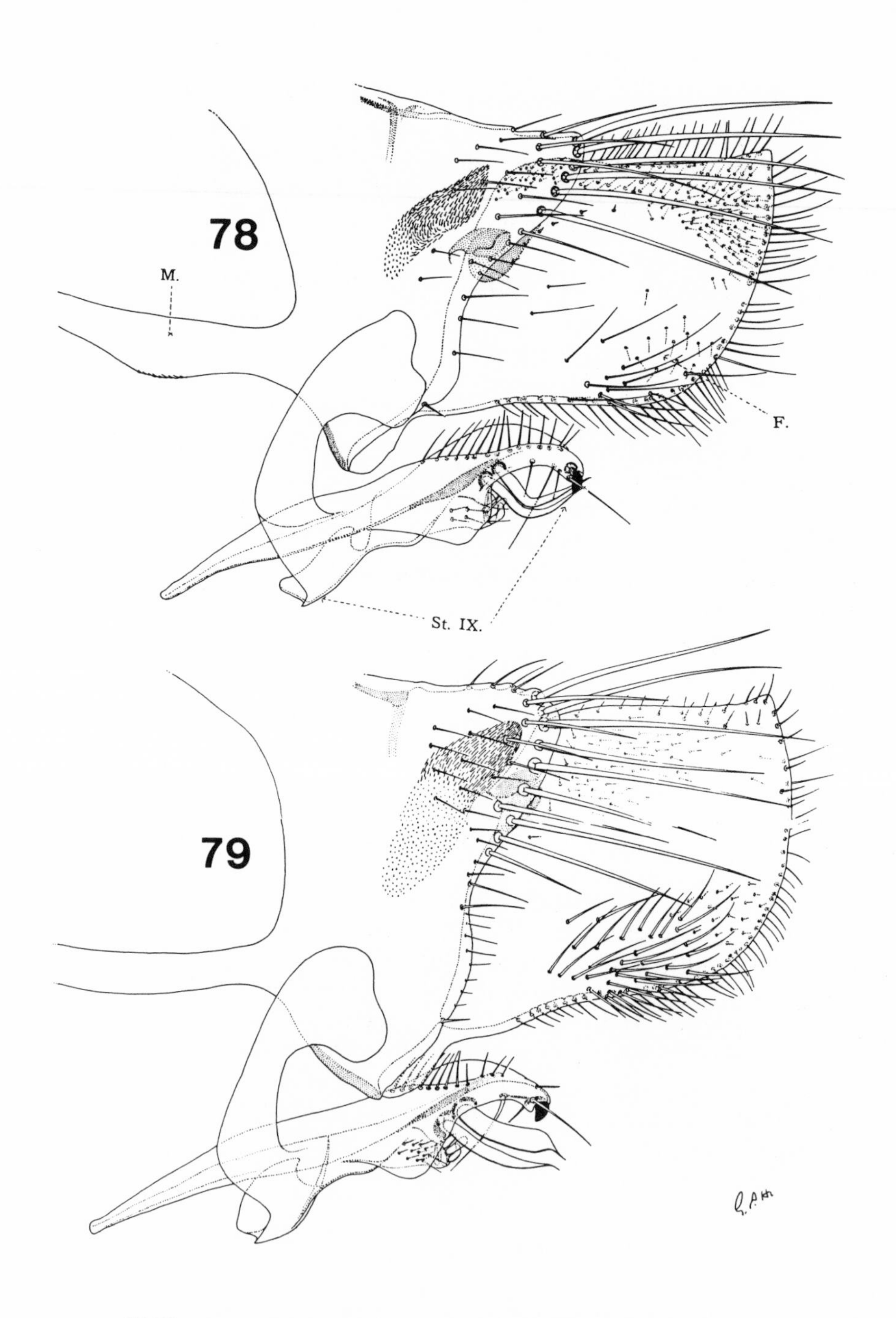

78-79 Male clasper and sternum IX in *Delotelis* species. **78** *D. telegoni* (Rothschild). **79** *D. hollandi* Smit. After Holland, 1985.

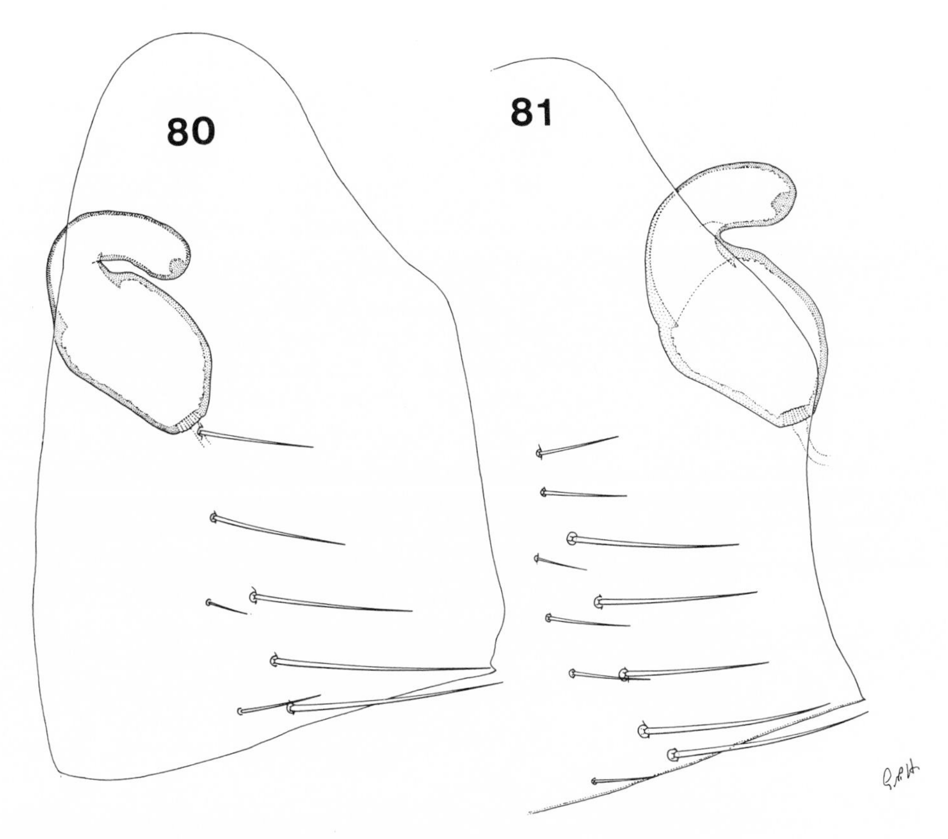

80-81 Female sternum VII and spermatheca in *Delotelis* species. **80** *D. telegoni* (Rothschild). **81** *D. hollandi* Smit. After Holland, 1985.

GENUS **Phalacropsylla** Rothschild

Phalacropsylla. Rothschild, 1915, Ectoparasites I: 39 (type species by designation: *P. paradisea* Rothschild, 1915).

The most recent review of this genus is that of Lewis and Maser (1978) which also provided a key to the six known species. These range from the northwestern United States to central Mexico. Although many collections have been made from various species of *Neotoma,* species of *Peromyscus* also seem to be good hosts, and a few collections have been made from *Ochotona.* The following species are reported from the area under consideration.

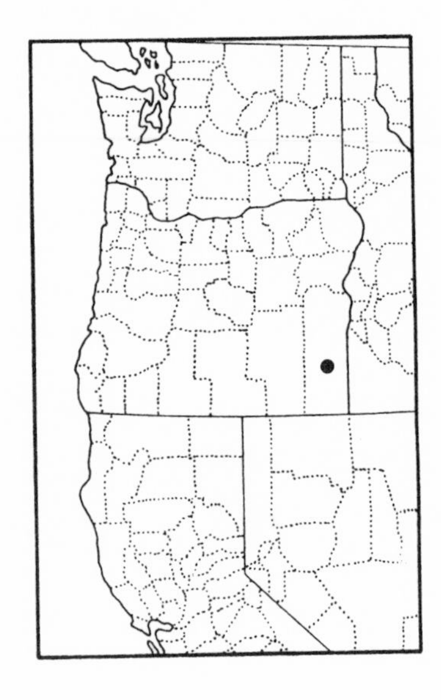

Phalacropsylla oregonensis Lewis and Maser

[97813]

(Figs. 82, 83)

Phalacropsylla oregonensis. Lewis and Maser, 1978, J. Parasitol. 64: 147-150, figs. 1, 2 (Succor Creek State Park, Malheur Co., Oregon, from *Neotoma lepida nevadensis, Peromyscus maniculatus sonoriensis,* and *P. crinitus crinitus*).

No further specimens have come to hand since the description of this species. As suggested in the description, this is evidently a nest flea and likely occurs in the adult stage mainly during the colder months. Those working with *Neotoma* species could contribute to our knowledge of these fleas by actively searching for ectoparasites in the nesting material of wood rats.

The specimen reported from Benton County, Washington, by O'Farrell (1975) probably belongs to this taxon.

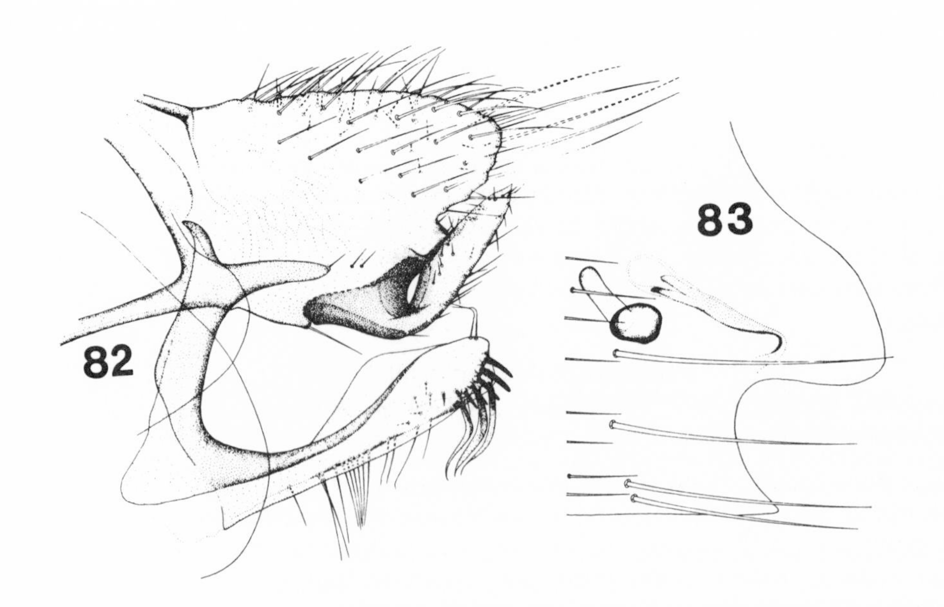

82-83 *Phalacropsylla oregonensis* Lewis and Maser. **82** Modified abdominal segments of male. **83** Female sternum VII and spermatheca. Original.

Phalacropsylla allos Wagner

[93637]

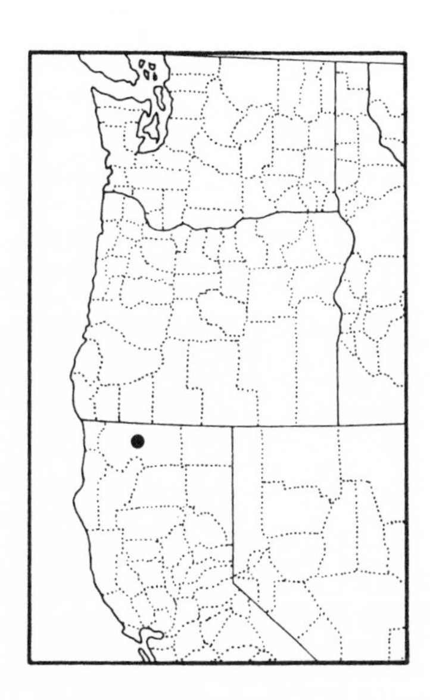

Phalacropsylla allos. Wagner, 1936, Z. Parasitenk. 8: 657 (Logan, Utah, from *Neotoma c. cinerea*).
Phalacropsylla monticola. Augustson, 1941, Bull. S. Calif. Acad. Sci. 40: 144 (Tully's Hole, Fresno Co., California, from *Ochotona schisticeps muiri).*
Phalacropsylla allos Wagner. Hubbard, 1947, Fleas of Western North America, pp. 338, 340.

Stark and Kinney (1969) report 66 specimens of this flea from the nests of *Neotoma cinerea* from the Lava Beds National Monument in northern California. An additional single specimen was taken on *Peromyscus maniculatus.* Hopkins and Rothschild (1962) list collections from Utah, Wyoming, and Montana, and Holland (1985) lists two records from British Columbia. The species has not yet been reported from Washington or Oregon, but possibly occurs in the eastern, more arid, portion of these states.

GENUS **Meringis** Jordan

Meringis. Jordan, 1937, Novit. Zool. 40: 332 (type species by designation: *M. parkeri* Jordan, 1937).
Atheropsylla. Stewart, 1940, Pan-Pacific Ent. 16: 18 (type species by designation: *A. bakeri* Stewart, 1940 [= *M. cummingi* (C. Fox, 1926)]).

Of the eighteen currently recognized species assigned to this genus, five are known to occur in the area under consideration. Three of these are mainly associated with *Dipodomys* species, while the other two parasitize *Perognathus* species, as well as other small rodents.

KEY TO THE NORTHWESTERN SPECIES OF *MERINGIS*

1 Distal arm of male st. IX not divided into dorsal and ventral lobes, its apex with two heavy spiniform bristles, with three antepygidial bristles per side; margin of female st. VII undulate-convex or with a deep incision **2**

Distal arm of male st. IX divided into dorsal and ventral lobes, each usually bearing a heavy spiniform seta, with two or three antepygidial bristles per side; margin of female st. VII straight or lobed **3**

2 Movable process of male clavate, about half as wide as long at widest point, apex of st. IX tapered to a sharp point; caudal margin of female st. VII with deep incision .... **dipodomys**

Movable process of male digitate, no more than one-third as wide as long at widest point, apex of st. IX bluntly rounded; caudal margin of female st. VII shallowly convex, subtended by a shallow ventral depression .... **cummingi**

3 Males and females both with three antepygidial bristles per side; apical margin of female st. VII straight or almost so .... **parkeri**

Males with only two antepygidial bristles per side; apical margin of female st. VII not straight **4**

4 Subapical spiniform of dorsal lobe of male st. IX subtended by two subspiniform bristles, movable process pyriform; apical margin of female st. VII relatively straight above and below dorsal lobe .... **shannoni**

Bristles subtending subapical spiniform not subspiniform, movable process clavate; apical margin of female st. VII as in *shannoni* .... **hubbardi**

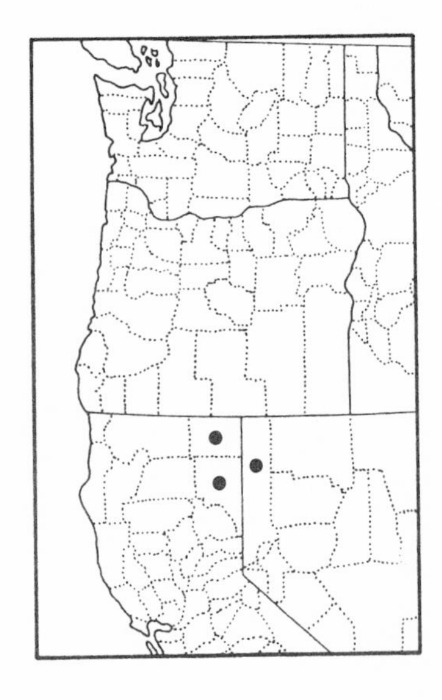

Meringis dipodomys Kohls

[93808]

Meringis dipodomys. Kohls, 1938, Publ. Hlth. Rep., Wash. 53: 1219 (Imperial Co., California, from *Dipodomys* sp.).

As the name implies, this is typically a parasite of kangaroo rats and the northwestern part of its range extends into northern California and western Nevada. It is also known from Utah and Arizona, and it probably extends into northern Mexico.

Meringis cummingi (C. Fox)

[92614]

(Figs. 84, 85)

Phalacropsylla cummingi. C. Fox, 1926, Pan-Pacific Ent. 2: 182, figs. 1, 2 (Los Angeles, California, from *Dipodomys agilis*).
Meringis cummingi Fox. Jordan, 1937, Novit. Zool. 40: 268.
Atheropsylla bakeri. Stewart, 1940, Pan-Pacific Ent. 16: 19, figs. 4-6 (near Jamesburg, Monterey Co., California, from *Dipodomys venustus* ssp.).
Meringis cummingi C. Fox, 1926. Hubbard, 1947, Fleas of Western North America, pp. 317, 321, fig. 190 (*A. bakeri* synonymized).

No additional specimens of this species have, to our knowledge, been reported from Oregon since those listed in Hubbard (1947). According to Hall and Kelson (1959), the range of *Dipodomys heermanni californicus* includes only portions of Jackson and Klamath counties in Oregon and it is evident that the range of this flea is quite restricted in the state. Hubbard (1949, 1961) and Stark and Kinney (1969) report the species from California and Nevada from *D. heermanni* and *D. microps.* Additional studies should be conducted in southern Oregon to determine the actual range of this species.

Meringis hubbardi Kohls

[93807]

(Figs. 86, 87)

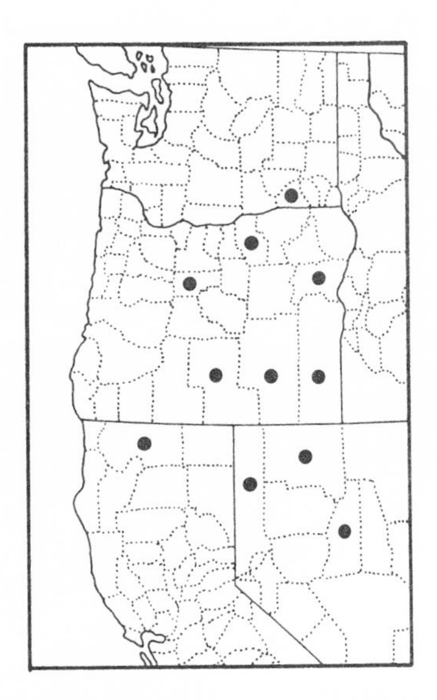

Meringis hubbardi. Kohls, 1938, Publ. Hlth. Rep., Wash. 53: 1217, figs. 1-3 (Mayfield, Idaho, from *Sylvilagus* sp.).
Meringis walkeri. Hubbard, 1940, Pacific Univ. Bull. 37(5): 2 (Central Oregon desert, 15 miles S. of Boardman, from *Perognathus parvus parvus*).
Meringis jewetti. Hubbard, 1940, *op. cit.*, p. 3 (Baker, Oregon, from *Peroganthus parvus parvus*).
Meringis hubbardi Kohls. Smit, 1953, Ent. Ber., Amst. 14: 397, figs. 7, 10, 11 (*walkeri & jewetti* synonymized).
Meringis hubbardi. Hansen, 1964, Great Basin Nat. 24: 81.

Hubbard (1947) states that this species is relatively scarce, and our collection data would seem to confirm this. However, Hansen (1964) reported 85 specimens from his collections in the Steens

Mountain area of Harney County, Oregon. While Hopkins and Rothschild (1962) state that *Perognathus* species are the preferred hosts, we can find no host preference for this species in Hubbard, Hansen, or our own data. About all that can be said is that small rodents in general are preferred over other types of hosts. The 25 specimens reported here were collected in Jefferson and Malheur counties, Oregon, from *Spermophilus townsendi, Thomomys talpoides, Peroganthus parvus, Peromyscus maniculatus, Onychomys leucogaster, Microtus montanus* and *Lagurus curtatus.* Seasonal occurrence is shown below.

	J	F	M	A	M	J	J	A	S	O	N	D
Males	0	0	3	0	0	0	0	2	3	1	0	3
Females	1	0	2	0	0	0	0	4	1	1	0	4

The known range of this species includes northern California, Oregon, Washington, Nevada, and Idaho.

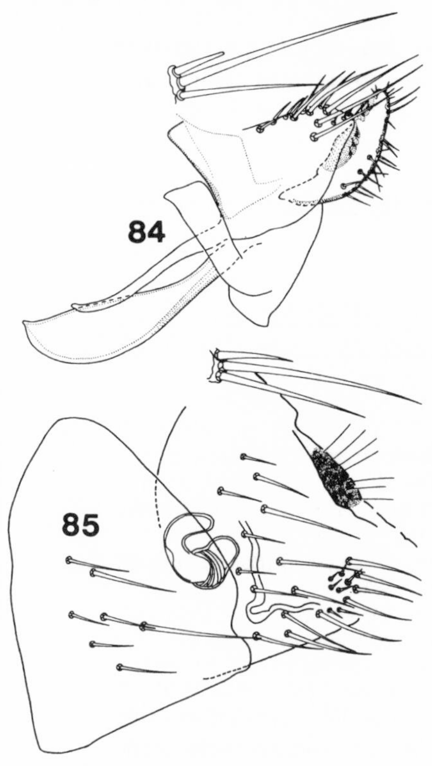

84-85 *Meringis cummingi* (C. Fox). **84** Male clasper. **85** Modified abdominal segments of female. Original.

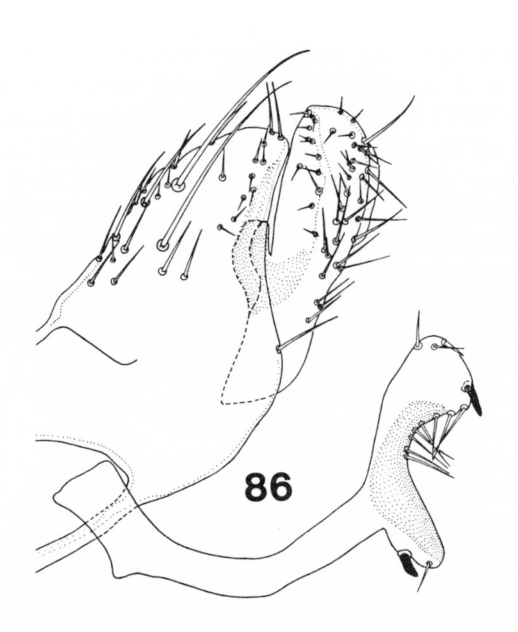

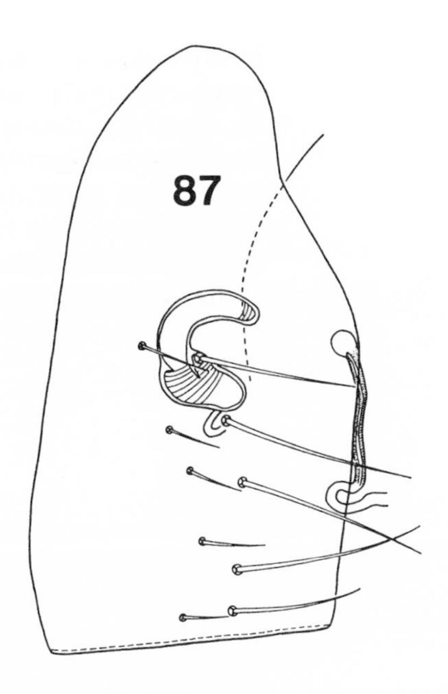

86-87 *Meringis hubbardi* Kohls. **86** Male clasper and sternum IX. **87** Female sternum VII and spermatheca. Original.

Meringis parkeri Jordan

[93710]

(Figs. 88, 89)

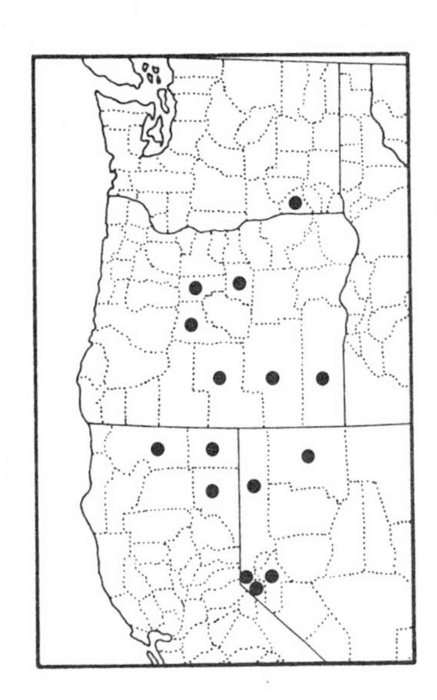

Meringis parkeri. Jordan, 1937, Novit. Zool. 40: 269, figs. 54, 55 (Powderville, Powder River Co., Montana, from *Dipodomys* sp.).

Meringis parkeri Jordan. Hubbard, 1940, Pacific Univ. Bull. 37: 4.

Meringis parkeri Jordan. Hubbard, 1947, Fleas of Western North America, pp. 318, 323, fig. 193.

Meringis walkeri Hubbard. Hubbard, 1947, (*nec* 1940) *op cit.,* pp. 318, 322, fig. 191 (*partim*).

Meringis parkeri Jordan. Smit, 1953, Ent. Ber., Amst. 14: 397, figs. 9, 12 (*walkeri* Hubbard, 1947, *nec* 1940, considered equivalent).

Meringis parkeri. Hansen, 1964, Great Basin Nat. 24: 81.

This is certainly the commonest member of the genus in Oregon. Its preferred host would seem to be *Dipodomys ordi* and 68 percent of the 47 specimens reported here were taken from this kangaroo rat. It is also a common parasite of

Perognathus sp. and sometimes is taken on *Peromyscus maniculatus*. Hubbard (1947) indicated that adults of this species could be collected at any time of the year. Known range of the species includes Washington south to California and east to Nevada, Utah, and Idaho. Jellison and Senger (1973) also list collections from Powder River, Custer, and Bighorn counties of Montana. Part of these made up the original type series and were collected by the late R. R. Parker in the extreme southeastern part of the state.

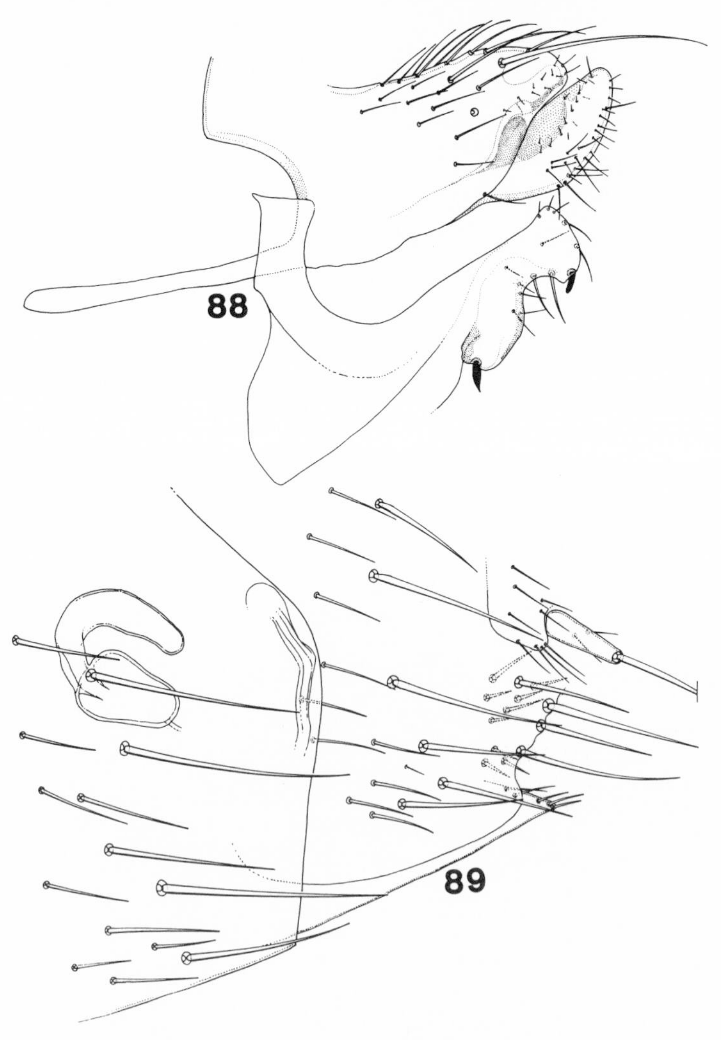

88-89 *Meringis parkeri* Jordan. **88** Male clasper and sternum IX. **89** Modified abdominal segments of female. After Holland, 1985.

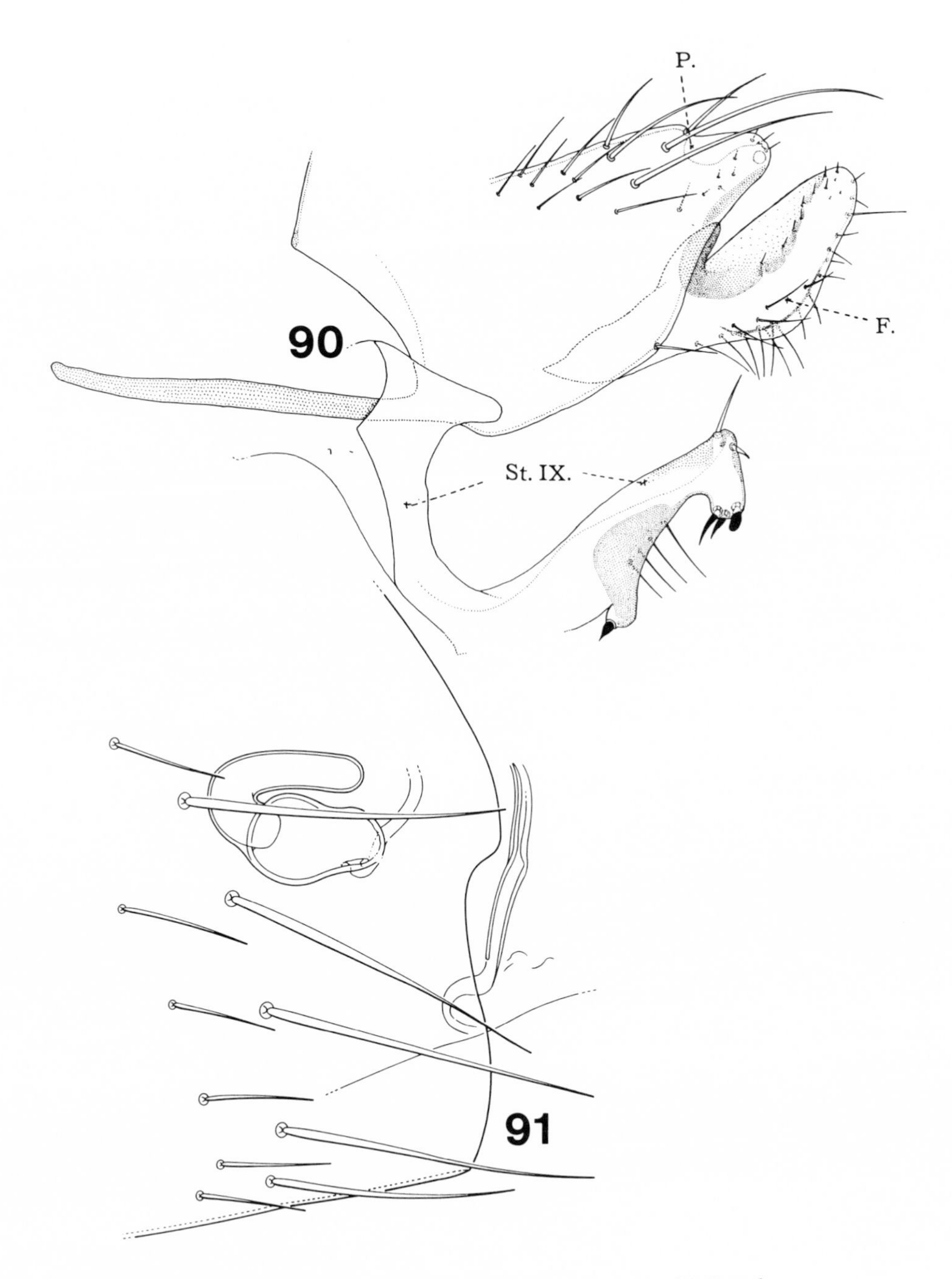

90-91 *Meringis shannoni* (Jordan). **90** Male clasper and sternum IX. **91** Female sternum VII and spermatheca. After Holland, 1985.

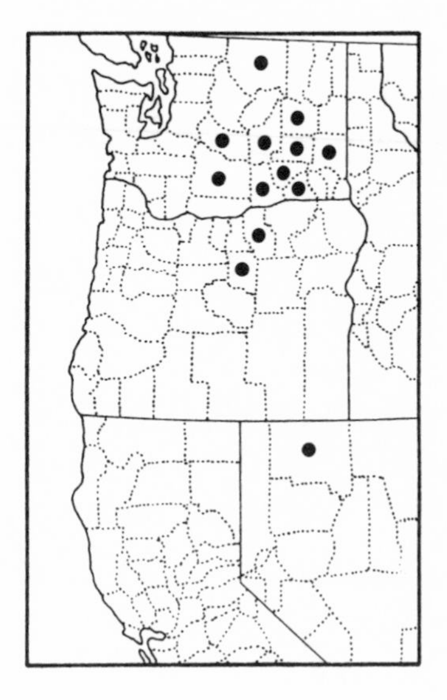

Meringis shannoni (Jordan)

[92920]

(Figs. 90, 91)

Phalacropsylla shannoni. Jordan, 1929, Novit. Zool. 35: 38, pl. 2, figs. 28, 29 (Ritzville, Washington, from field mice).
Meringis shannoni Jordan. Jordan, 1937, Novit. Zool. 40: 269.

Although this species has not been collected during this study, it is included here since Hubbard (1947) cites a number of collections from Washington and Oregon. He states that *M. shannoni* is mainly found on pocket mice (*Perognathus* sp.), their probable predators, and grasshopper mice (*Onychomys* sp.) but that they may also be taken as strays on a number of other small desert mice. Lewis (1972) reported this species from *Lagurus curtatus* collected in the Crooked River National Grassland in Jefferson County, Oregon. This was based on a single female that looked more like this than related species. No males that might confirm this determination have subsequently been taken. Known distribution for the species includes southern British Columbia, Washington, and Oregon.

SUBFAMILY **Anomiopsyllinae**

This subfamily has recently received attention from Tipton et al. (1979) and Barnes et al. (1977). With respect to the North American genera, excluding *Eopsylla* Argyropulo, 1946, and *Wagnerina* Ioff and Argyropulo, 1934, both of which are Palaearctic in distribution, this subfamily contains five genera that are found exclusively in the western Nearctic Region. One of these, *Conorhinopsylla* Stewart, 1930, has not been collected in the Pacific Northwest. Males of the remaining four genera may be separated using the following modification of the key by Tipton et al. (1979).

KEY TO THE NORTHWESTERN GENERA OF *ANOMIOPSYLLINAE*

1 Pronotal comb absent; chaetotaxy greatly reduced; caudoventral margin of coxae II and III with an acuminate spur .... **Anomiopsyllus**

Pronotal comb present; chaetotaxy variable but conspicuous; eye vestigial or absent; caudoventral margin of coxae II and III without spur **2**

2 Eye vestigial but present; st. VIII covering more than half the distal arm of st. IX; apex of movable process of clasper level with or extending only slightly beyond apex of fixed process .... **Megarthroglossus**

Eye absent; st. VIII covering less than half of distal arm of st. IX; apex of movable process of clasper extending well beyond apex of fixed process **3**

3 Frons evenly convex; dorsal margin of aedeagal apodeme not extending into a long coiled rod; preantennal area and abdominal sterna lacking enlarged bristles; mesotibia without dorsolateral comb ...... **Callistopsyllus**

Frons subacuminate; dorsal margin of aedeagal apodeme extending into a long, coiled rod; preantennal area and abdominal sterna possessing slightly or greatly enlarged bristles; mesotibia with dorsolateral comb .... **Stenistomera**

GENUS **Stenistomera** Rothschild

Stenistomera. Rothschild, 1915, Novit. Zool. 22: 307 (type species by designation: *S. alpina* Baker [1895, as *Typhlopsylla*]).

Stenistomera (Stenistomera). Good, 1942, Proc. ent. Soc. Wash. 44: 133 (declaration of nominate subgenus for *S. alpina* (Baker, 1895)).

Stenistomera (Miochaeta). Good, 1942, *op. cit.* (erection of new subgenus for *S. macrodactyla* Good, 1942).

Miochaeta Good. Stark, 1958, The Siphonaptera of Utah, U.S.D.H.E.W., Publ. Hlth. Ser., CDC, p. 103 (elevation of *Miochaeta* to full generic status).

Stenistomera Rothschild. Egoscue, 1968, Bull. S. Calif. Acad. Sci. 6: 138 (sinks *Stenistomera* and *Miochaeta* as subgenera).

Stenistomera Rothschild. Tipton et al., 1979, Great Basin Nat. 39: 403.

The most recent treatment of this genus is that of Tipton et al. (1979). They review the status of the species, host preferences and distribution, and provide a key to the males only. Following is a modification of their key, incorporating characters whereby females may be determined with a fair degree of accuracy.

KEY TO THE SPECIES OF *STENISTOMERA*

1 Frons with five rows of spiniform bristles in both sexes; distal arm of male st. IX truncated, with one subapical spiniform bristle; st. VII of female either straight or almost imperceptibly concave .... **alpina**

Frons with four rows of bristles which are unmodified or only slightly spiniform in both sexes; male and female modified segments not as above **2**

2 Frons with four rows of unmodified bristles in both sexes; spiniforms on posterior margin of male movable process widely separated; distal arm of male st.IX subacuminate with one subapical spiniform; st. VII of female with smoothly rounded lobe subtended by a shallow ventral sinus .... **macrodactyla**

Frons with at least anterior row of bristles subspiniform in both sexes; spiniforms on posterior margin of male movable process approximate; distal arm of male st. IX subacuminate, with two subapical spiniforms; st. VII of female with distinct, straight-edged dorsal lobe subtended by a deep sinus .... **hubbardi**

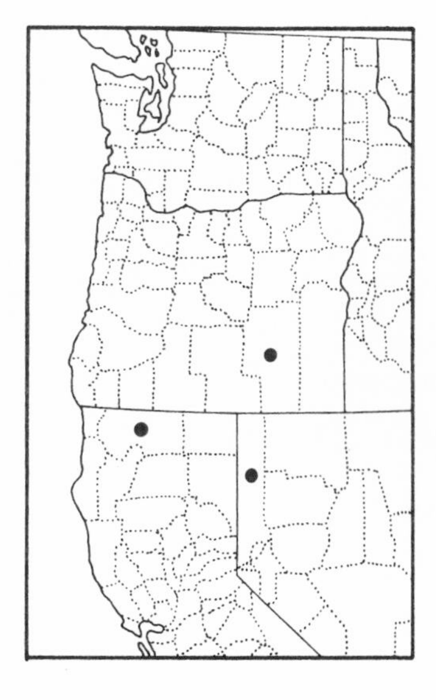

Stenistomera alpina (Baker)

[89513]

(Figs. 92, 93)

Typhlopsylla alpina. Baker, 1895, Canad. Ent. 27: 189, 191 (Georgetown, Colorado, from "mountain rat").

Ctenopsyllus alpinus Baker. Baker, 1904, Proc. U. S. Nat. Mus. 27: 427, 452 (host *Neotoma* sp.).

Stenistomera alpina Baker. Rothschild, 1915, Novit. Zool. 22: 307.

Delotelis mohavensis. Augustson, 1942, Bull. S. Calif. Acad. Sci. 40: 138, figs. 1-3 (Mojave, Kern Co., California, from *Neotoma lepida lepida*).

Stenistomera (Stenistomera) alpina (Baker). Good, 1942, Proc. ent. Soc. Wash. 44: 133, figs. 1-3.

Stenistomera alpina Baker. Hubbard, 1947, Fleas of Western North America, p. 305, fig. 181 (*mohavensis* synonymized).

Stenistomera (Stenistomera) alpina (Baker). Hopkins and Rothschild, 1962, An Illustrated Catalogue... Vol. III, pp. 350-353.

Stenistomera alpina (Baker). Tipton et al., 1979, Great Basin Nat. 39: 408-410, figs. 30, 38, 102, 103, 106, 111, 112, 115.

This is by far the commonest species of this genus. It has a broad range from Arizona, New Mexico, and California north to Montana, Idaho, and British Columbia. It appears to be primarily associated with species of *Neotoma* but is not uncommon on *Peromyscus* species also. The latter association doubtless accounts for the frequent predator records. Tipton et al. (1979) reported that over 90 percent of their specimens were taken from *N. lepida* and that most of them were collected during the winter months. Although this species has been associated with plague transmission, its lack of catholicity with respect to hosts certainly must mitigate its efficiency as a vector.

This report includes a single male taken from *N. cinerea* collected in Harney County, Oregon, in August of 1974, but the species is doubtless much more common than this single record suggests.

Stenistomera hubbardi Egoscue

[96818]

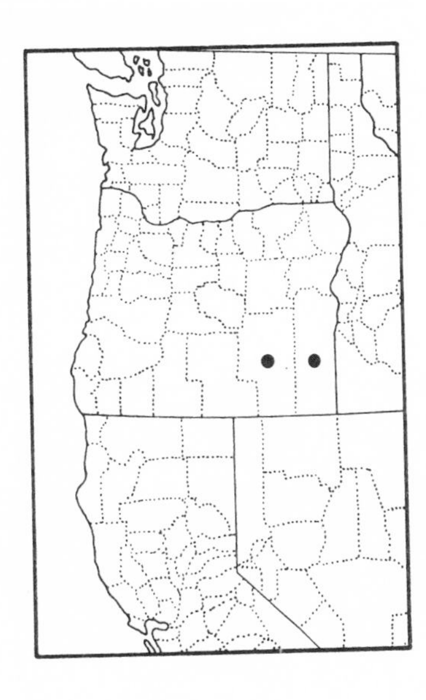

Stenistomera hubbardi. Egoscue, 1968, Bull. S. Calif. Acad. Sci. 67: 138-142 (5 mi. S. of Crane, Harney Co., Oregon, from *Peromyscus maniculatus).*

Stenistomera hubbardi Egoscue. Tipton et al., 1979, Great Basin Nat. 39: 406, figs. 28, 105, 108, 109, 114, 115.

Until now this species has been known only from the type series, consisting of one male, one female, and a broken, unsexed specimen. Through the courtesy of Dr. Egoscue, I have been able to examine these specimens and compare them with collections from other localities in Oregon. Two females from *P. maniculatus* collected 3 miles south of Vale, Malheur County (about 75 miles northeast of the type locality), on 21 March 1975 are identical to the female allotype in the configuration of the spermatheca and st. VII, as well as in their chaetotaxy. An additional male and two females taken 35 miles south of this locality 2 months later are clearly *S. macrodactyla*.

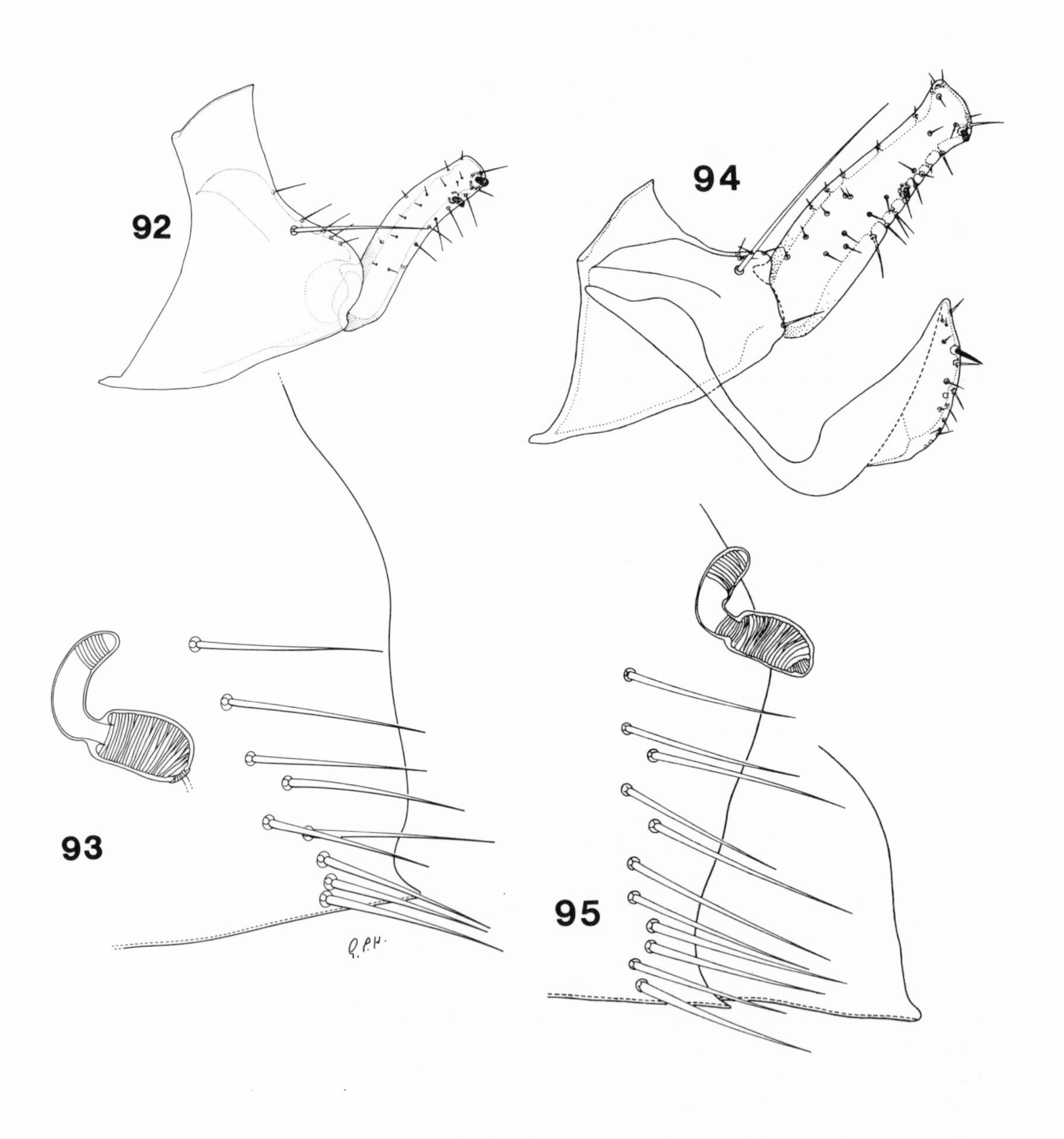

92-93 *Stenistomera alpina* (Baker). **92** Male clasper. **93** Female sternum VII and spermatheca. After Holland, 1985.
94-95 *Stenistomera macrodactyla* Good. **94** Male clasper and sternum IX. **95** Female sternum VII and spermatheca. Original.

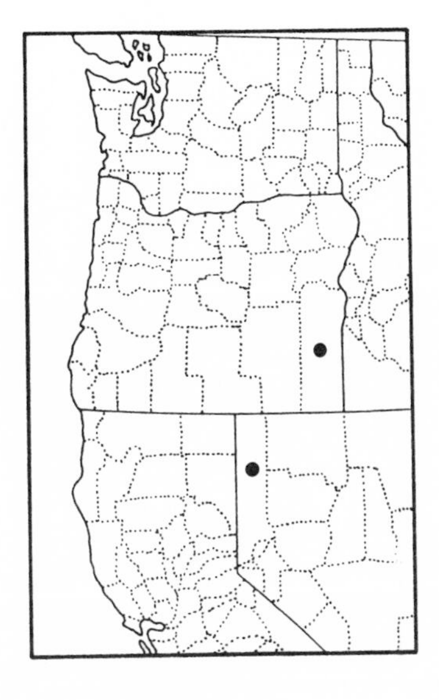

Stenistomera macrodactyla Good

[94227]

(Figs. 94, 95)

Stenistomera (Miochaeta) macrodactyla. Good, 1942, Proc. ent. Soc. Wash. 44: 135, figs. 4, 5 (Mojave Co., Arizona, 21 miles S. of St. George, Utah, from *Peromyscus eremicus*).
Stenistomera macrodactyla Good. Hubbard, 1947, Fleas of Western North America, p. 306, fig. 182.
Miochaeta macrodactyla (Good). Stark, 1958, The Siphonaptera of Utah, U.S.D.H.E.W., Publ. Hlth. Ser., CDC, p. 103.
Stenistomera (Miochaeta) macrodactyla Good. Hopkins and Rothschild, 1962, An Illustrated Catalogue... Vol. III, pp.353-354.
Stenistomera macrodactyla Good. Tipton et al., 1979, Great Basin Nat. 39: 406-407, figs. 29, 104, 107, 110, 113, 115.

This is also an uncommon species among the collections reported here, being represented by only the three specimens mentioned under the account of *S. hubbardi.* These specimens were collected on *Peromyscus crinitus.* Material mentioned by Tipton et al. (1979) came mainly from *P. maniculatus,* but they do report one pair from *P. crinitus* as well. The species is broadly distributed from Oregon, Idaho, and Wyoming south through Utah and Colorado to Nevada and New Mexico. As is true of the other two species, it is most frequently encountered in the winter.

GENUS **Callistopsyllus** Jordan and Rothschild

Callistopsyllus. Jordan and Rothschild, 1915, Ectoparasites I: 46 (type species by designation: *C. terinus* Roths. [1905, as *Ceratophyllus*]).
Callistopsyllus Jordan and Rothschild. Tipton et al., 1979, Great Basin Nat. 39: 353-354.

Again, the most recent treatment of this genus is that of Tipton et al. (1979). These authors reduced the three recognized species to subspecific status and made a case for relegating all named taxa to the nominate species. Traub (*in lit.*) questioned that a "nest flea" could have such a broad distribution, but Tipton et al. (1979) state that individual variation within isolated populations more than transcends the characters by which the currently recognized taxa are separated. While there are numerous records from various small rodents and squirrels, species of *Peromyscus* seem to be the preferred hosts.

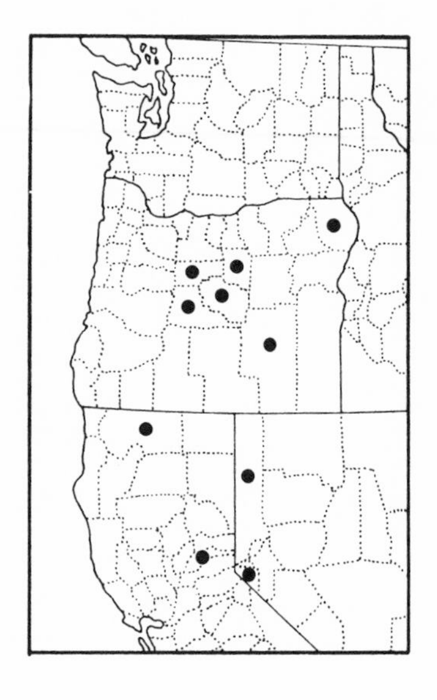

Callistopsyllus terinus terinus (Rothschild)

[90504]

(Figs. 96, 97)

Ceratophyllus terinus. Rothschild, 1905, Novit. Zool. 12: 158, pl. 8, fig. 26, pl. 9, fig. 29 (Mable Lake, British Columbia, from *Spermophilus columbianus*).

Callistopsyllus terinus Rothschild. Jordan and Rothschild, 1915, Ectoparasites I: 46.

Callistopsyllus paraterinus. Wagner, 1940, Z. Parasitenk. 11: 465, figs. 5, 6 (North Fork of Eagle River, British Columbia, from *Peromyscus maniculatus*).

Callistopsyllus terinus (Rothschild). Holland, 1949, The Siphonaptera of Canada, p. 97, pl. 17, figs. 116-119, map 14 (*paraterinus* synonymized).

Callistopsyllus terinus (Rothschild). Hopkins and Rothschild, 1962, An Illustrated Catalogue... Vol. III, pp. 355-357.

Callistopsyllus terinus (Fox). Tipton et al., 1979, Great Basin Nat. 39: 354-355.

Callistopsyllus terinus terinus Rothschild. Tipton et al., 1979, *op. cit.*, 356-359, figs. 4, 7, 10, 13, 14, 99.

Callistopsyllus terinus deuterus Jordan. Tipton et al., 1979, *op. cit.*, 359, figs. 1, 3, 6, 8, 11, 14, 36, 97.

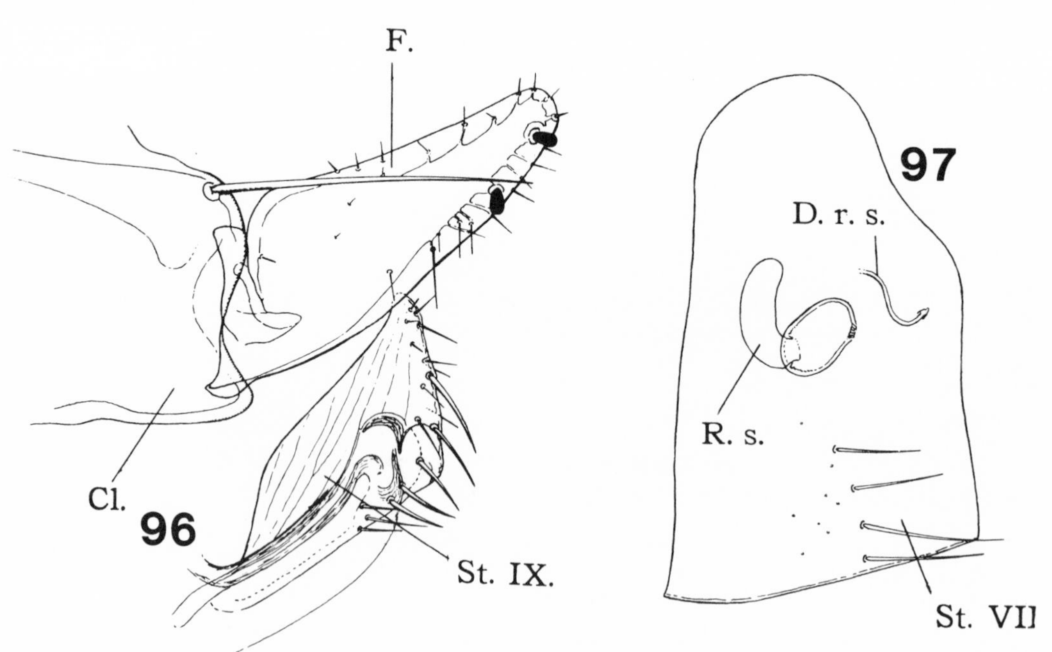

96-97 *Callistopsyllus terinus terinus* (Rothschild). **96** Male clasper and apex of sternum IX. **97** Female sternum VII and spermatheca. After Holland, 1985.

Under the current treatment of this genus a single species is represented containing three subspecies. Its range extends from British Columbia, Idaho, and Montana south to California, Arizona, and New Mexico. The easternmost published record for the species is from Custer County, in extreme southwestern South Dakota. The fact that this is a nest flea probably accounts for its rarity in collections. Tipton et al. (1979) list the nominate subspecies from Deschutes, Harney, and Jefferson counties in Oregon. They also show one record of *C. t. deuterus* from Jefferson County, Oregon. New records reported here are from Crook, Wallowa, and Malheur counties of Oregon and involve only three pairs, all but one specimen taken on *Peromyscus* sp.

GENUS **Megarthroglossus** Jordan and Rothschild

Megarthroglossus. Jordan and Rothschild, 1915, Ectoparasites I: 46 (type species by designation: *M. procus* J.& R., 1915).

Megarthroglossus Jordan and Rothschild. Méndez, 1956, Univ. Calif. Publ. Ent. 11: 159-192, 14 figs.

Megarthroglossus Jordan and Rothschild. Tipton et al., 1979, Great Basin Nat. 39: 371-377.

This difficult genus currently contains thirteen recognized taxa, variously distributed throughout the western United States. Species of *Neotoma* seem to be their preferred hosts, but many of the species have been reported from other small mammals, including chipmunks, tree and flying squirrels, wood mice, and pikas, as well as scattered predator records.

KEY TO THE SPECIES OF *MEGARTHROGLOSSUS*

(Modified from Tipton et al. (1979))

1 Males **2**

Females **5**

2 Height of hump on dorsal margin of aedeagus greater than 10 microns **3**

Height of hump on dorsal margin of aedeagus less than 10 microns **4**

3 Anterior margin of movable process with angular denticle, inner fovea of immovable process exceeding 50 microns below dorsal margin; ventrolateral lobe of st. VIII evenly convex .... **spenceri**

Anterior margin of movable process without angular denticle, inner fovea of immovable process less than 50 microns below dorsal margin; ventrolateral lobe of st. VIII truncate .... **jamesoni**

4	Ventrolateral lobe of st. VIII divided into an upper convex lobe and a lower acuminate lobe, spur of crochet less than 10 microns long	**procus**
	Ventrolateral lobe of st. VIII not divided, spur of crochet less than 10 microns long	**divisus**
5	Bulga of spermatheca not compressed	**6**
	Bulga of spermatheca compressed	**7**
6	Posterior margin of st. VII with a sinus	**spenceri**
	Posterior margin of st. VII without a sinus	**jamesoni**
7	Posterior margin of st. VII with sinus more than 10 microns deep	**divisus**
	Posterior margin of st. VII usually without a sinus but, if present, less than 10 microns deep	**procus**

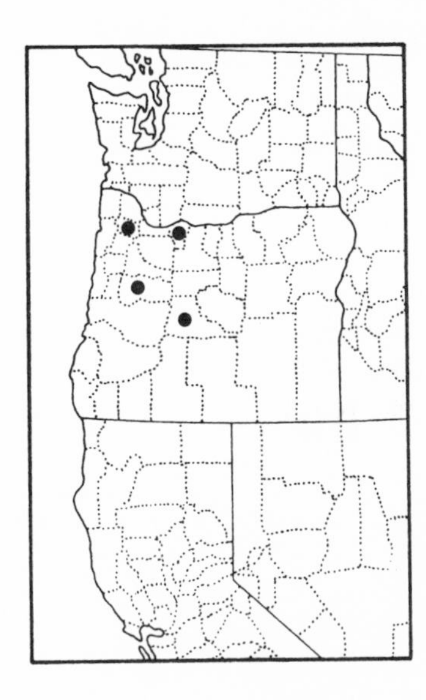

Megarthroglossus procus Jordan and Rothschild

[91531]

(Figs. 98, 99)

Megarthroglossus procus. Jordan and Rothschild, 1915, Ectoparasites I: 47, figs. 50-52 (Chilliwack, British Columbia, from *Spilogale* [*gracilis olympica*]).

Megarthroglossus similis. Wagner, 1936, Canad. Ent. 68: 196, fig. 1 (Beaverdell, British Columbia, from *Neotoma cinerea occidentalis*).

Megarthroglossus procus procus Jordan and Rothschild. Hubbard, 1947, Fleas of Western North America, p. 297, fig. 171.

Megarthroglossus procus oregonensis. Hubbard, 1947, *op. cit.,* p. 299, fig. 172 (lectotype from Parkdale, Oregon, from *Tamiasciurus douglasi cascadensis*).

Megarthroglossus procus Jordan and Rothschild. Hopkins, 1952, J. Wash. Acad. Sci. 42: 365 (*oregonensis* synonymized).

Megarthroglossus muiri. Augustson, 1953, Bull. S. Calif. Acad. Sci. 52: 122-125, pl. 23, figs. 1, 2 (Tully's Hole, N. E. Fresno Co., California, from *Tamiasciurus douglasi albolimbatus*).

Megarthroglossus procus muiri Augustson. Hopkins and Rothschild, 1962, An Illustrated Catalogue... Vol. III, pp. 374-376.

Megarthroglossus procus Jordan and Rothschild. Smit and Wright, 1965, Mitt. Hamburg. Zool. Mus. Inst. 62: 41 (*similis* and *muiri* synonymized).

Megarthroglossus procus muiri Augustson. Tipton et al., 1979, Great Basin Nat. 39: 384-387, figs. 49, 61, 78, 90, 101 (*muiri* resurrected).

Megarthroglossus procus procus Jordan and Rothschild. Tipton et al., 1979, *op. cit.,* 387-390, figs. 50, 62, 76, 77, 88, 89, 101.

The status of the subspecies of this species seems debatable at best, *muiri* having been synonymized by Smit and Wright (1965), but treated as a valid subspecies by Tipton et al. (1979). The latter authors do not cite the Smit and Wright work in their references, nor do they indicate that they are resurrecting *muiri* from the status of a synonym of *procus*. Based on distribution, host preference and morphology, we are convinced that the characters used to distinguish the two from each other are well within the range of individual variation and consider the two taxa to be conspecific.

The preferred host of this species would appear to be *Tamiasciurus hudsonicus,* but it has been taken on a broad range of other hosts, suggesting that its preferences are not all that specific. Additional hosts include other species of tree squirrels, wood mice, and pikas.

The species is distributed from central California and Nevada north to British Columbia, and east to Wyoming. Recent Oregon collections include one male and five females from various hosts, all collected in Linn County.

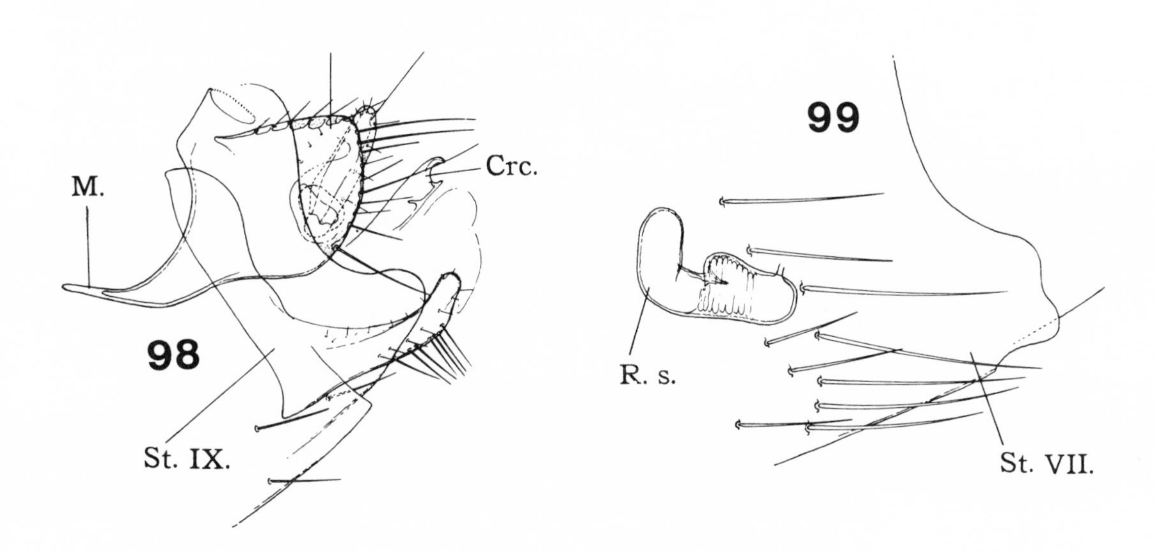

98-99 *Megarthroglossus procus* J. & R. **98** Modified abdominal segments of male. **99** Female sternum VII and spermatheca. After Holland, 1985.

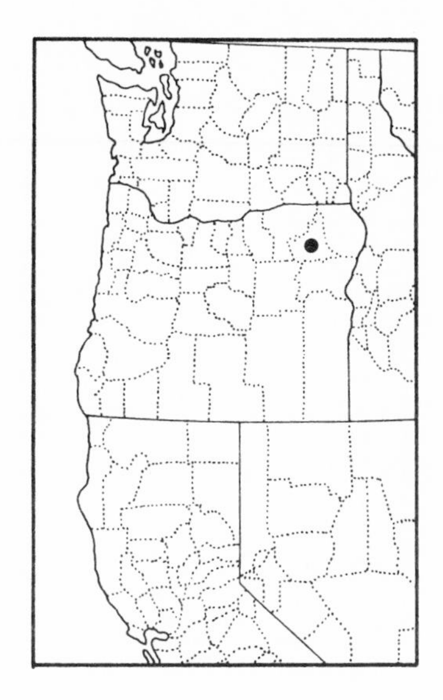

Megarthroglossus spenceri Wagner

[93641]

(Figs. 100, 101)

Megarthroglossus spenceri. Wagner, 1936, Canad. Ent. 68: 196, fig. 2 (Nicola, British Columbia, from *Ochotona princeps*).

Megarthroglossus pygmaeus. Wagner, 1936, *op. cit.*, 196, fig. 4 (Nicola, British Columbia, from *Neotoma cinerea*).

Megarthroglossus pygmaeus Wagner. Méndez, 1956, Univ. Calif. Publ. Ent. 11: 163, 172, figs. 3B, 6C, 8B, 10B.

Megarthroglossus spenceri Wagner. Méndez, 1956, *op. cit.*, 176, figs. 12K, 14A.

Megarthroglossus spenceri Wagner. Hopkins and Rothschild, 1962, An Illustrated Catalogue... Vol. III, p. 383-384.

Megarthroglossus spenceri Wagner. Smit and Wright, 1978a, A List of Code Numbers..., p. 22 (*pygmaeus* synonymized).

Megarthroglossus spenceri Wagner. Tipton et al., 1979, Great Basin Nat. 39: 395-398, figs. 52, 63, 80, 94, 101.

This species has been reported previously from California and British Columbia, although there was little doubt that it occurred in the intervening area. The single pair reported here was taken on *Neotoma cinerea* collected in Union County, Oregon, in September 1975. The few collection records for this species indicate that wood rats are the preferred hosts, and, as with other members of the genus, this is a nest flea.

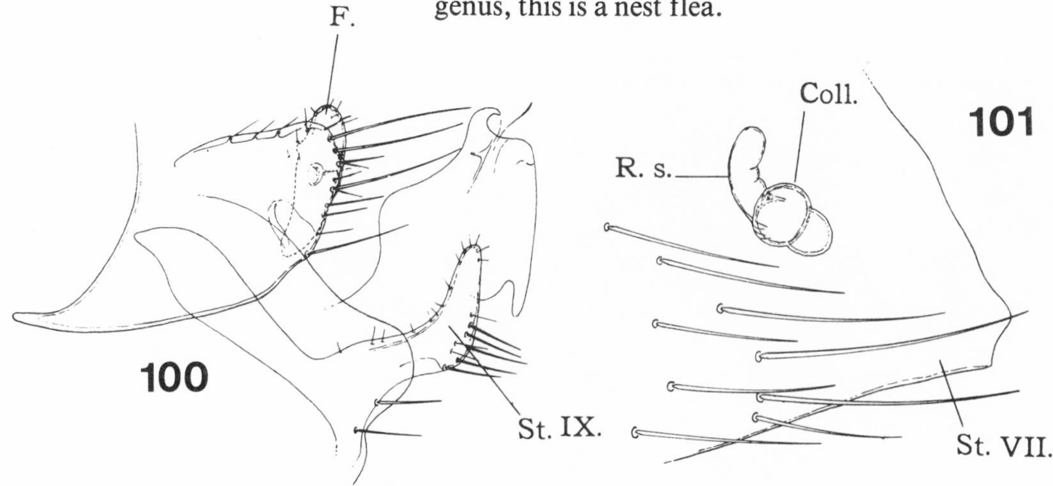

100-101 *Megarthroglossus spenceri* Wagner. **100** Modified abdominal segments of male. **101** Female sternum VII and spermatheca. After Holland, 1985.

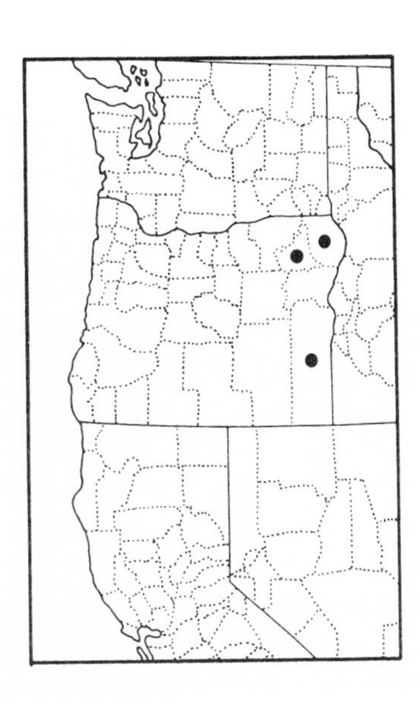

Megarthroglossus divisus (Baker)

[89820]

(Figs. 102, 103)

Pulex longispinus. Baker, 1898, (*nec Pulex longispinus* Wagner, 1893), Canad. Ent. 27: 131, 132 (Georgetown, Colorado, from Fremont's Chickaree).

Pulex divisus. Baker, 1898, J. N. Y. ent. Soc. 6: 54 (*nomen novum* for name given above).

Ceratophyllus divisus Baker. Baker, 1904, Proc. U. S. Nat. Mus. 27: 388, 416, 441, pl. 21, figs. 7-10 (host given on p. 441 as *Sciurus fremonti*).

Megarthroglossus divisus Baker. Jordan and Rothschild, 1915, Ectoparasites I: 52, figs. 55, 56.

Megarthroglossus longispinus var. *exsecatus.* Wagner, 1936, Canad. Ent. 68: 196 (Avola, British Columbia, from *Sciurus duglasii* [= *Tamiasciurus hudsonicus streatori*]).

Megarthroglossus divisus divisus (Baker). Jellison and Good, 1942, Bull. U. S. Natl. Inst. Hlth. No. 178, p. 83.

Megarthroglossus divisus exsecatus Wagner. Jellison and Good, 1942, *op. cit,* p. 83.

Megarthroglossus divisus wallowensis. Hubbard, 1947, Fleas of Western North America, p. 301, fig. 174 (Wallowa Lake, Wallowa Co., Oregon, from *Tamiasciurus hudsonicus richardsoni*).

Megarthroglossus divisus divisus (Baker). Traub, in: Hopkins, 1952, J. Wash. Acad. Sci. 42: 365 (*wallowensis* synonymized).

Megarthroglossus divisus divisus (Baker). Hopkins and Rothschild, 1962, An Illustrated Catalogue... Vol. III, pp. 387-389.

Megarthroglossus divisus exsecatus Wagner. Hopkins and Rothschild, 1962, *op. cit.,* pp. 389-390.

Megarthroglossus divisus (Baker). Tipton et al., 1979, Great Basin Nat. 39: 381-384 (*exsecatus* treated as a synonym of *divisus*).

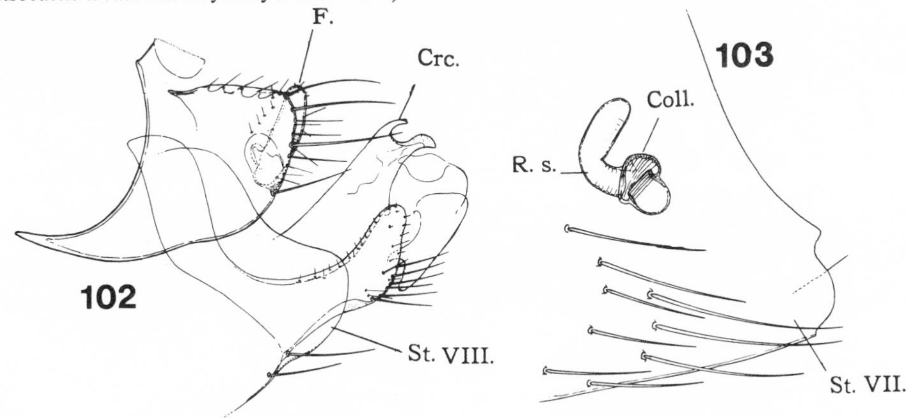

102-103 *Megarthroglossus divisus* (Baker). **102** Modified abdominal segments of male. **103** Female sternum VII and spermatheca. After Holland, 1985.

This species has the broadest range of any member of the genus, being reported from most of the western states east to Texas and Nebraska. *Tamiasciurus hudsonicus* is the type host but it is frequently taken on *Peromyscus* and *Neotoma* species, and eight of the nine specimens reported here came from *Glaucomys sabrinus.*

Tipton et al. (1979) noted that there seemed to be three populations represented in the material that they examined. One of these, commonly referred to as the nominate subspecies, was associated with *T. hudsonicus.* The second, referred to as *M. d. exsecatus,* was associated with *Neotoma* species. The third, an intermediate form, was taken on *Peromyscus* species. The three populations are sympatric in parts of their range and these authors could find no reliable character that could be used in separating them from each other.

Material reported here was collected in Malheur, Union, and Wallowa counties, Oregon. The single specimen not taken from flying squirrels was a female from *Peromyscus crinitus.*

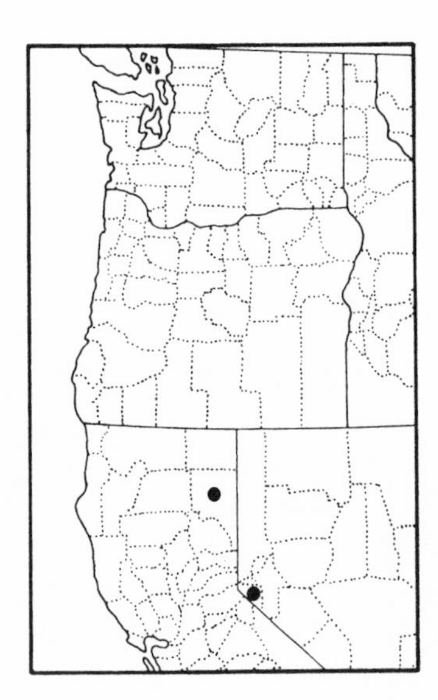

Megarthroglossus jamesoni Smit

[95346]

Megarthroglossus jamesoni. Smit, 1953, Bull. Brit. Mus. (nat. Hist.), Ent. 3: 202 (Pine Nut Mountains, Douglas Co., Nevada, from *Neotoma cinerea* nest).

Megarthroglossus jamesoni Smit. Tipton et al., 1979, Great Basin Nat. 39: 384.

This species is known from very few specimens, all taken from the wood rat, *Neotoma cinerea.* Hopkins and Rothschild (1962) suggest that *M. sierrae* Augustson, 1953, may be a junior synonym. If so, this would extend the range of this taxon south to Fresno County, California. Méndez (1956) examined two males and one female collected in Malheur County, Oregon, but failed to assign them to this species because of the limited material. It is quite likely that the species will be collected in Oregon since the range of its preferred host extends into the state.

In passing, attention should be called to the fact that the publications bearing the descriptions of *M. jamesoni* and *M. sierrae* are both assumed to have been released on 31 December 1953, in accordance with Article 21(a)(ii) of the International Code of Zoological Nomenclature. Since the former was published in London and the latter in California, *M. jamesoni* has 8 hours priority over *M. sierrae* should the two turn out to be synonyms.

GENUS **Anomiopsyllus** Baker

Anomiopsyllus. Baker, 1904, Proc. U. S. Nat. Mus. 27: 377, 425 (type species: *Typhlopsylla nudata* Baker).
Anomiopsyllus Baker. Barnes et al., 1977, Great Basin Nat. 37: 138-206.

This genus currently contains fifteen named taxa that are collectively restricted to the more arid parts of the western United States, south to southern Mexico. They are true parasites of wood rats, but are frequently taken on other small mammals that associate with wood rats. A number of taxa are known from very limited material, but one of the species reported here is known from hundreds of collections from Arizona, Nevada and Utah, as well as a record from Wyoming and another from Oregon.

Males of the two species reported here may be separated with the following key.

KEY TO THE NORTHWESTERN SPECIES OF *ANOMIOPSYLLUS*

1 Triangular movable process bearing one distal and one proximal spiniform .... **amphibolus**

Triangular movable process bearing three distal and one proximal spiniform (proximal spiniform may be missing in a few specimens) .... **falsicalifornicus congruens**

Anomiopsyllus amphibolus Wagner

[93631]

(Figs. 104, 105)

Anomiopsyllus amphibolus. Wagner, 1936, Z. Parasitenk. 8: 654, fig. I (Salina, Utah, from *Neotoma desertorum*).
Anomiopsyllus amphibolus Wagner. Barnes et al., 1977, Great Basin Nat. 37: 156-163, figs. 1-6, 30, 52, 80, 95.
Anomiopsyllus amphibolus Wagner. Egoscue, 1980, Great Basin Nat. 40: 361.

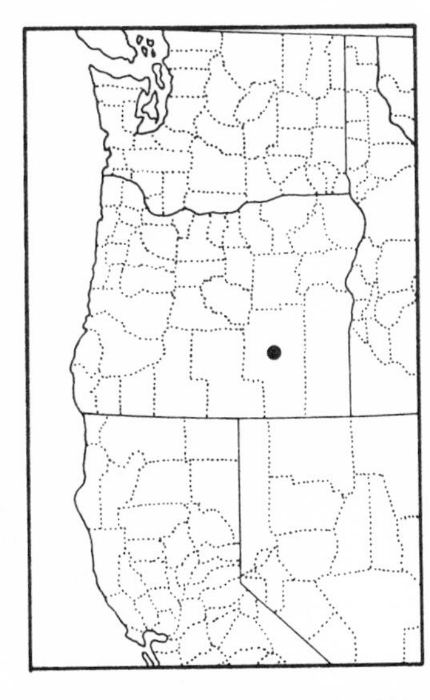

This species was first reported from Oregon by Egoscue (1980) from two specimens collected in Harney County. His specimens, one male and one female, were collected from *Neotoma cinerea* and *Peromyscus maniculatus,* respectively. Collection records from Oregon would thus suggest that this is a rare faunal element. In reality, wood rats occur commonly in the state and examination of their nests is almost certain to produce this species in large numbers. Such an effort would be particularly appropriate during the winter when other "rare" species are most likely to occur.

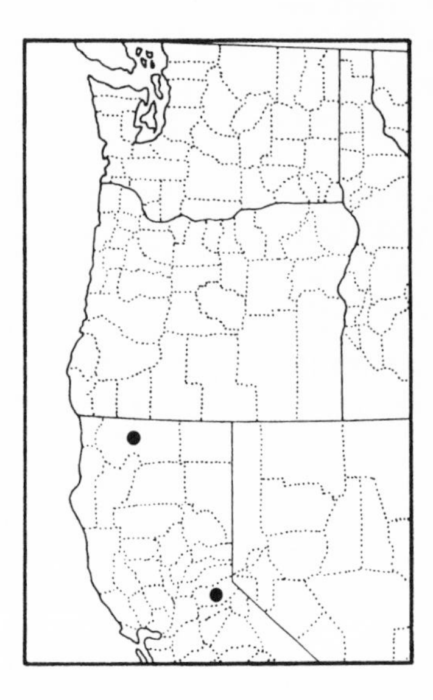

Anomiopsyllus falsicalifornicus congruens Stewart

[94001]

Anomiopsyllus congruens. Stewart, 1940, Pan-Pacific Ent. 16: 17-18 (Hastings Natural History Reservation near Jamesburg, Monterey Co., California, from *Spilogale gracilis*).

Anomiopsyllus falsicalifornicus congruens Stewart. Hopkins, 1952, J. Wash. Acad. Sci. 42: 365.

Anomiopsyllus falsicalifornicus congruens Stewart. Barnes et al., 1977, Great Basin Nat. 37: 184-192.

This taxon evidently has a broader distribution than the nominate form, although both are restricted to California. Historically, various authors have disagreed as to the rank accorded this name, since the two subspecies are easily separated on the basis of male characters. Barnes et al. (1977) recently concluded, after examining many specimens from various localities, that subspecific rank was appropriate. The species shows some clinal variation from north to south, with this subspecies being northernmost. The preferred host is *Neotoma fuscipes,* but there are many records from other rodents.

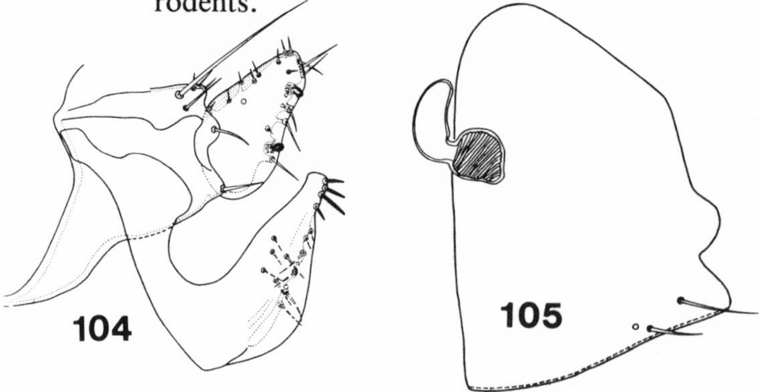

104-105 *Anomiopsyllus amphibolus* Wagner. **104** Modified abdominal segments of male. **105** Female sternum VII and spermatheca. Original.

SUBFAMILY **Rhadinopsyllinae**

There are at present seven recognized genera belonging to this subfamily, members of which are all restricted to the Northern Hemisphere. The two species of *Wenzella* Traub, 1953, occur in Central America where they parasitize heteromyid rodents. The nine species of *Stenischia* Jordan, 1932, are found in southern and eastern Asia as parasites of small rodents and insectivores. Two more genera are Holarctic in distribution, while the remaining three are unique to North America.

The five genera with representatives in the Northwest may be separated with the following key, modified from Hopkins and Rothschild (1962).

KEY TO THE NORTHWESTERN GENERA OF THE *RHADINOPSYLLINAE*

1 Genal comb strongly reflexed, forming an acute angle with the ventral margin of the head and composed of spatulate spines 2

Genal comb not reflexed, either absent completely or composed of peg-like spines 3

2 Genal comb of six spines; abdominal terga at least slightly (usually strongly) incrassate dorsally .... **Corypsylla**

Genal comb of five spines; abdominal terga not more strongly sclerotized dorsally than elsewhere .... **Nearctopsylla**

3 Genal comb vestigial or absent .... **Trichopsylloides**

Genal comb well developed, of at least three spines 4

4 Spines of genal comb subequal, confined to ventral margin of head; no subapical patch of bristles on inner side of hind coxae .... **Paratyphloceras**

Genal comb including an ocular spine which is usually smaller than other spines; with a subapical patch of spiniform or subspiniform bristles on inner side of hind coxae .... **Rhadinopsylla**

GENUS **Corypsylla** C. Fox

Corypsylla. C. Fox, 1908, Ent. News 19: 452 (type species by designation: *C. ornatus* C. Fox, 1908).

Corypsylloides. Hubbard, 1940, Pacific Univ. Bull. 37: 7 (type species by designation: *Corypsylloides (Corypsylla) kohlsi* Hubbard).

Corypsylla C. Fox. Hopkins, 1952, J. Wash. Acad. Sci. 42: 365 (*Corypsylloides* synonymized).

This genus is unique to the Californian subregion of North America where three species are parasites of insectivores. It is separable from *Nearctopsylla* in having six spines in the genal comb and slight to heavy incrassations in the abdominal tergites.

KEY TO THE SPECIES OF *CORYPSYLLA*

(Modified from Hopkins and Rothschild, 1962)

1 Tergal incrassations poorly developed; third spine from ventral end of genal comb not longer than other spines and not drawn out to a fine point .... **kohlsi**

Tergal incrassations well developed; third spine in genal comb the longest, drawn out to a fine point **2**

2 Third spine of genal comb with apical half almost symmetrical .... **jordani**

Third spine of genal comb strongly asymmetrical, its ventral margin nearly straight .... **ornata**

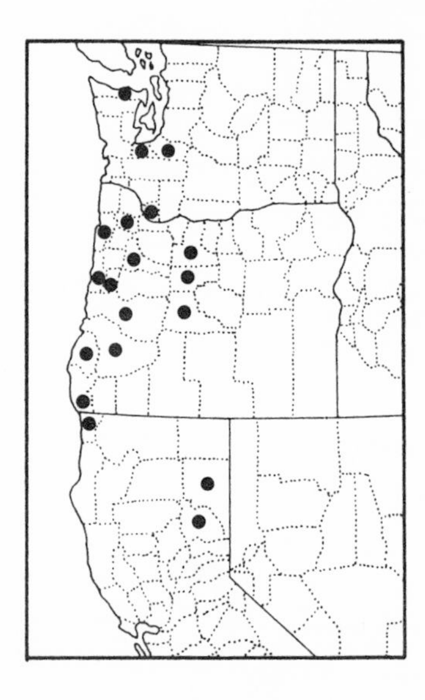

Corypsylla ornata C. Fox

[90836]

(Figs. 106, 107)

Corypsylla ornatus. C. Fox, 1908, Ent. News 19: 452 (San Francisco, California, from *Scapanus californicus* [= *latimanus*]).

Corypsylla setosifrons. Stewart, 1940, Pan-Pacific Ent. 16: 23, fig. 12 (near Jamesburg, Monterey Co., California, from *Scapanus latimanus* ssp.).

Corypsylla ornata C. Fox. Hubbard, 1947, Fleas of Western North America, pp. 363, 364.

Corypsylla setosifrons Stewart. Hubbard, 1947, *op. cit.*, pp. 363, 366.

Corypsylla ornata C. Fox. Jameson, in: Hopkins, 1952, J. Wash. Acad. Sci. 42: 365 (*setosifrons* synonymized).

Of the 139 males and 244 females reported here, 100 percent of the males and 97.9 percent of the females were collected on the moles *Scapanus townsendi* and *S. orarius* in the western part of Oregon. Three of the remaining females came from shrews and one from *Peromyscus maniculatus*. The species showed no seasonal peak and specimens were taken throughout the year. Most collections were from the coastal tier of counties.

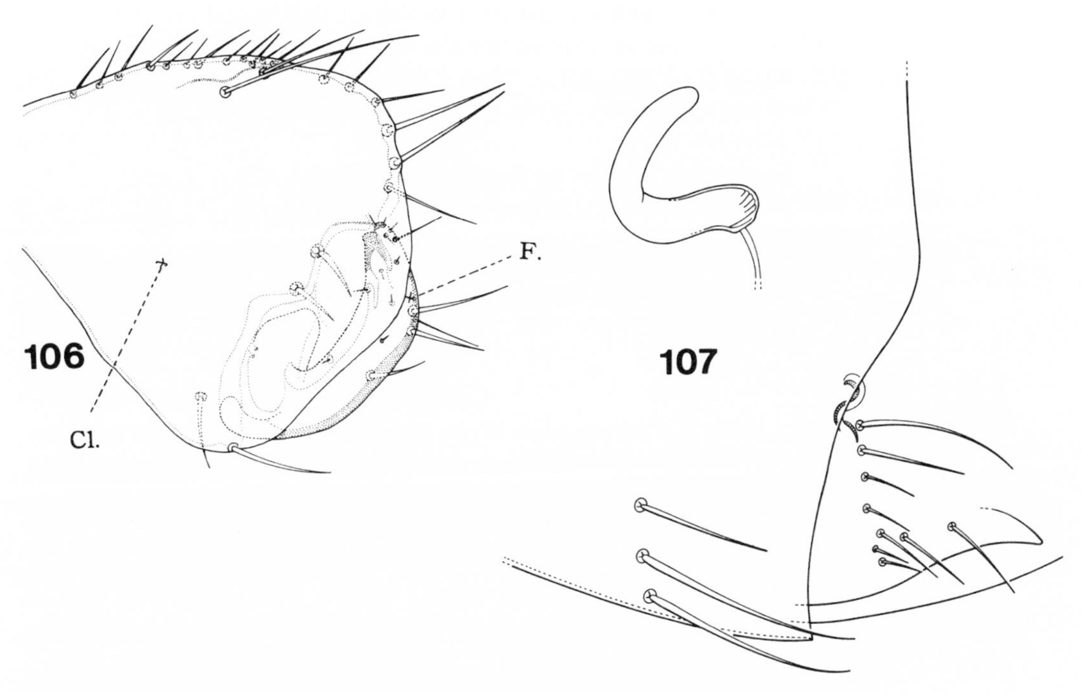

106-107 *Corypsylla ornata* C. Fox. **106** Male clasper. **107** Modified abdominal segments of female. After Holland, 1985.

Corypsylla kohlsi Hubbard

[94024]

(Figs. 108, 109)

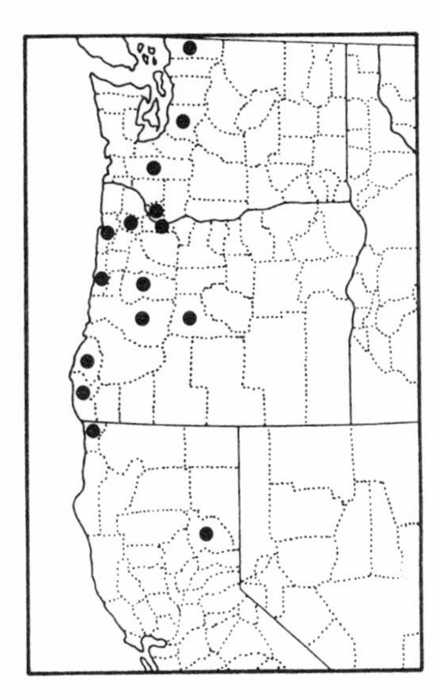

Corypsylla kohlsi. Hubbard, March, 1940, Pacific Univ. Bull. 37(1): 10 (Cannon Beach, Oregon, from *Sorex obscurus bairdi*).

Corypsylloides kohlsi Hubbard. Hubbard, April, 1940, Pacific Univ. Bull. 37(2): 7.

Corypsylloides spinata. I. Fox, June, 1940, J. Wash. Acad. Sci. 30: 273, figs. 1, 2, 4 (Portland, Oregon, from *Microtus townsendi*).

Corypsylloides kohlsi (Hubbard). Ewing and Fox, 1943, Miscl. Publ. U. S. Dept. Agric. No. 500, p. 96 (*spinata* synonymized).

Corypsylloides kohlsi Hubbard. Hubbard, 1947, Fleas of Western North America, p. 367.

Corypsylla kohlsi Hubbard. Hopkins and Rothschild, 1962, An Illustrated Catalogue... Vol. III, pp. 408-410.

Of the 31 males and fifty females of this species reported here, all but one male and three females were collected from shrews. The remaining four specimens came from a mole and small rodents. The following table shows the distribution of occurrence.

Host associations of *Corypsylla kohlsi*

Host	M	(%)	F	(%)	%
Sorex bendirei	1	(1)	0	(0)	1.2
Sorex pacificus	7	(8)	15	(17)	27.1
Sorex trowbridgei	15	(17)	22	(27)	45.6
Sorex vagrans	6	(6)	10	(11)	18.7
Neurotrichus gibbsi	1	(1)	0	(0)	1.2
Accidental hosts	1	(1)	3	(3)	4.9

Again, most specimens came from the coastal tier of Oregón counties; records from the interior were limited to Linn and Deschutes counties. Specimens were taken in all months of the year except January, February, and July. Incidence during the summer was low.

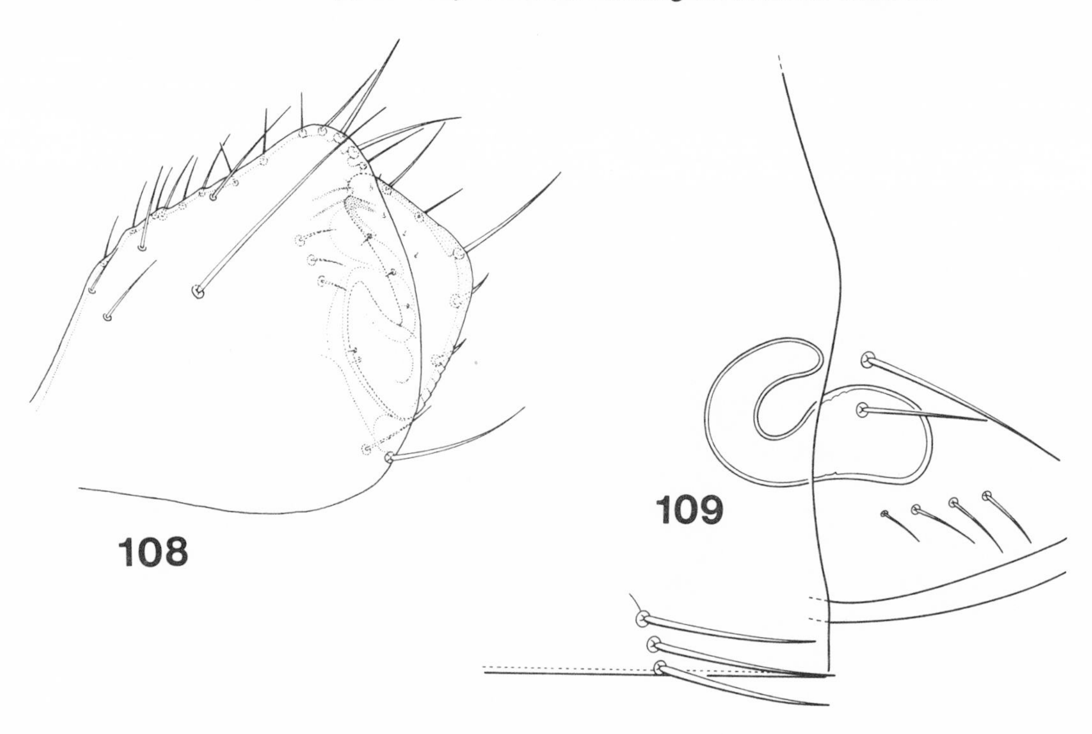

108-109 *Corypsylla kohlsi* Hubbard. **108** Male clasper. **109** Female sternum VII and spermatheca. After Holland, 1985.

Corypsylla jordani Hubbard

[94023]

(Figs. 110, 111)

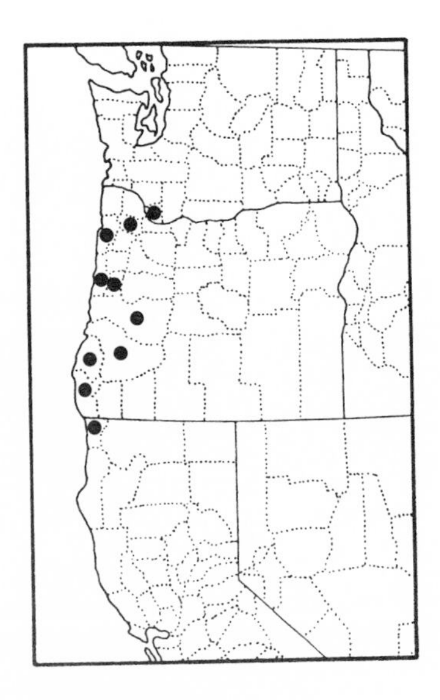

Corypsylla jordani. Hubbard, 1940, Pacific Univ. Bull. 37: 9 (Gaston, Oregon, from *Neurotrichus gibbsi gibbsi* and *Peromyscus maniculatus rubidus*).

Corypsylla 'ornata C. Fox.' Holland, 1949, (*partim, nec* C. Fox, 1908), The Siphonaptera of Canada, p. 103.

Corypsylla jordani Hubbard. Hopkins and Rothschild, 1962, An Illustrated Catalogue... Vol. III, pp. 410-412.

Thirty-seven males and 44 females of this species were collected. Thirty-three (40.7 percent) of the males and 37 (45.6 percent) of the females were taken on *Neurotrichus gibbsi.* The remaining records came from various other insectivores. All collections came from the coastal tier of Oregon counties.

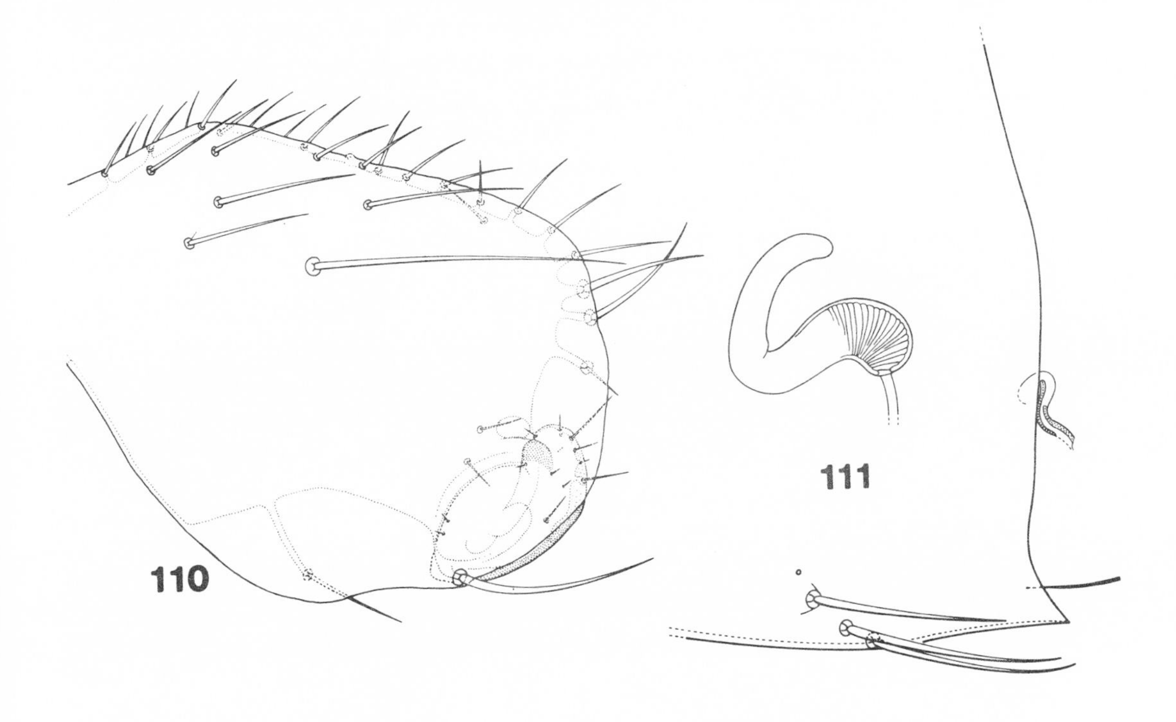

110-111 *Corypsylla jordani* Hubbard. **110** Male clasper. **111** Female sternum VII and spermatheca. After Holland, 1985.

GENUS **Nearctopsylla** Rothschild

Nearctopsylla. Rothschild, 1915, Novit. Zool. 22: 307 (type species by designation: *N. brooksi* Roths., 1904 [as *Ctenopsyllus*]).

Nearctopsylla subgenus *Beringiopsylla*. Ioff, 1950, (in: Ioff et al.) Med. Parasitol., Moscow 19: 272 (type species by designation: *N. ioffi* Sychevskiy, 1950).

Nearctopsylla subgenus *Hollandiana*. Hopkins, 1951, Ann. Mag. Nat. Hist. (12)4: 544 (type species by designation: *N. princei* Holland and Jameson, 1950).

Nearctopsylla subgenus *Chinopsylla*. Ioff, 1950 (in: Ioff et al.) Med. Parasitol., Moscow 19: 272 (type species by designation: *N. beklemischevi* Ioff, 1950).

Nearctopsylla subgenus *Beringiopsylla* Ioff, 1950. Hopkins and Rothschild, 1962, An Illustrated Catalogue ... Vol. III, p. 417 (subgenus *Hollandiana* included as a synonym).

Nearctopsylla subgenus *Neochinopsylla*. Wu, Wang and Liu, 1965, Acta zootax. Sinica 2: 201-205 (type species by monotypy: *N. brevidigita* Wu, Wang and Liu, 1965).

As the name suggests, this is primarily, though not exclusively, a Nearctic genus. Five of the seventeen currently recognized taxa are native to the Palaearctic Region. The remaining known taxa are restricted to North America, with the greatest variety of species concentrated in the west. They are associated mainly with insectivores, but predator records are common and one taxon is evidently a parasite of mustelids. Eight species are known to occur in the area under consideration. The species may be separated with the following key.

KEY TO THE NORTHWESTERN SPECIES OF *NEARCTOPSYLLA*

(Modified from Hopkins and Rothschild, 1962)

1 Pronotum and mesonotum with three irregular rows of bristles; pseudosetae present under the metanotal collar .... **(Nearctopsylla s.s.)—brooksi**

Pronotum and mesonotum with only one row of bristles; metanotum without pseudosetae .... **(Beringiopsylla)—2**

2 Third spine of genal comb extending much further caudad than others **3**

Second spine extending further back than others **5**

3 Pronotal comb of more than forty sharp spines in males, about thirty in females; distal arm of st. IX of male almost forming a right angle, its posterior half bearing a thick ventral fringe of bristles; female st. VII with lobe above sinus pointed or broadly rounded .... **hamata**

Pronotal comb of 32 or fewer spines; male st. IX much straighter; dorsal lobe of female st. VII acute **4**

4 Pronotal comb of 24 spines in both sexes; distal arm of male st. IX bent strongly dorsad; apical margin of female st. VII with short to moderately long, sharp, dorsal lobe with margin below it a sinus; spermatheca large .... **jordani**

Pronotal comb of 28-32 spines in males, 26-27 in females; distal arm of male st. IX less deflected dorsally; apical margin of female st. VII with long, sharp dorsal projection, subtended by a marked angular projection; spermatheca much smaller .... **traubi**

5 Apex of male st. IX sharp, forming an angle of no more than 60 degrees; with not more than twenty small bristles in a clump on inner side of fixed process of clasper; female with bulga of spermatheca sub-cylindrical and slightly broader than the hilla .... **genalis hygini**

Apex of male st. IX blunt; clump of thirty or more small bristles on inner side of fixed process of clasper; female with bulga of spermatheca globular **6**

6 Pronotal comb only feebly curved forward at ventral end; no bristles on male st. IX nearly as long as its maximum width .... **princei**

Pronotal comb strongly curved forward at ventral end; male with numerous bristles on st. IX as long or longer than its maximum width **7**

7 Pronotal comb of not more than 34 spines, usually fewer; distal arm of male st. IX widest just before apex, its apical half rather narrow and with subparallel sides .... **hyrtaci**

Pronotal comb of 39 spines; distal arm of male st. IX widest considerably before the apex, much broader with sides not subparallel...... **martyoungi**

Nearctopsylla (Nearctopsylla) brooksi (Rothschild)

[90475]

(Figs. 112, 113)

Ctenopsyllus brooksi. Rothschild, 1904, Novit. Zool. 11: 649, pl. 15, figs. 86, 88, pl. 16, fig. 89 (Calgary, Alberta, from *Putorius longicaudatus* [= *Mustela frenata longicauda*]).
Nearctopsylla brooksi Rothschild. Rothschild, 1915, Novit. Zool. 22: 307.
Nearctopsylla (Nearctopsylla) brooksi (Rothschild). Hopkins and Rothschild, 1962, An Illustrated Catalogue... Vol. III, pp. 415-417.

Although this species has not yet been taken in Washington or Oregon, there are numerous records from Alberta and British Columbia, and Hopkins and Rothschild (1962) list one collection from Mono County, California. Jellison and Senger (1973) list two females from Flathead County, Montana, and it is safe to assume that the range of the species includes both Oregon and Washington. It is evidently specific to mustelids and most collections have been made during the cooler months.

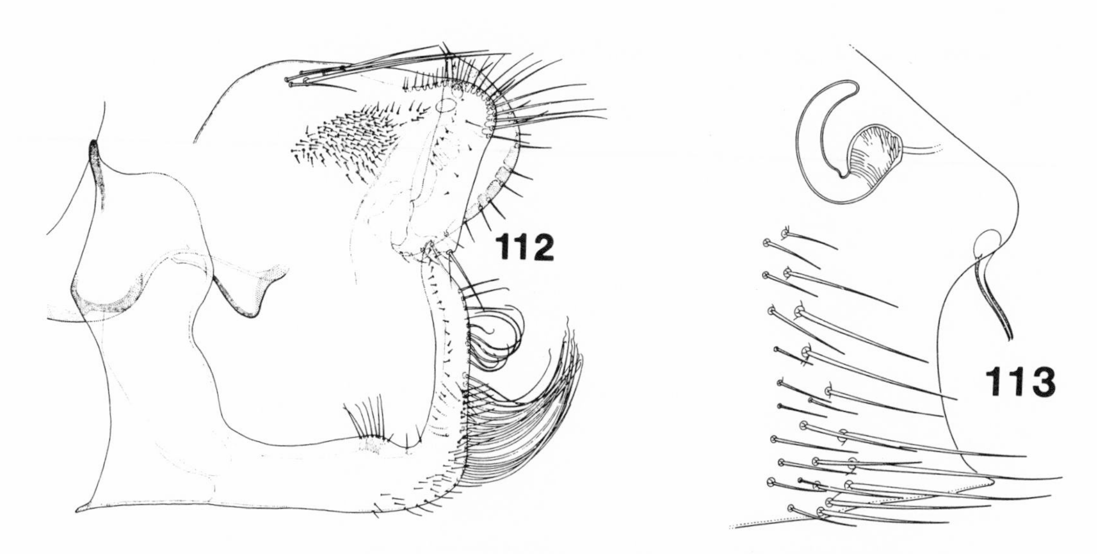

112-113 *Nearctopsylla brooksi* (Rothschild). **112** Modified abdominal segments of male. **113** Female sternum VII and spermatheca. After Holland, 1985.

Nearctopsylla (Beringiopsylla) hamata Holland and Jameson

[95001]

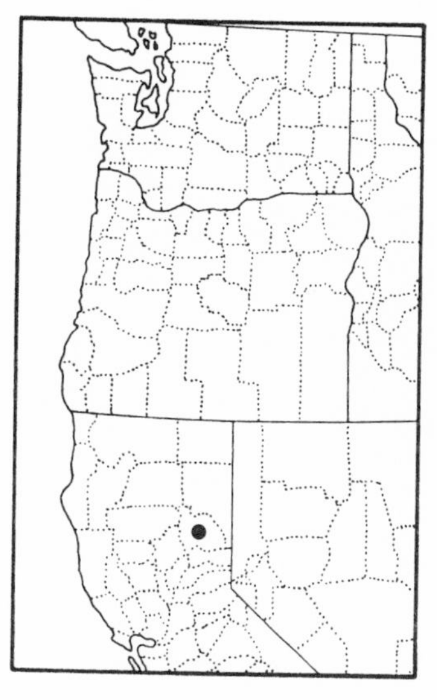

Nearctopsylla hamata. Holland and Jameson, 1950, Canad. Ent. 81: 250, 253, figs. 1-3 (Quincy, Plumas Co., California, from *Neurotrichus gibbsi*).

This is a relatively little-known species that Hopkins and Rothschild indicated was a close relative of *N. jordani* and *N. traubi.* In addition to anatomical similarities, these three taxa are the only members of the genus having moles as their preferred hosts. The species is known only from a limited number of collections from California.

Nearctopsylla (Beringiopsylla) jordani Hubbard

[94022]

(Figs. 114, 115)

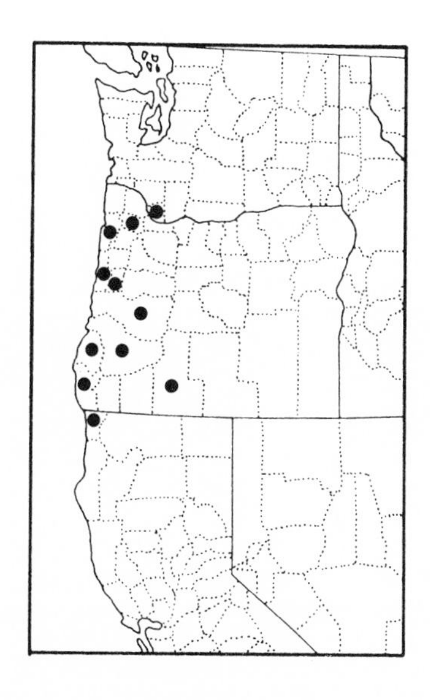

Nearctopsylla jordani. Hubbard, March, 1940, Pacific Univ. Bull. 37(1): 5 (Forest Grove, Oregon, from *Scapanus townsendi*).

Nearctopsylla hygini columbiana. Wagner, May, 1940, Z. Parasitenk. 11: 467, Fig. 7 (Vancouver, British Columbia, from *Scapanus orarius schefferi*).

Nearctopsylla jordani Hubbard. Holland, 1942, Canad. Ent. 74: 158 (*columbiana* synonymized).

Nearctopsylla (Beringiopsylla) jordani Hubbard. Hopkins and Rothschild, 1962, An Illustrated Catalogue... Vol. III, pp. 422-424.

This species, typically a parasite of moles of the genus *Scapanus,* is mainly restricted to the coastal areas. Reported here are 21 males and 34 females from Oregon, mostly taken during the cooler months. All but four were taken from either *S. townsendi* or *S. orarius.* The species also has been reported from southern British Columbia from *Neurotrichus gibbsi.* It is very similar to *N. traubi,* and earlier records may represent mixed collections since these two flea species are sympatric.

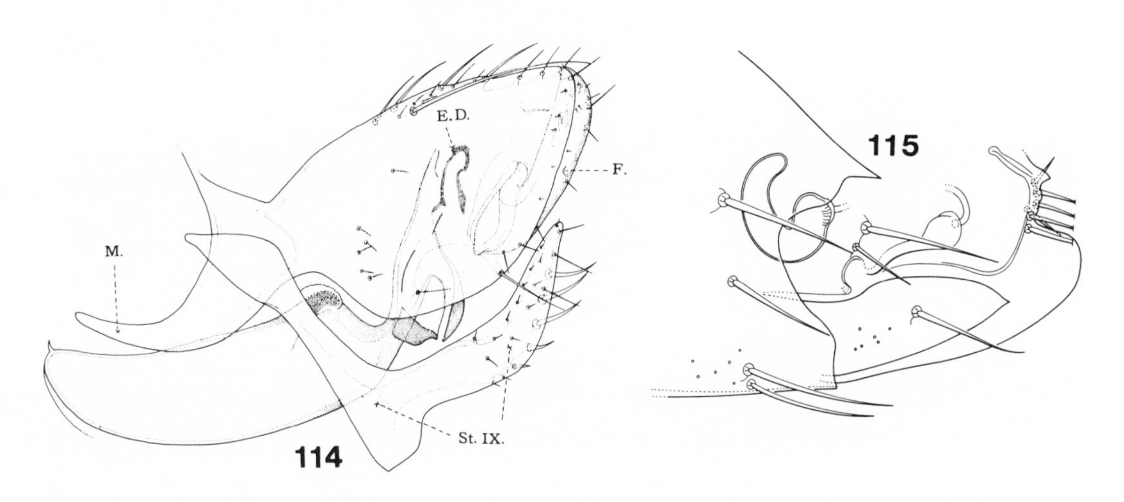

114-115 *Nearctopsylla jordani* Hubbard. **114** Modified abdominal segments of male. **115** Modified abdominal segments of female. After Holland, 1985.

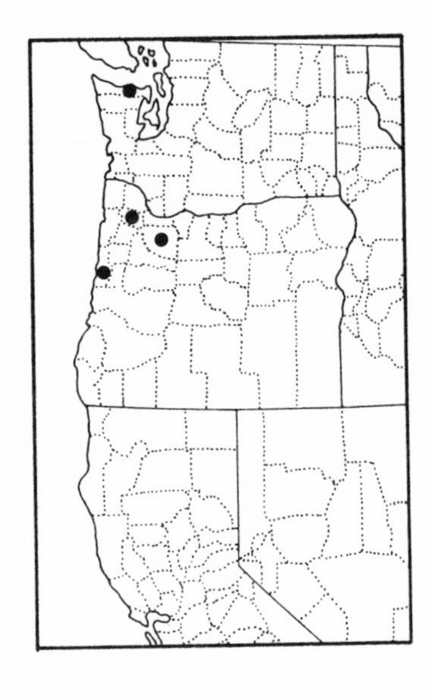

Nearctopsylla (Beringiopsylla) traubi Hubbard

[94926]

(Figs. 116, 117)

Nearctopsylla jordani traubi. Hubbard, 1949, Bull. S. Calif. Acad. Sci. 48: 49 (Sandy, Clackamas Co., Oregon, from *Scapanus townsendi*).
Nearctopsylla traubi Hubbard. Smit, 1952, Ann. Mag. Nat. Hist. (12)5: 340-346, figs. 2, 4, 6, 9, 10, 12, 14, 16.
Nearctopsylla (Beringiopsylla) traubi Hubbard. Hopkins and Rothschild, 1962, An Illustrated Catalogue... Vol. III, pp. 424-425.

Our collection contains one male and six females belonging to this species, taken by M. L. Johnson in Clallam County, Washington, and there are numerous records in the literature of collections from Oregon, Washington, and British Columbia. According to Hopkins and Rothschild (1962), it is said to be restricted to moles of the genus *Scapanus*.

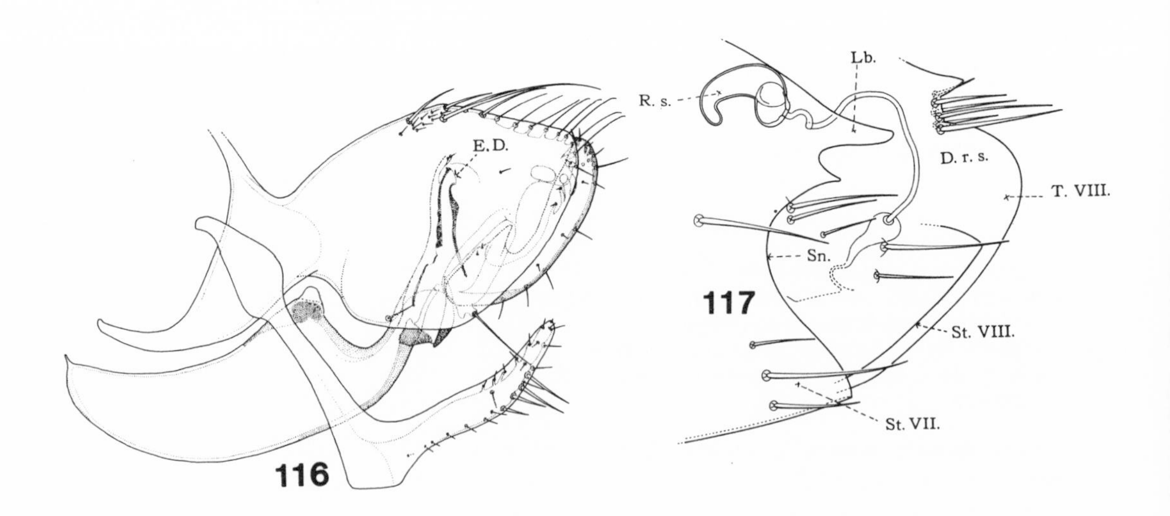

116-117 *Nearctopsylla traubi* Hubbard. **116** Modified abdominal segments of male. **117** Modified abdominal segments of female. After Holland, 1985.

Nearctopsylla (Beringiopsylla) princei H. & J.

[95002]

(Figs. 118, 119)

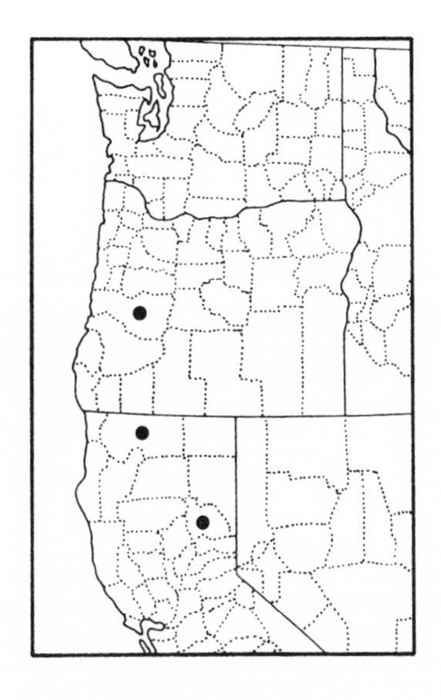

Nearctopsylla princei. Holland and Jameson, 1950, Canad. Ent. 81: 251, 253, figs. 4, 5, 7 (Quincy, Plumas Co., California, from *Sorex trowbridgei*).

Nearctopsylla (Beringiopsylla) princei Holland and Jameson. Hopkins and Rothschild, 1962, An Illustrated Catalogue... Vol. III, pp. 426-428.

This species was described from Plumas County, California, and does not seem to have been reported beyond the confines of the northern part of the state. We have two females from Lane County, Oregon, which have been tentatively assigned to this taxon, as well as a pair collected in Plumas County, California, by E. W. Jameson. Females of many of the species are so similar that they cannot be determined with confidence in the absence of accompanying males. Comparison with California material suggests that the two females from Oregon are closer to *N. princei* than to related species. *Sorex trowbridgei* seems to be its sole host.

Nearctopsylla (Beringiopsylla) martyoungi Hubbard

[95421]

(Figs. 120, 121)

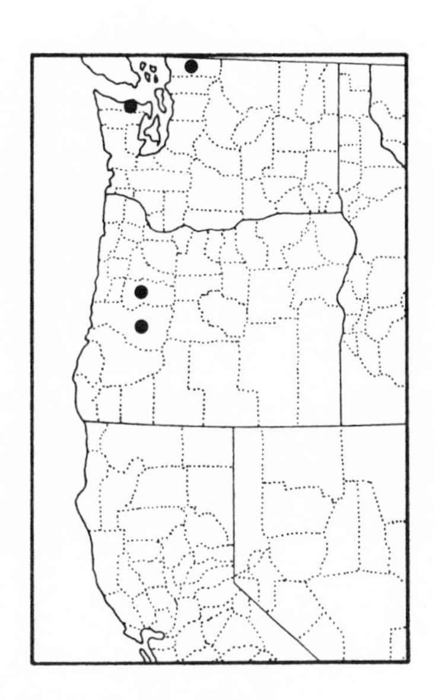

Nearctopsylla martyoungi. Hubbard, 1954, Ent. News 65: 173 (Hurricane Ridge, Olympic National Park, Clallam Co., Washington, from *Scapanus townsendi*).

Nearctopsylla (Beringiopsylla) martyoungi Hubbard. Hopkins and Rothschild, 1962, An Illustrated Catalogue... Vol. III, pp. 428-429.

This species shows a strong preference for shrews as hosts and, of the seven males and twelve females reported here, twelve came from shrews. The remaining specimens were taken from *Mustela erminea,* a common predator of shrews. Collection dates range from April through September and collections were made only in Lane and Linn counties in Oregon.

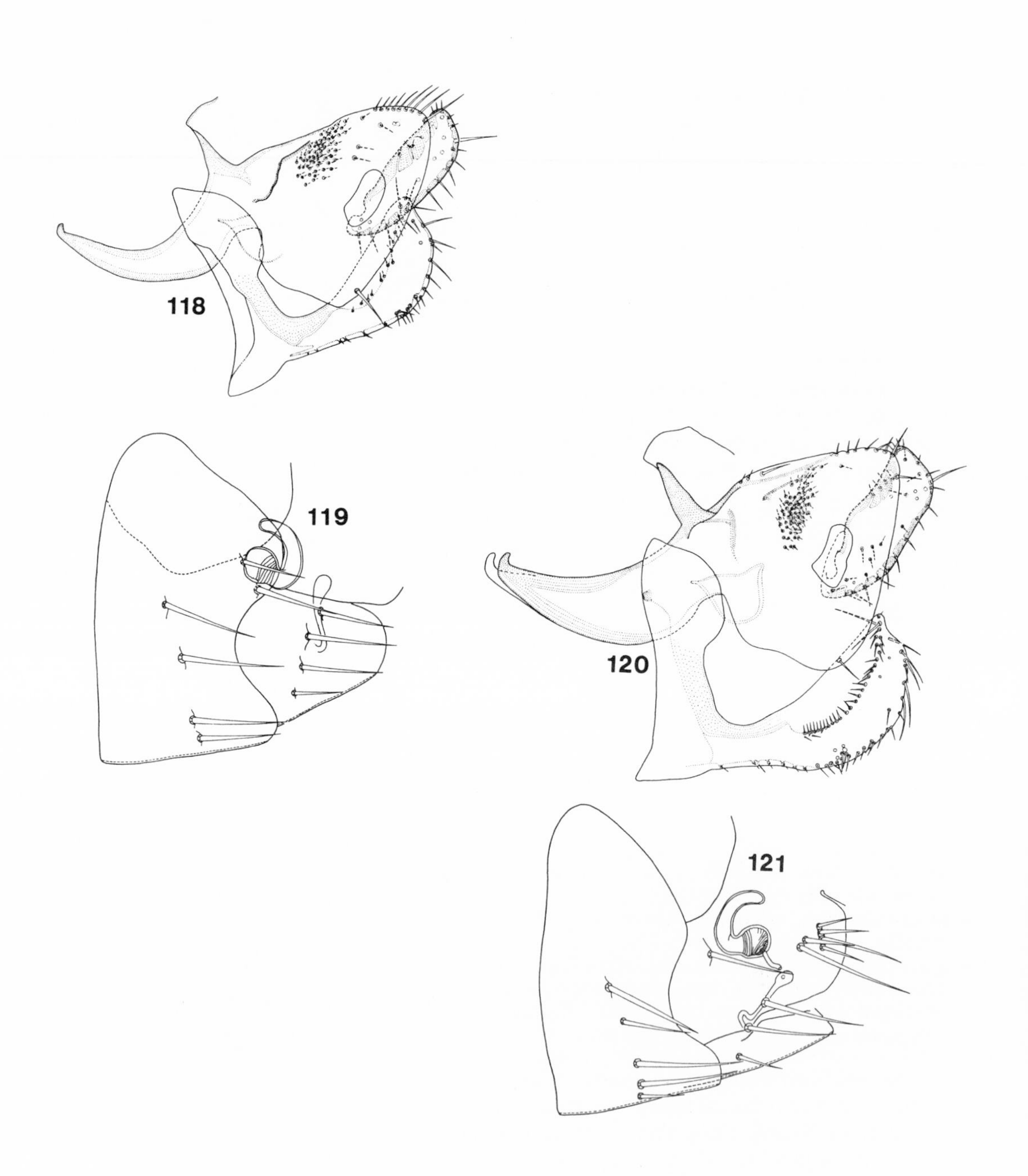

118-119 *Nearctopsylla princei* Holland and Jameson. **118** Modified abdominal segments of male. **119** Modified abdominal segments of female. Original.

120-121 *Nearctopsylla martyoungi* Hubbard. **120** Modified abdominal segments of male. **121** Modified abdominal segments of female. Original.

Nearctopsylla (Beringiopsylla) hyrtaci (Rothschild)

[90477]

(Figs. 122, 123)

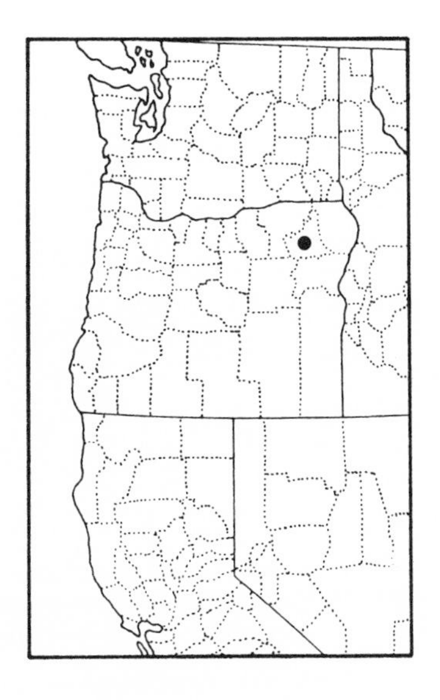

Ctenopsyllus hyrtaci. Rothschild, 1904, Novit. Zool. 11: 652, pl. 16, figs. 92, 95 (British Columbia, from *Sorex obscurus*).
Nearctopsylla hyrtaci Roths. Rothschild, 1915, Novit. Zool. 22: 307.
Nearctopsylla hyrtaci (Rothschild). Holland, 1949, The Siphonaptera of Canada, p. 107.

Considered rare by Holland (1949), this species has been reported from British Columbia and Alberta by Hopkins and Rothschild (1962) and from Montana by Jellison and Senger (1973). Recent Oregon records include one male and four females from *Sorex vagrans* and three males and two females from *Mustela frenata.* Both collections came from Union County in October and November. There is some speculation as to the preferred hosts of this species. Many collections come from mustelids, but Hopkins and Rothschild (1962) suggest that the true host is more likely a shrew belonging to the genus *Sorex.*

Nearctopsylla (Beringiopsylla) genalis hygini (Rothschild)

[90476]

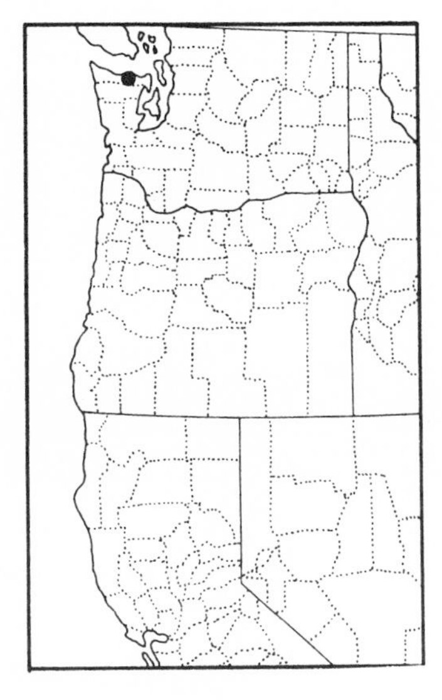

Ctenopsyllus hygini. Rothschild, 1904, Novit. Zool. 11: 650 (red deer, Alberta, Canada, from *Putorius richardsoni* [= *Mustela erminea richardsoni*]).
Nearctopsylla hygini Roths. Rothschild, 1915, Novit. Zool. 22: 307.
Nearctopsylla genalis (Baker). I. Fox, 1940, Fleas of Eastern United States, p. 91 (*N. g. hygini* erroneously synonymized).
Nearctopsylla genalis hygini (Rothschild). Holland, 1949, The Siphonaptera of Canada, p. 104-105.

This western subspecies ranges from Manitoba west in Canada, although it has not been reported from British Columbia according to Holland (1985). In the United States it is known from Minnesota and Iowa west to the Pacific Ocean. As Hopkins and Rothschild (1962) indicate, it is probably a shrew flea in spite of the many collections from mustelids.

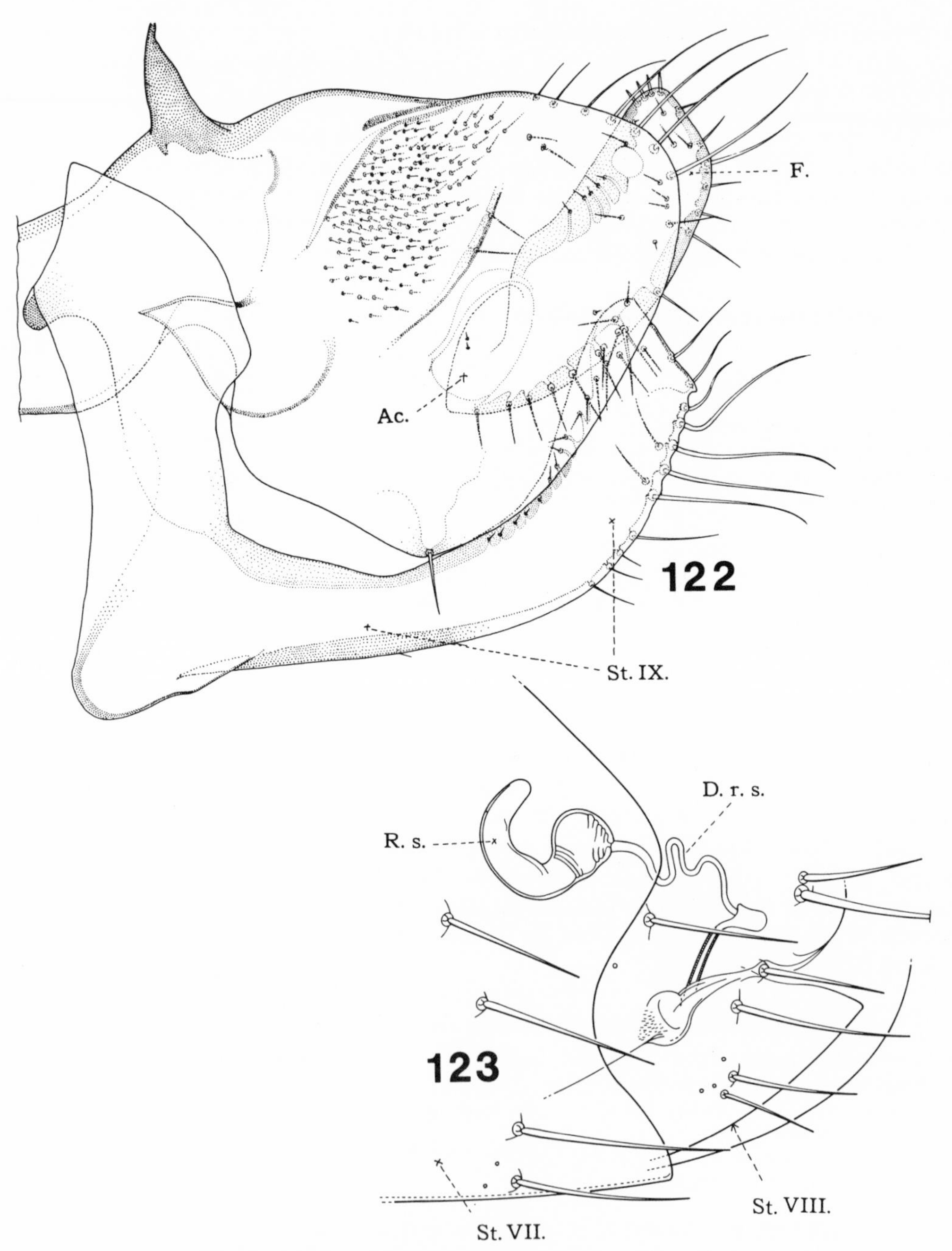

122-123 *Nearctopsylla hyrtaci* (Rothschild). **122** Modified abdominal segments of male. **123** Modified abdominal segments of female. After Holland, 1985.

GENUS **Paratyphloceras** Ewing

Paratyphloceras. Ewing, 1940, Proc. biol. Soc. Wash. 53: 35 (type species by designation: *P. oregonensis* Ewing, 1940).

This is a monotypic genus, one of three that are specific and unique to the archaic rodent, *Aplodontia rufa.* It is rare in collections because its host has a very restricted range in northwestern North America from northern California to southern British Columbia. This is the largest member of the Rhadinopsyllinae.

Paratyphloceras oregonensis Ewing
[94028]
(Figs. 124, 125)

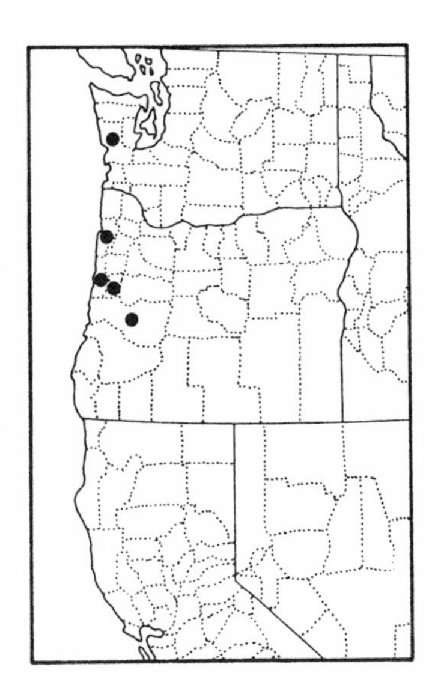

Paratyphloceras oregonensis. Ewing, 1940, Proc. biol. Soc. Wash. 53: 35 (Mercer Lake, Oregon, from mink).
Paratyphloceras oregonensis Ewing. Hubbard, 1954, Ent. News 65: 172 (description of allotype male).

Sixty-five males and 75 females of this species were collected from mountain beavers taken in Benton, Lincoln, and Tillamook counties in Oregon. Three females were collected in February and the remainder of the specimens in April. Although conclusions concerning seasonal occurrence cannot be accurately drawn from these records alone, species belonging to this subfamily typically build up in numbers during the winter and occur in the greatest abundance in the spring. The large number taken in April suggests that this is probably the pattern in this species as well. Fifty-eight specimens were taken from a single host collected in the Cascade Head Experimental Forest in Oregon on 17 April 1971.

GENUS **Trichopsylloides** Ewing

Trichopsylloides. Ewing, 1938, Proc. ent. Soc. Wash. 40: 94 (type species by designation: T. oregonensis Ewing, 1938).
Phaneris. Jordan, 1939, Novit. Zool. 41: 317 (type species by designation: *P. hubbardi* Jordan, 1939).
Trichopsylloides hubbardi (Jordan). Ewing and Fox, 1943, Miscl. Publ. U. S. Dept. Agric. No. 500, p. 19 (*Phaneris* synonymized).

This is another genus unique to *Aplodontia rufa.* Individuals are much smaller than in the preceding genus, although still large for members of this subfamily.

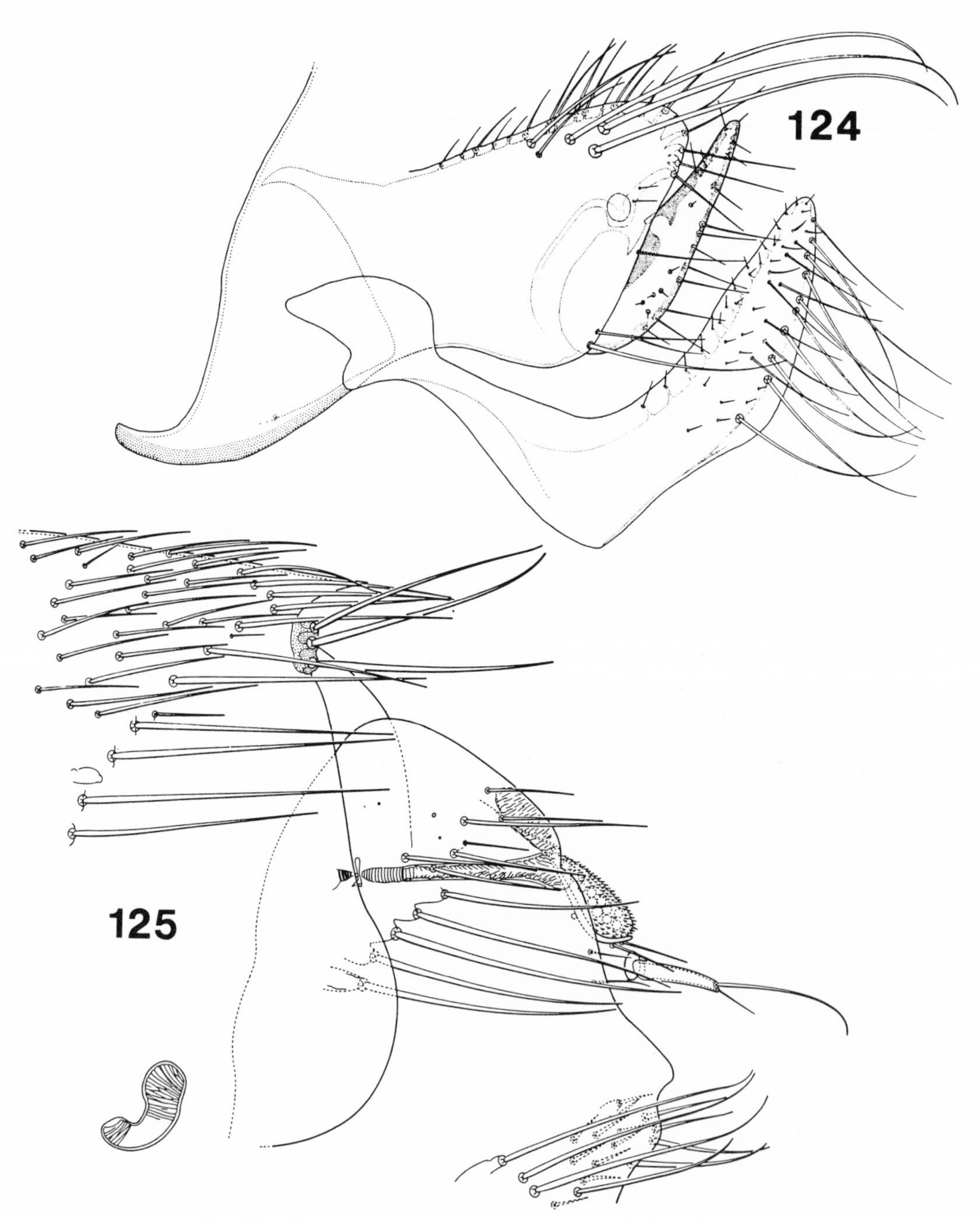

124-125 *Paratyphloceras oregonensis* Ewing. **124** Modified abdominal segments of male. **125** Modified abdominal segments of female. After Holland, 1985.

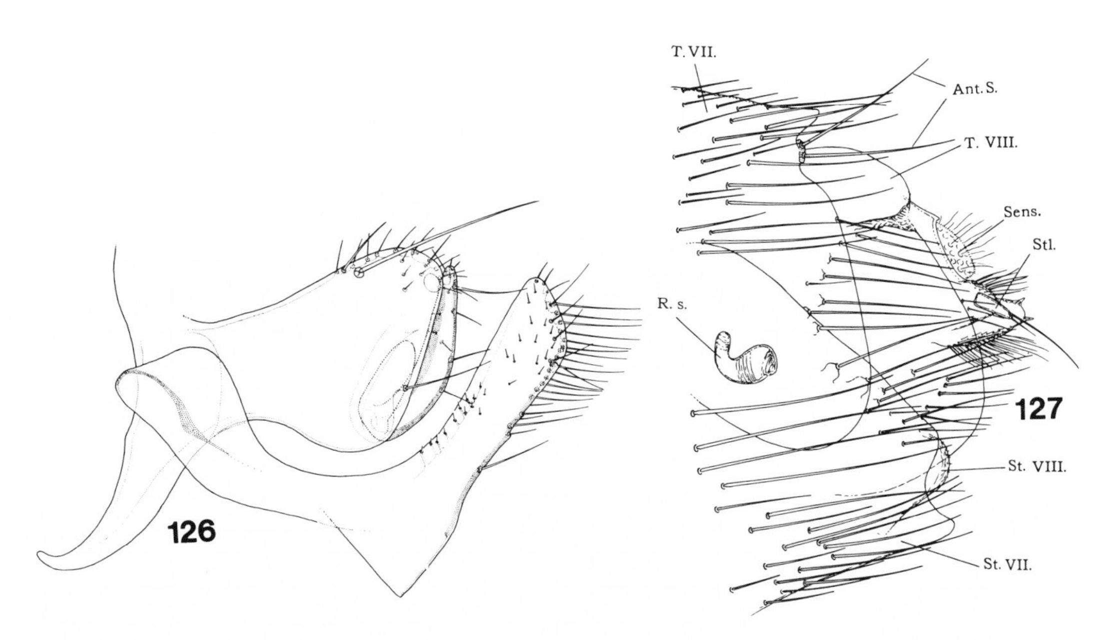

126-127 *Trichopsylloides oregonensis* Ewing. **126** Modified abdominal segments of male. **127** Modified abdominal segments of female. After Holland, 1985.

Trichopsylloides oregonensis Ewing

[93804]

(Figs. 126, 127)

Trichopsylloides oregonensis. Ewing, 1938, Proc. ent. Soc. Wash. 40: 94 (Delake [Devils Lake?], Lincoln Co., Oregon, from *Aplodontia rufa pacifica*).

Phaneris hubbardi. Jordan, 1939, Novit. Zool. 41: 318, figs. 268, 269 (Springwater, Oregon, from *Aplodontia rufa*).

Trichopsylloides hubbardi (Jordan). Ewing and Fox, 1943, Miscl. Publ. U. S. Dept. Agric. No. 500, p. 19.

Trichopsylloides oregonensis Ewing. Ewing and Fox, 1943, *op. cit.*, p. 19, fig. 4A.

Trichopsylloides oregonensis Ewing. Hubbard, 1947, Fleas of Western North America, p. 308, fig. 183 (*hubbardi* synonymized).

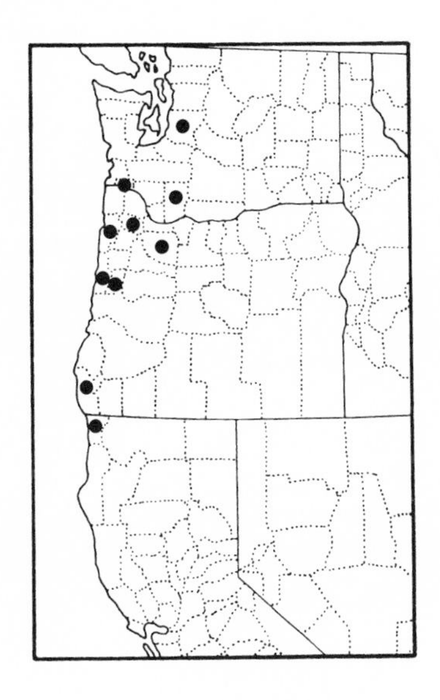

This interesting flea is also specific to the mountain beaver, but may be taken on mink, predators of this rodent. Holland (1949) cites records of this species from British Columbia, Washington, and Oregon. Additional records from these states and province are cited by Hubbard (1947), who adds records from California. He also comments that the species reaches a population peak in April and May but is rare at other times of the year. Our records would certainly confirm this observation. Of the 77 males and 82 females reported here, 145 were collected in April, and one host collected during this month in the Cascade Head Experimental Forest in Oregon had the incredible number of 89 fleas. Remaining collections include four males and five females taken in August and one male and four females collected in December. The December records include a male from *Sorex pacificus* and two pairs from *Spilogale putorius* and do not reflect prevalence on the preferred host.

GENUS **Rhadinopsylla** Jordan and Rothschild

Rhadinopsylla. Jordan and Rothschild, 1912, Novit. Zool. 19: 367 (type species by designation: *R. masculana* Jordan and Rothschild, 1912).

Actenophthalmus. C. Fox, 1925, Ent. News 36: 121 (type species by designation: *Actenophthalmus heiseri* McCoy, 1911, as *Ctenophthalmus*). Here treated as a subgenus.

Micropsylla. Dunn, 1923, (in: Dunn and Parker), Publ. Hlth. Rep., Wash. 38: 2767 (type species by designation: *M. peromyscus* Dunn, 1923 [=*sectilis*]). Here treated as a subgenus.

Five named taxa belonging to this large Holarctic genus are known from the Pacific Northwest. They are seldom of common occurrence and seem to reach a peak population only during the cooler months. All are associated with small rodents and may be separated with the following key.

KEY TO THE NORTHWESTERN SPECIES OF *RHADINOPSYLLA*

(Modified from Hopkins and Rothschild, 1962)

1 Without a ridge separating the metepisternum from the metanotum .... **(Micropsylla)—2**

Metepisternum separated from the metanotum by a strong ridge .... **(Actenophthalmus)—3**

2 Genal comb normally of five spines; male with apex of movable process of clasper only slightly notched; female with sinus in margin of st. VII hardly deeper than wide (west of the Cascade Mountains) .... **sectilis goodi**

Genal comb normally with four spines; movable process slightly longer with a more marked apical notch; sinus in female st. VII deeper (east of Cascade Mountains) .... **sectilis sectilis**

3 Six pairs of lateral plantar bristles on fifth tarsal segment .... **heiseri**

Four or five pairs of lateral plantar bristles on fifth tarsal segment **4**

4 Normally with two bristles below spiracle on t. III-VI and two on inner side of hind tibia; male with a bristle on t. VIII and apex of distal arm of st. IX truncate; female with spiracle of t. VIII rather small and square, tubular portion short .... **fraterna**

Only one bristle below spiracle on t. III-VI and usually one or none on inside of hind tibia; male without bristle on t. VIII and apex of distal arm of st. IX obliquely pointed; female with spiracle of t. VIII much larger, tubular portion short .... **difficilis**

Rhadinopsylla (Micropsylla) sectilis sectilis J. & R.

[92314]

(Figs. 128, 129)

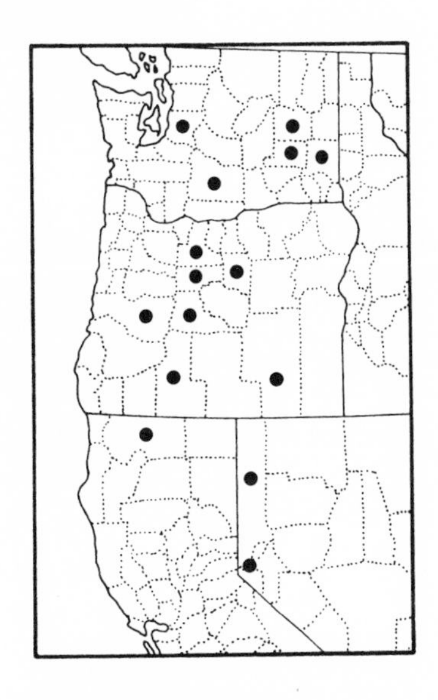

Rhadinopsylla sectilis. Jordan and Rothschild, 20 November 1923, Ectoparasites I: 314, fig. 318 (Kelowna, British Columbia, from *Peromyscus* sp.).

Micropsylla peromyscus. Dunn, 23 November 1923, (in: Dunn and Parker) Publ. Hlth. Rep., Wash. 38: 2767, 2775 (Lolo Canyon, Bear Creek, 3 mi. W. of Woodman, Montana, from *Peromyscus maniculatus artemisiae*).

Micropsylla sectilis Jordan and Rothschild. Jordan, 1937, Novit. Zool. 40: 270 (*peromyscus* synonymized).

Micropsylla sectilis Jordan and Rothschild. Hubbard, 1947, Fleas of Western North America, p. 349, fig. 214.

Rhadinopsylla (Micropsylla) sectilis sectilis Jordan and Rothschild. Hopkins and Rothschild, 1962, An Illustrated Catalogue... Vol. III, pp. 548-549.

and

Rhadinopsylla (Micropsylla) sectilis goodi (Hubbard)

[94111]

(Figs. 130, 131)

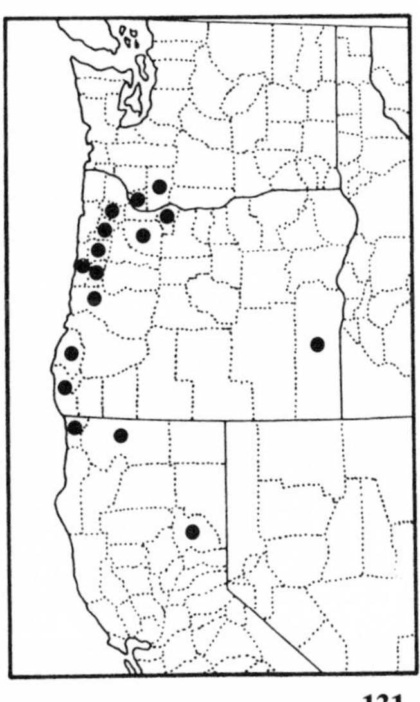

Rectofrontia 'sectilis J. & R'. Wagner, 1936 (*nec* Jordan and Rothschild, 1923), Canad. Ent. 68: 203, figs. 8, 9 (Vancouver, B. C., from *Peromyscus maniculatus austerus*).

Micropsylla goodi. Hubbard, 1941, Pacific Univ. Bull. 37(10): 2 (Forest Grove, Oregon, from *Peromyscus maniculatus rubidus*).

Micropsylla sectilis goodi Hubbard. Holland, 1949, The Siphonaptera of Canada, pp. 89, 90.

Rhadinopsylla (Micropsylla) sectilis goodi (Hubbard). Smit, 1957, Bull. Brit. Mus. (nat. Hist.), Ent. 6: 69, 75.

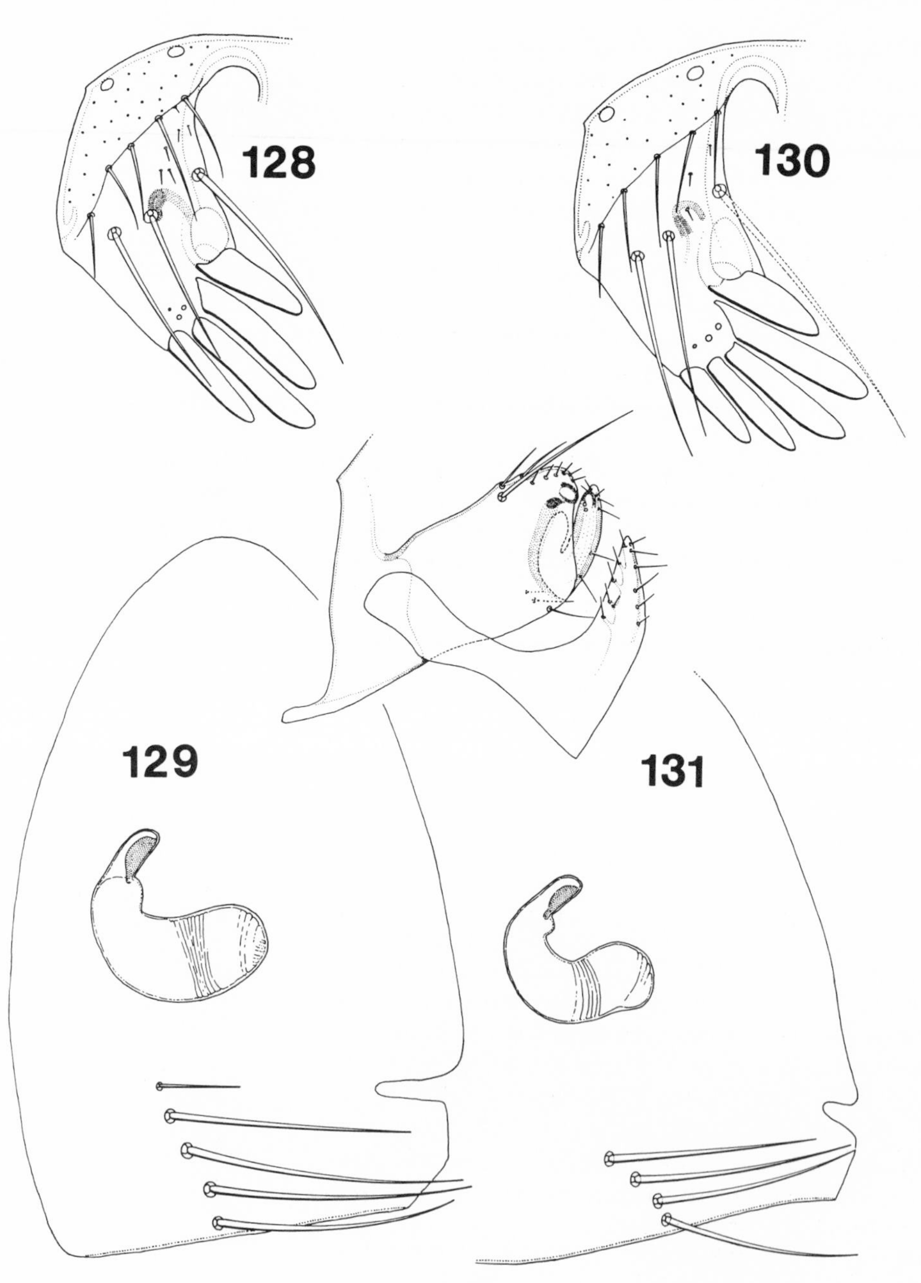

128-131 *Rhadinopsylla sectilis* subspecies. **128** *R. sectilis sectilis* J. & R., preantennal region of head. **129** Female sternum VII and spermatheca. **130** *R. sectilis goodi* Hubbard, preantennal region of head. **131** Female sternum VII and spermatheca. Subspecies inseparable on basis of male genitalia. After Holland, 1985.

These two subspecies are very similar in most respects and a strong case could be made for considering them to be conspecific. However, the Cascade Mountains do seem to be the dividing line between the western (*R. s. goodi*) and the eastern (*R. s. sectilis)* populations. The nominate subspecies is reported here from six males and eleven females taken in Jefferson, Lane, and Klamath counties in Oregon. The other subspecies is reported here from fourteen males and fourteen females taken in Benton, Lincoln, Coos, Curry, and Lane counties in Oregon. Two males included here were taken in Malheur County in the southeastern part of the state, but this record may be in error. All other records correspond to the typical east-west distribution. Svihla (1941) reports both subspecies from Washington.

Most collections came from microtines (i.e., *Microtus, Lagurus, Clethrionomys, Arborimus*) but *Peromyscus, Eutamias, Spermophilus* and *Spilogale* species also yielded a few specimens. Again, adults appear to be more common during the cooler months.

Rhadinopsylla (Actenophthalmus) heiseri (McCoy)

[91129]

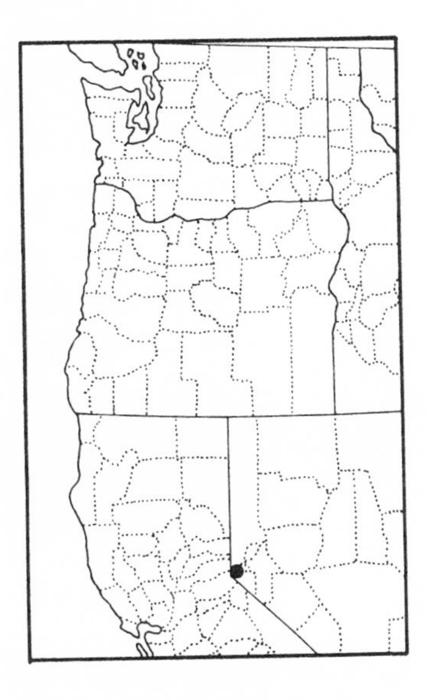

Ctenophthalmus heiseri. McCoy, 1911, Ent. News 22: 445 (Mojave, California, from unknown host).

Actenophthalmus heiseri McCoy. C. Fox, 1925, Ent. News 36: 121.

Rhadinopsylla (Actenophthalmus) heiseri (McCoy). Morlan and Prince, 1955, Texas Rep. Biol. Med. 12: 1039.

This, the type species of the subgenus *Actenophthalmus,* is a large flea that is associated with ground squirrels in the Great Basin and other arid regions of the northwest. While not a common faunal element in the area under consideration, it has been collected in California and Nevada, The allotype is from Carson City, Nevada. The bulk of the material in our collection is from Utah. The species is unique in having six pairs of lateral plantar bristles on the fifth tarsal segment, rather than the usual four or five pairs characteristic of other members of the subgenus.

Rhadinopsylla (Actenophthalmus) fraterna (Baker)

[89512]

(Figs. 132, 133, 134)

Typhlopsylla fraterna. Baker, 1895, Canad. Ent. 27: 189, 190 (Brookings, South Dakota, host not given).

Ctenophthalmus fraternus Baker. Baker, 1904, Proc. U. S. Nat. Mus. 27: 420, 423, 450.

Typhlopsylla fraterna Baker. Rothschild, 1913, Entomologist 46: 297 (referred to *Rhadinopsylla*).

Neopsylla hamiltoni. Dunn, 1923 (in: Dunn and Parker) Publ. Hlth. Rep., Wash. 38: 2770, 2775 (Spoon Creek, southwest of Darby, Montana, from *Neotoma cinerea*).

Rectofrontia fraterna Baker. Jordan, 1937, Novit. Zool. 40: 270 (*hamiltoni* synonymized).

Rhadinopsylla (Rectofrontia) fraterna (Baker). Morlan and Prince, 1955, Texas Rep. Biol. Med. 12: 1045, fig. 3.

Rhadinopsylla (Actenophthalmus) fraterna (Baker). Smit, 1957, Bull. Brit. Mus. (nat. Hist.), Ent. 6: 57, 71, 75.

This is mainly an eastern species and is included here on the strength of a record of a single female reported by Hubbard (1947) from the summit of McKenzie Pass, Oregon. Here he also quotes Holland's records from British Columbia and Alberta. In view of the rarity of members of this genus in collections, and the fact that a number of additional taxa have been described since 1947, the record of this species seems subject to some question. The whole North American complex of species in this genus is in much need of scrutiny, but such a study must await the collection of additional material.

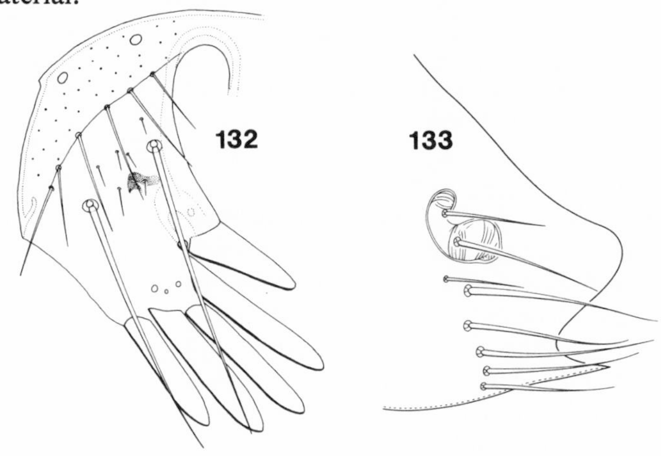

132-133 *Rhadinopsylla fraterna* (Baker). **132** Preantennal region of head of male. **133** Female sternum VII and spermatheca. After Holland, 1985.

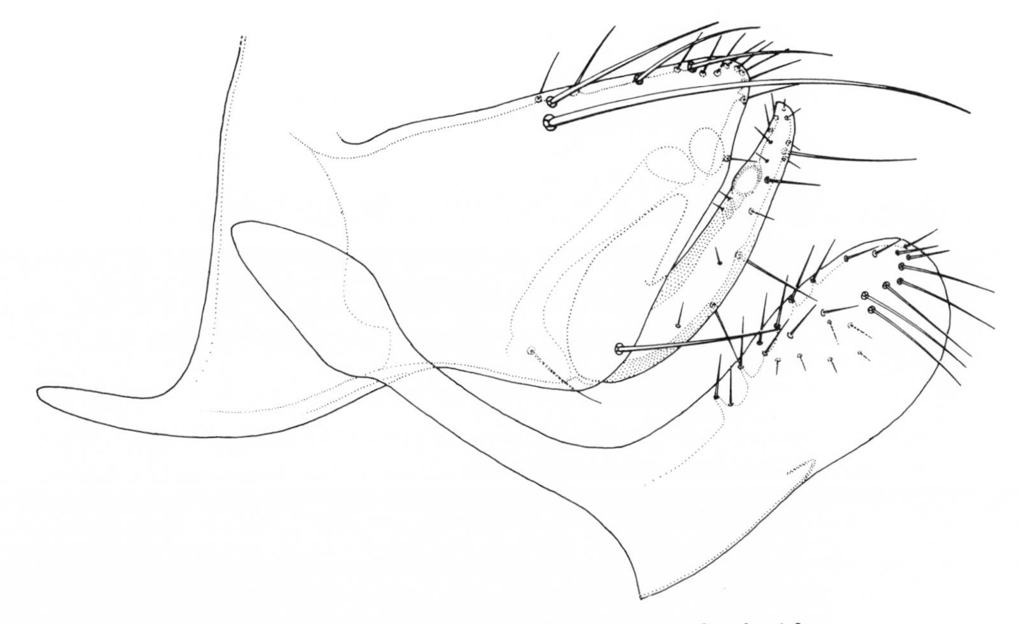

134 *Rhadinopsylla fraterna* (Baker). Modified abdominal segments of male. After Holland, 1985.

Rhadinopsylla (Actenophthalmus) difficilis Smit

[95765]

(Figs. 135, 136)

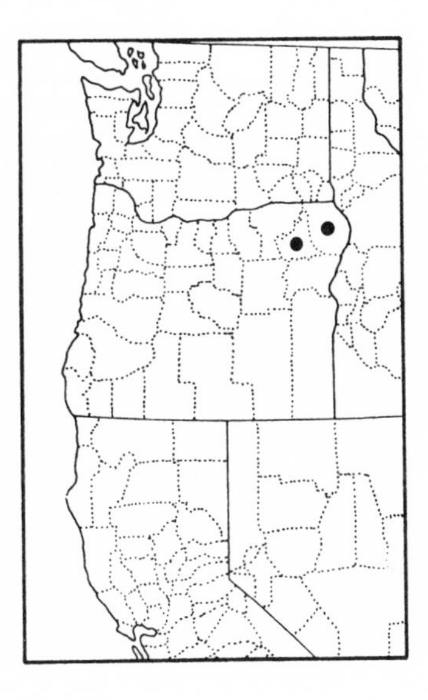

Rhadinopsylla (Actenophthalmus) difficilis. Smit, 1957, Bull. Brit. Mus. (nat. Hist.), Ent. 6: 65, 75, figs. 41-43, 47 (Kelowna, British Columbia, from *Mustela* sp.).

Rhadinopsylla (Actenophthalmus) difficilis Smit. Hopkins and Rothschild, 1962, An Illustrated Catalogue... Vol. III, pp. 541-542.

This species is reported on the strength of two specimens. A single male was taken on *Mustela frenata* in Union County, Oregon, in November, and a female was found on *Martes americana* in Wallowa County, Oregon, in January. Both locations are in the extreme northeastern part of the state. Although Hopkins and Rothschild (1962) state that *Mustela* is not likely the true host, these two records suggest differently.

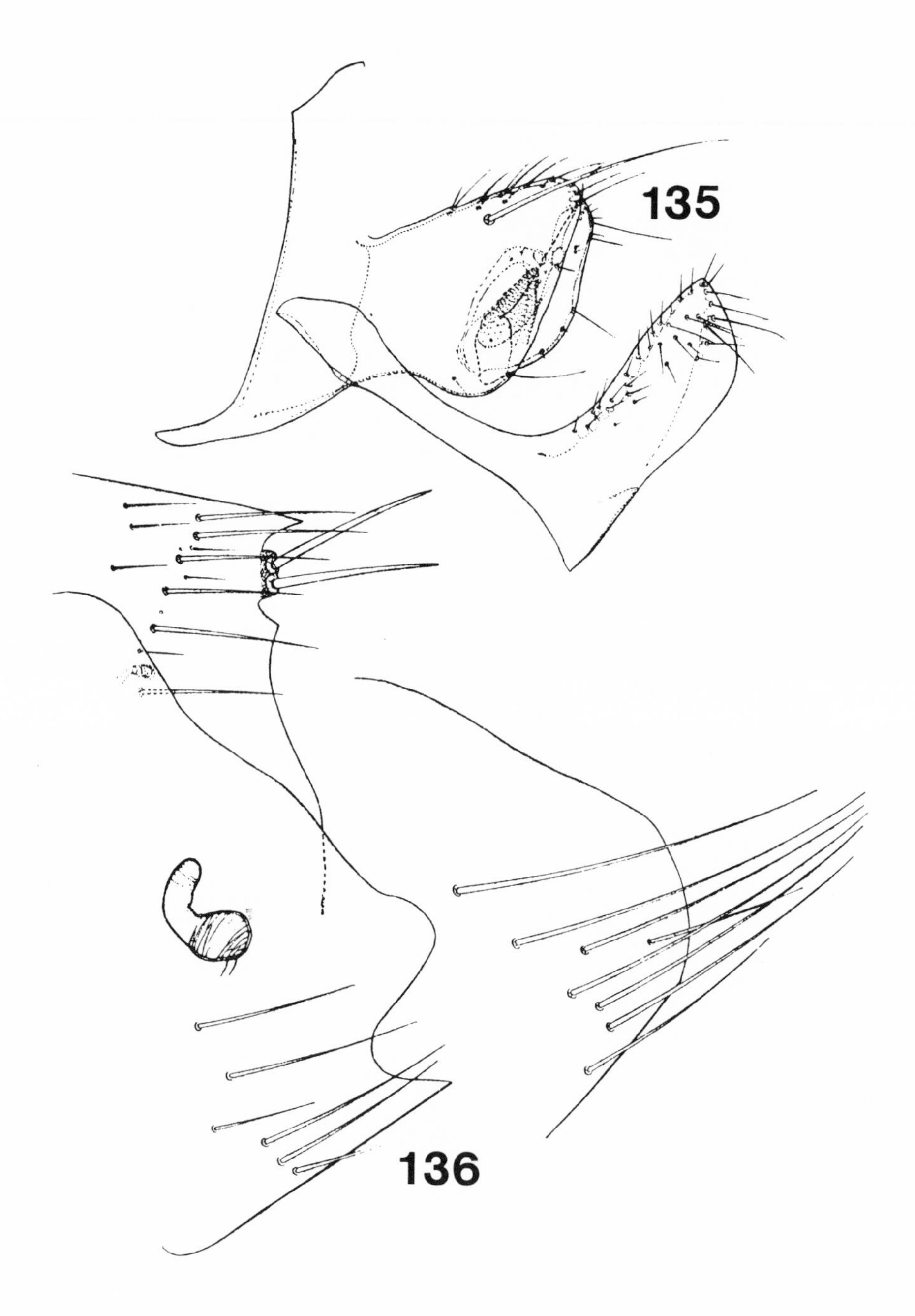

135-136 *Rhadinopsylla difficilis* Smit. **135** Modified abdominal segments of male. **136** Modified abdominal segments of female. After Holland, 1985.

SUBFAMILY **Doratopsyllinae**

There are six genera assigned to this subfamily. One of these, *Xenodaeria* Jordan, 1932, is known from the Himalayas and adjacent mountain systems where it is a parasite of small insectivores. The ten recognized taxa of *Doratopsylla* J. & R., 1912, except one, are distributed through the Palaearctic Region as parasites of shrews and moles. The exception, *D. blarinae* C. Fox, 1914, occurs in the eastern United States. Both *Acedestia* Jordan, 1937, and *Idilla* Smit, 1957, are Australian, parasitizing small marsupials. *Adoratopsylla* Ewing, 1925, species are parasites of marsupials in South America. The only northwestern representative of this subfamily belongs to the genus *Corrodopsylla.*

GENUS **Corrodopsylla** Wagner

Doratopsylla subgenus *Corrodopsylla.* Wagner, 1929, Konowia 8: 317 (type species by subsequent designation: *Doratopsylla curvata* Rothschild, 1915).
Corrodopsylla Wagner. Wagner, 1936, Canad. Ent. 68: 205.

This small genus contains four species, one of which occurs in Asia. The remaining three occur in the Western Hemisphere. They are mainly parasites of insectivores of the family Soricidae but there are many published records from rodents as well. Only one species occurs in the area under consideration.

Corrodopsylla curvata obtusata (Wagner)
[92930]
(Figs. 137, 138)

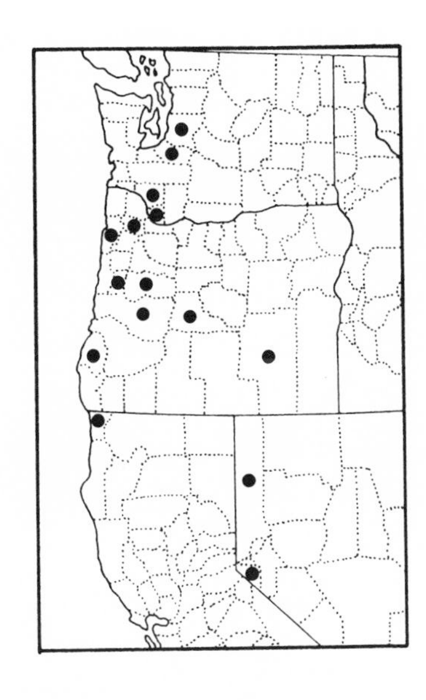

Doratopsylla curvata obtusata. Wagner, 1929, Konowia 8: 318, fig. 2 (Abbotsford, British Columbia, from a shrew).
Doratopsylla jellisoni. Hubbard, 1940, Pacific Univ. Bull. 37: 8 (3 miles N.E. Forest Grove, Oregon, from Trowbridge shrew).
Doratopsylla curvata obtusata Wagner. Holland, 1942, Canad. Ent. 74: 157 (*jellisoni* synonymized).
Corrodopsylla curvata obtusata (Wagner). Smit and Wright, 1965, Mitt. Hamburg. Zool. Mus. Inst. 62: 30 (selection of lectotype).

Although this is not a common element of the fauna, four males and six females are reported here from Benton, Coos, Deschutes, Lane, and Linn counties in Oregon. All were taken from *Sorex vagrans* and they show no particular peak in seasonal abundance. Hansen (1964) lists this species from Oregon but does not indicate which subspecies. There is no current evidence that the nominate subspecies occurs in the area under consideration. Holland (1949) suggests that the differences between both sexes of *C. curvata curvata* and *C. curvata obtusata* were so clear and consistent that perhaps the latter taxon should be given the status of a full species. Jellison and Senger (1973) list both from Montana, and the fact that they seem to be sympatric would support Holland's contention. Material from the eastern part of the state might help to resolve this question. For the present, subspecific status is maintained.

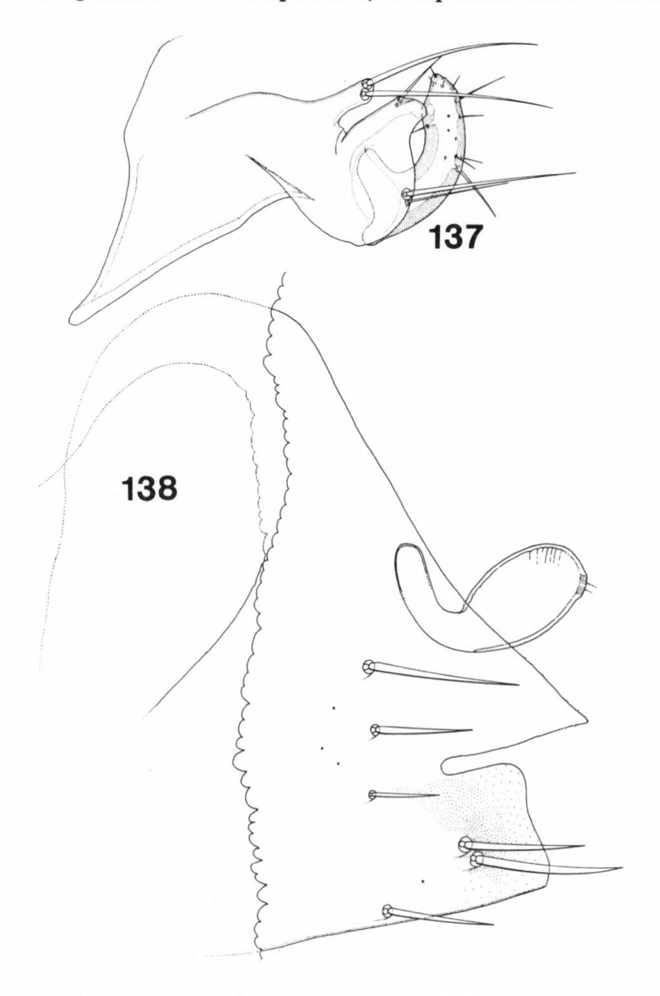

137-138 *Corrodopsylla curvata obtusata* (Wagner). **137** Male clasper. **138** Female sternum VII and spermatheca. After Holland, 1985.

SUBFAMILY **Ctenophthalminae**

According to the system followed by Hopkins and Rothschild (1966), this subfamily contains four tribes. The nominate tribe contains the genera *Palaeopsylla* with 53 taxa and *Ctenophthalmus* with 269 described forms. Both genera are extralimital to this study. The remaining three tribes contain four genera, three of which are restricted to South America. *Neotyphloceras* has four taxa, *Chiliopsylla* one, and *Agastopsylla* five, respectively. Only *Carteretta* might occur in the area under consideration.

Carteretta carteri C. Fox

[92706]

Carteretta carteri. C. Fox, 1927, Trans. Amer. Ent. Soc. 53: 209 (Los Angeles, California, from *Neotoma fuscipes*).

Trirarchipsylla digitiformis. Stewart, 1940, Pan-Pacific Ent. 16: 24, figs. 13-15 (near Jamesburg, Monterey Co., California, from *Perognathus californicus* ssp.).

Carteretta carteri clavata C. Fox. Good, 1942, Ann. ent. Soc. Amer. 35: 110, 112 (*T. digitiformis* synonymized and *C. carteri clavata* described).

Carteretta carteri C. Fox, 1927. Hopkins and Rothschild, 1966, An Illustrated Catalogue... Vol. IV, pp. 149-150 (*C. clavata* elevated to full species).

This species is probably extralimital to the area treated here, but there is one marginal record from California and the possibility exists that more collecting in the northern part of the state will produce it. It would seem to be a winter flea with a preference for *Perognathus californicus* as a host, although there are records from a few other rodents as well.

FAMILY **LEPTOPSYLLIDAE**

Derivation: This family name is derived from the Greek words *lepto* (fine, slender) and *psylla* (a flea) and the Latin suffix *idae,* and alludes to the slender, graceful appearance of most members of the family. The type genus is *Leptopsylla* Jordan and Rothschild, 1911.
General Characters: Tentorial arm present, complete; with or without genal ctenidium; pronotal ctenidium always present; eye usually heavily pigmented and well developed, sinuate ventrally; dorsal bristle of ocular row arising above upper margin of eye; marginal or submarginal spinelets present (except in *Ornithophaga);* dorsal surface of sensilium practically straight in both sexes; acetabular bristles absent or weakly developed; Wagner's gland absent; anal stylet of female with one or two long, sublateral bristles in addition to apical bristle.
World Distribution: Distributed worldwide except for the Neotropical Region. Poorly represented except for the Holarctic Region.
Number of Northwestern Taxa: 9
General References: Hopkins and Rothschild (1971), Holland (1985).

Genera belonging to both subfamilies occur in North America, but only *Peromyscopsylla* is well represented by a number of species. For the purpose of identifying the fauna of the Northwest, subfamilies may be separated with the following key.

KEY TO THE SUBFAMILIES OF *LEPTOPSYLLIDAE*

(Modified from Hopkins and Rothschild, 1971)

1 Head always fracticipit; genal comb always present .... **Leptopsyllinae**

Head always integricipit; genal comb never present .... **Amphipsyllinae**

SUBFAMILY **Leptopsyllinae**

Only two genera belonging to this subfamily are known from the Northwest. Both belong to the tribe Leptopsyllini and are associated with small rodents, both domestic and feral. They may be separated with the following key.

KEY TO THE NORTHWESTERN GENERA OF *LEPTOPSYLLINAE*

1 Genal comb of two spines; trabecula centralis absent; fore coxae arising at apex of prosternosome .... **Peromyscopsylla**

Genal comb of four spines; frons with two spiniforms; trabecula centralis present; fore coxae arising below apex of prosternosome .... **Leptopsylla**

GENUS **Peromyscopsylla** I. Fox

Peromyscopsylla. I. Fox, 1939, Proc. ent. Soc. Wash. 41: 47 (type species by designation: *Ctenopsyllus hesperomys* Baker, 1904).
Peromyscopsylla I. Fox. Johnson and Traub, 1954, Smithson. miscl. Coll. 123: 2.
Peromyscopsylla I. Fox. Hopkins and Rothschild, 1971, An Illustrated Catalogue... Vol. V, p. 105.

Recent workers divide this genus into two species groups based on whether or not the genal process is hidden by the genal comb. The four taxa known from the area fall into the *"sylvatica"* group in which the genal process is visible. They may be separated with the following key.

KEY TO THE NORTHWESTERN SPECIES OF *PEROMYSCOPSYLLA*

(Modified from Hopkins and Rothschild, 1971)

1 Dorsal spine of genal comb extending farther caudad than ventral spine; usually with four spiniforms in submarginal row of frons; movable process broadest before apex; female with three antepygidial bristles per side .... **selenis**

Dorsal spine of genal comb about the same length as ventral spine; with three or four spiniforms in submarginal row of frons; movable process triangular; female with four antepygidial bristles per side .... **hesperomys—2**

2 Four spines in submarginal row of frons; males with dense bristles on dorsal metanotum; sinus in female st. VII not very deep and often extremely narrow; usually found on *Neotoma* species ..... **hesperomys ravalliensis**

Three spines in submarginal row of frons; with fewer metanotal bristles; associated with *Peromyscus* species 3

3 Movable process of male very short, its posterior margin semicircular; dorsal lobe of female st. VII short and relatively broad, far shorter than ventral lobe .... **hesperomys adelpha**

Movable process of male somewhat longer, its posterior margin less than semicircular; ventral lobe of female st. VII narrow, often smoothly rounded .... **hesperomys pacifica**

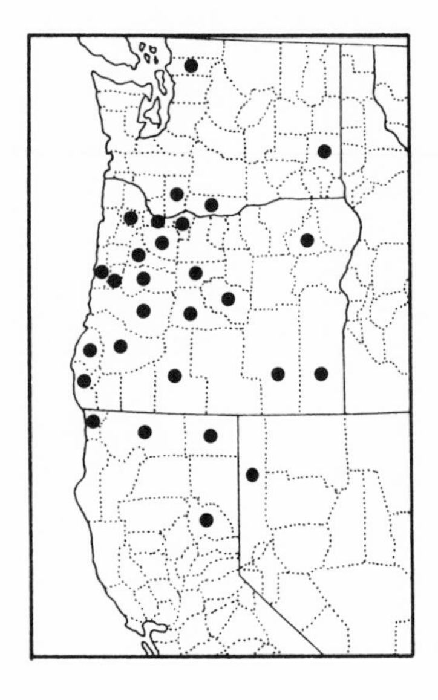

Peromyscopsylla selenis (Rothschild)

[90613]

(Figs. 139, 140)

Ctenopsyllus selenis. Rothschild, 1906, Canad. Ent. 38: 322, fig. 43 (Horse Creek, Upper Columbia Valley, Canada, from *Peromyscus canadiani* [= *P. maniculatus*]).

Leptopsylla selenis Rothschild. Jordan, 1928, Novit. Zool. 34: 186.

Peromyscopsylla selenis (Rothschild). Jellison and Good, 1942, Bull. U. S. Natl. Inst. Hlth. (178): 123.

Peromyscopsylla duma. Traub, 1944, Zool. Ser. Field Mus. 29: 217 (Logan Canyon, Cache Co., Utah, from *Microtus* sp.).

Peromyscopsylla selenis (Rothschild). Hubbard, 1947, Fleas of Western North America, pp. 329, 333, fig. 200 (*partim*).

Peromyscopsylla duma Traub. Hubbard, 1947, *op. cit.,* p. 334, fig. 200a (inseparable from *selenis*).

Peromyscopsylla selenis (Rothschild). Johnson and Traub, 1954, Smithson. miscl. Coll. 123: 5, 7, 31, figs. 53-58, 114 (*duma* synonymized).

This is one of the commoner faunal elements in Oregon and 104 males and 122 females are reported here from the state. As shown in the following table, this species is mainly a parasite of microtines, but also may be taken in fair numbers on their associates.

Host Associations in *Peromyscopsylla selenis*

Host Species	Males	Females
Microtus townsendi	0	8
Microtus montanus	8	14
Microtus oregoni	26	32
Microtus longicaudus	4	4
Microtus richardsoni	6	5
Microtus sp.	0	2
Clethrionomys californicus	32	35
Clethrionomys gapperi	4	1
Phenacomys intermedius	1	0
Lagurus curtatus	3	4
Peromyscus maniculatus	9	8
Glaucomys sabrinus	0	1
Sorex vagrans	5	2
Sorex trowbridgei	6	6
Total	104	122

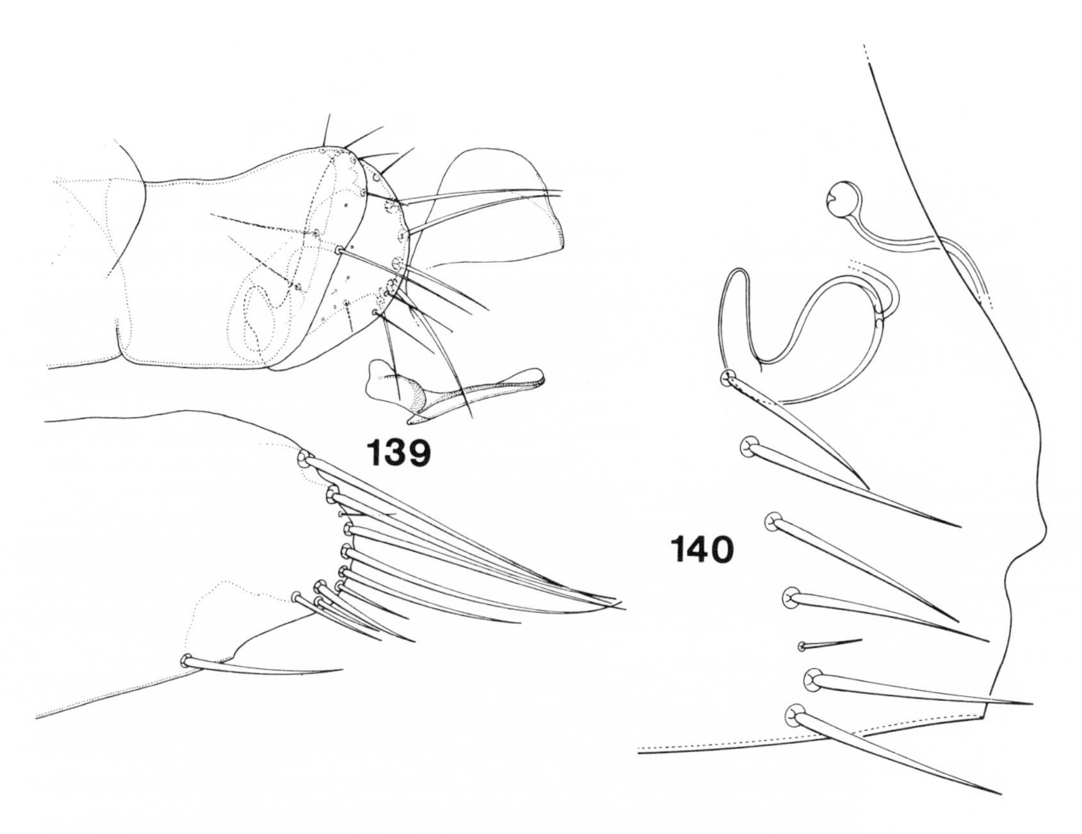

139-140 *Peromyscopsylla selenis* (Rothschild). **139** Male clasper and sternum VIII. **140** Female sternum VII and spermatheca. After Holland, 1985.

Slightly over 83 percent of the collections reported here came from microtine hosts. These collections also show that this species occurs as adults throughout the year, with peak populations in August and September. This is not in agreement with Hubbard (1947) who claimed it was more common during the fall, winter, and spring. More collections during the winter are required to resolve this conflict.

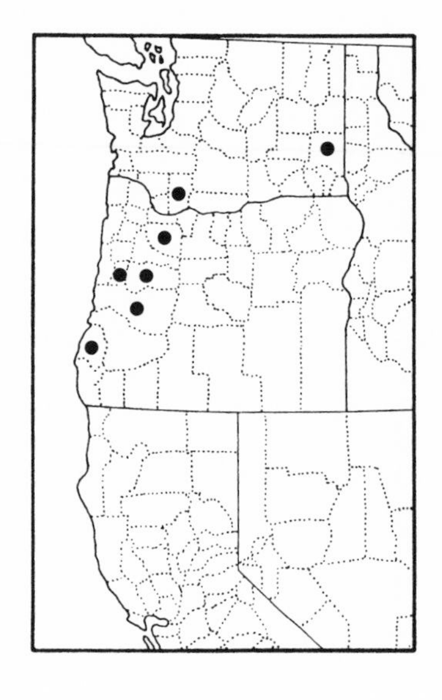

Peromyscopsylla hesperomys pacifica Holland

[94938]

(Figs. 141, 142)

Peromyscopsylla 'hesperomys (Baker).' Hubbard, 1947, (*nec* Baker, 1904), Fleas of Western North America, p. 329, fig. 197 (*partim*).

Peromyscopsylla hesperomys pacifica. Holland, 1949, The Siphonaptera of Canada, p. 176, pl. 40, figs. 319-321, 324 a-e, map 41 (Vancouver, British Columbia, from *Peromyscus maniculatus austerus*).

Peromyscopsylla hesperomys pacifica Holland. Hopkins and Rothschild, 1971, An Illustrated Catalogue... Vol. V, pp. 154-155.

This species is not as common as *P. selenis;* only seven males and twelve females are reported from our collections. All the males and nine of the females are from *Peromyscus maniculatus,* its preferred host. Collections came from Benton, Coos, Lane, and Linn counties in Oregon and, with the exception of one specimen taken in December, all were taken in August and September. Hubbard (1947) cites this as a common species, but our records do not reflect this abundance.

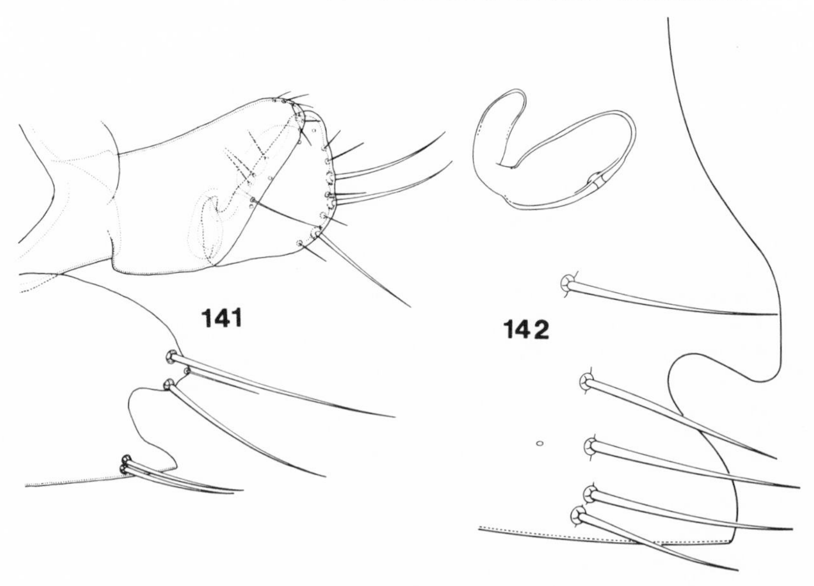

141-142 *Peromyscopsylla hesperomys pacifica* Holland. **141** Male clasper and sternum VIII. **142** Female sternum VII and spermatheca. After Holland, 1985.

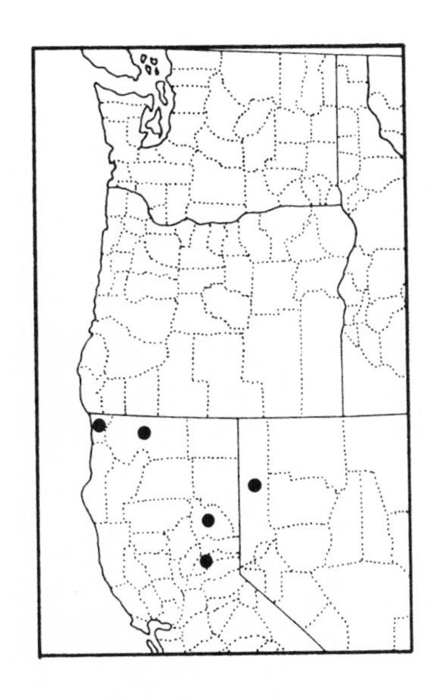

Peromyscopsylla hesperomys adelpha (Rothschild)

[91505]

Leptopsylla adelpha. Rothschild, 1915, Novit. Zool. 22: 304 (Paradise, Arizona, from *Mus* sp.).

Peromyscopsylla hemisphaerium. Stewart, 1940, Pan-Pacific Ent. 16: 25 (near Jamesburg, Monterey Co., California, from *Peromyscus truei*).

Peromyscopsylla hesperomys adelpha (Rothschild). Hopkins, 1951, Ann. Mag. Nat. Hist. (12)4: 541.

Peromyscopsylla hesperomys adelpha (Rothschild). Johnson and Traub, 1954, Smithson. miscl. Coll. 123: 7, 8, 17 (*hemisphaerium* synonymized).

Peromyscopsylla zempoalensis. Barrera, 1954, Ciencia, Mex. 14: 88 (Lagunas de Zempoala, Morelos, Mexico, from *Peromyscus maniculatus labecula*).

Peromyscopsylla hesperomys adelpha (Rothschild, 1915). Hopkins and Rothschild, 1971, An Illustrated Catalogue...Vol. V, pp. 155-157.

This is a widespread taxon ranging from Alberta and Saskatchewan in the north into Mexico in the south, and from Colorado and Wyoming west to California. As with many other members of the genus, this is almost exclusively a parasite of *Peromyscus*.

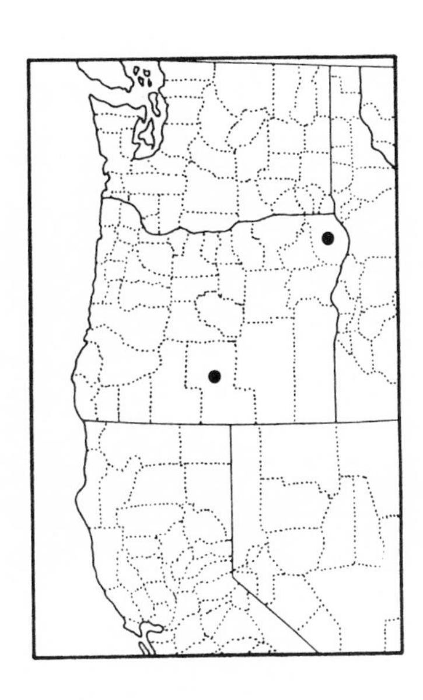

Peromyscopsylla hesperomys ravalliensis (Dunn)

[92340]

(Figs. 143, 144)

Ctenopsyllus ravalliensis. Dunn, 1923 (in: Dunn and Parker), Publ. Hlth. Rep., Wash. 38: 2760, 2775 (Tin Cup Creek, foothills of Bitterroot Mountains, S. W. of Darby, Montana, from *Neotoma cinerea*).

Peromyscopsylla ravalliensis (Dunn). Jellison and Good, 1942, Natl. Inst. Hlth. Bull. (178): 122.

Peromyscopsylla hesperomys ravalliensis (Dunn). Johnson and Traub, 1954, Smithson. miscl. Coll. 123: 6, 8, 14, figs. 12, 14, 18-21, 27, 28.

Peromyscopsylla ravalliensis (Dunn). Holland, 1985, Mem. Ent. Soc. Canada No. 130, pp. 227-230.

Although we have no new collections of this subspecies to report, it is included on the strength of the Oregon records noted by Hubbard (1947). As he points out, the species occurs mainly on the bushy-tailed wood rat, *Neotoma cinerea*, and is distributed throughout the northern portion of the Rocky Mountain Subregion, including southwestern Canada.

Hopkins (1985) reports finding this species in Deschutes County, Oregon, but does not assign a subspecific name. His collections were made in June, July, and September and were made from *Peromyscus maniculatus* and *Spermophilus lateralis*.

GENUS **Leptopsylla** Jordan and Rothschild

Ctenopsyllus. Kolenati, 1863 (*nec* 1856) Trudy Russk. Ent. Obshch. 2: 37 (type species by subsequent designation: (Baker, 1904) *musculi* Duges [= *quadridentatus*]).

Ctenopsylla. Wagner, 1893, Trudy Russk. Ent. Obshch. 27: 350 (an invalid emendation of *Ctenopsyllus* Kolenati, 1863).

Ctenophthalmus. Oudemans, 1907 (*nec* Kolenati, 1856), Ent. Ber., Amst. 2: 219 (type species designated: *Pulex musculi* [Duges, 1832, see Oudemans, 1914, Ent. Ber., Amst. 4: 136]; species not mentioned under *Ctenophthalmus* in Kolenati, 1856. Selection as type invalid).

Leptopsylla. Jordan and Rothschild, 1911, Novit. Zool. 18: 85 (*nomen novum* for *Ctenopsyllus* Kolenati, 1862 [*recte* 1863], *nec* Kolenati, 1856. Type species: *musculi* Duges, 1832).

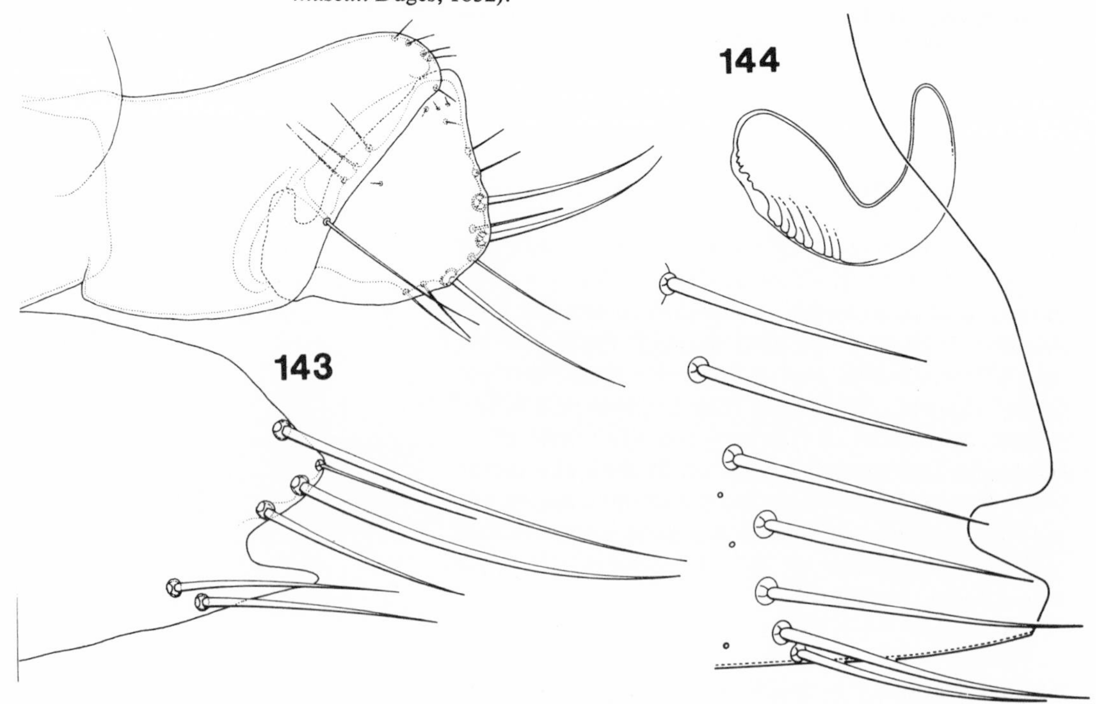

143-144 *Peromyscopsylla hesperomys ravalliensis* (Dunn). **143** Male clasper and sternum VIII. **144** Female sternum VII and spermatheca. After Holland, 1985.

Triainopsylla. Rosický, 1957, Čslká Parasit. 4: 298 (type species by designation: *Ctenopsylla taschenbergi* Wagner, 1898).
Leptopsylla. Smit, 1960, Proc. R. Ent. Soc. Lond. (B), 29: 12 (*Triainopsylla* treated as a synonym).

This is an Old World genus, the members of which are distributed throughout the Palaearctic Region and the northern portion of the Ethiopian Region. There are currently 21 recognized species and subspecies in the nominate subgenus. One of these, *L. segnis,* is worldwide in occurrence, having been transported via human agencies to all continents, on its preferred host, *Mus musculus.*

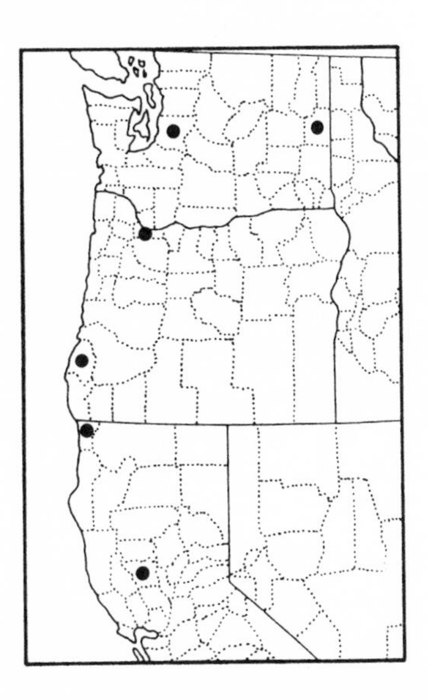

Leptopsylla (Leptopsylla) segnis (Schönherr)

[81101]

(Figs. 145, 146, 147)

Pulex segnis. Schönherr, 1811, K. svenska VetenskAcad. Handl.
(2)32: 99, pl. 5, figs. A, B (Sweden, from *Mus musculus*).
Leptopsylla musculi Duges. Jordan and Rothschild, 1911, Novit. Zool. 18: 85.
Leptopsylla segnis Schönh. Jordan, 1929, Novit. Zool. 35: 177.

For a more complete synonymy of this taxon, see Hopkins and Rothschild (1971).

No specimens of this species were collected during our Oregon studies and the taxon is included on the strength of records listed in Hubbard (1947) and those of other workers. This is a common flea on commensal rats and mice and has been introduced to all continents through their port cities. Where suitable murid hosts are an indigenous faunal element, the species may become established inland. As a result, it has been reported from all of the major landmasses of the world with the exception of Antarctica. The species has been associated with plague but evidently it is a poor vector at best and not considered to be of importance in the transmission of this disease.

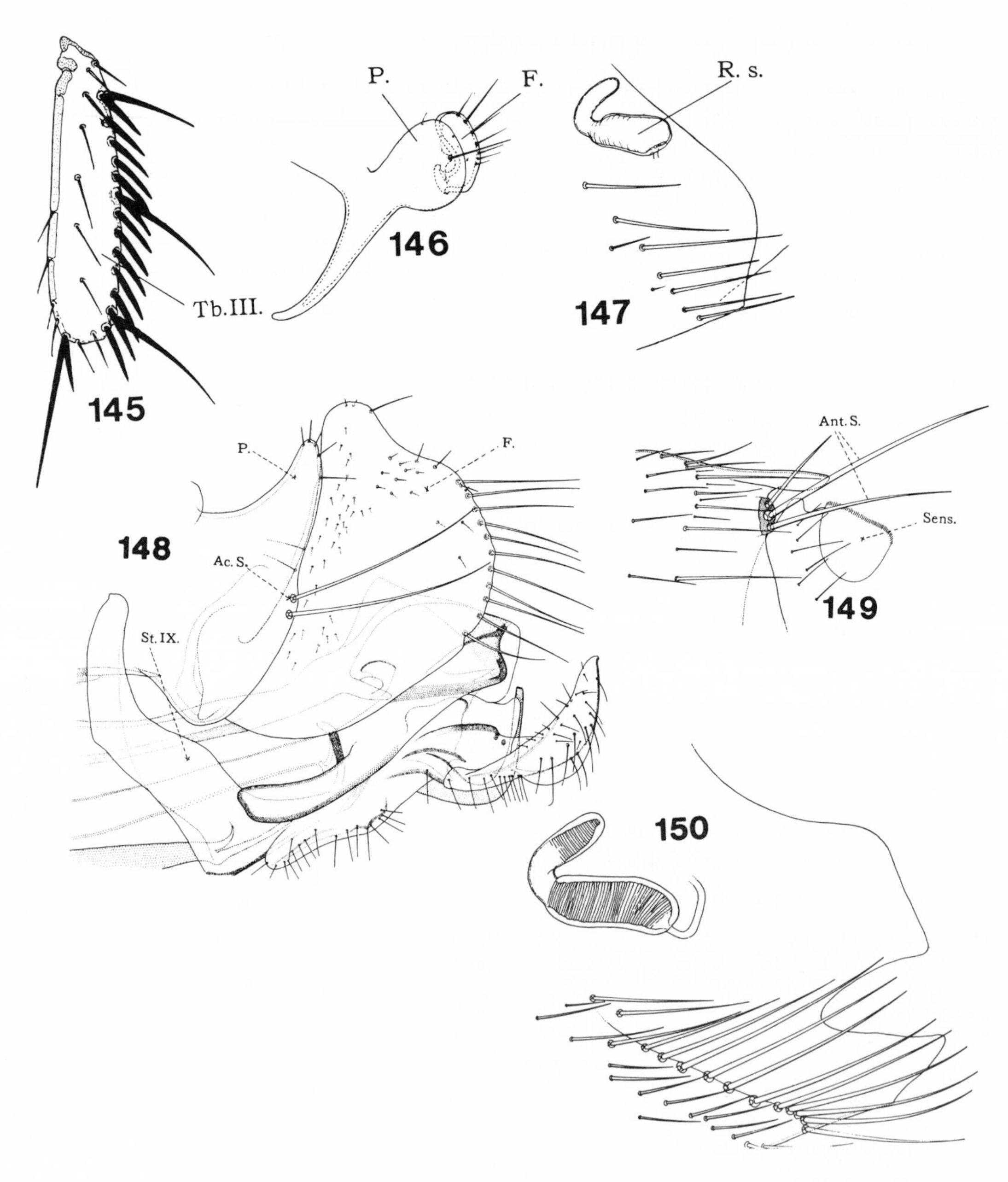

145-147 *Leptopsylla segnis* (Sch¨onherr). **145** Hind tibia showing false comb. **146** Male clasper. **147** Female sternum VII and spermatheca. After Holland, 1985.
148-150 *Dolichopsyllus stylosus* (Baker). **148** Male genitalia. **149** Apex of tergum VII of male. **150** Female sternum VII and spermatheca. After Holland, 1985.

SUBFAMILY **Amphipsyllinae**

Two of the four recognized tribes belonging to this subfamily are represented in the flea fauna of the Pacific Northwest. Indeed, the area is the center of distribution for the monotypic Dolichopsyllini. Following is a modification of the Hopkins and Rothschild (1971) key for separating the two tribes.

1 Large fleas (up to 6 mm), with no eye but with well developed pronotal comb .... **Dolichopsyllini**

Much smaller, eye present and usually large .... **Amphipsyllini**

TRIBE **DOLICHOPSYLLINI**

This tribe contains the single genus *Dolichopsyllus.* Its only species is specific to the primitive rodent *Aplodontia rufa.*

GENUS **Dolichopsyllus** Baker

Dolichopsyllus. Baker, 1905, Proc. U. S. Nat. Mus. 29: 127, 135, 155, 168 (type species by monotypy: *Ceratophyllus stylosus* Baker, 1904).

This is another monotypic genus associated with the mountain beaver. Like most other flea species associated with this host, individuals tend to be larger and more setose than most fleas. As might be expected, the range of the genus coincides with that of the host, from northern California to southern British Columbia.

Dolichopsyllus stylosus (Baker)

[90433]

(Figs. 148, 149, 150)

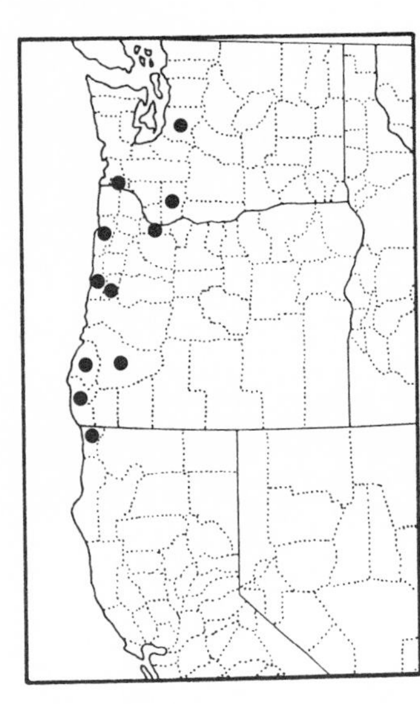

Ceratophyllus stylosus. Baker, 1904, Proc. U. S. Nat. Mus. 27: 388, 418, 420, 447, pl. 14, figs. 1-7, pl. 15, figs. 1, 2 (Astoria, Oregon, from *Aplodontia rufa*).
Dolichopsyllus stylosus Baker. Baker, 1905, *op. cit.* 29: 155.
Dolichopsylla stylosa (Baker). Ewing and Fox, 1943, Misc. Publ. U. S. Dept. Agric. No. 500: 29, figs. 5E, 5F.
Dolichopsyllus stylosus (Baker). Hopkins and Rothschild, 1971, An Illustrated Catalogue... Vol. V, pp. 229-230.

Twenty-six males and 48 females are reported here, mostly from *A. rufa* collected in Benton, Coos, Curry, Douglas, Lincoln, and Tillamook counties in Oregon. Although the species can be taken throughout the year, our records suggest that the peak population of adults probably occurs during the winter and early spring. The sex ratio is almost exactly two females to one male. A few of our records are from *Spilogale putorius, Mustela vison,* and *Felis rufus* but there is no evidence that these are anything other than chance associations.

TRIBE **AMPHIPSYLLINI**

Genera belonging to this tribe are found mainly in the Palaearctic Region, where their species are usually parasites of small rodents. Three genera have representatives in North America and all of them are known from the area under consideration.

KEY TO THE NORTHWESTERN GENERA OF AMPHIPSYLLINI

1 First pair of plantar bristles of tarsal segment V displaced mesad and arising between the bases of the second pair .... **Amphipsylla**

First pair of lateral plantar bristles of tarsal segment V not arising between bases of second pair **2**

2 Frontal row of bristles distinctly spiniform; bulga of spermatheca with orifice at the apex of a conspicuous cone .... **Ctenophyllus**

Frontal row of bristles not spiniform; with patch of small, spiniform bristles on inner surface of hind coxae .... **Odontopsyllus**

GENUS **Odontopsyllus** Baker

Odontopsyllus. Baker, 1905, Proc. U. S. Nat. Mus. 29: 129, 131 (type species by designation: *Pulex multispinosus* Baker, 1898).

This genus is known from three species, two from North America and one from the Iberian Peninsula. They are typically parasites of lagomorphs, but seem to be equally at home on rabbit and hare predators. *Odontopsyllus multispinosus* is an eastern species. The western species is a common faunal element in the Pacific Northwest.

Odontopsyllus dentatus (Baker)

[90412]

(Figs. 151, 152)

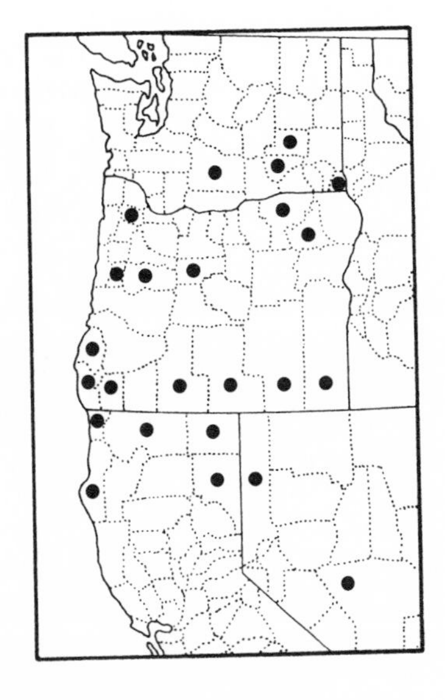

Ceratophyllus dentatus. Baker, 1904, Proc. U. S. Nat. Mus. 27: 386, 390, 441 (Moscow, Idaho, from *Lynx canadensis*).
Odontopsyllus dentatus Baker. Baker, 1905, *op. cit.* 29: 131, 145.
Ceratophyllus ponerus. Rothschild, 1909, Novit. Zool. 16: 54, pl. 8, fig. 5 (Palo Alto, California, from a fox).
Odontopsyllus spenceri. Dunn, 1923 (in: Dunn and Parker), Publ. Hlth Rep., Wash. 38: 2765, 2775 (foothills of Bitterroot Mountains west of Hamilton, Montana, from *Lepus americanus bairdi*).
Odontopsyllus dentatus Baker. Kohls, 1940, Natl. Inst. Hlth. Bull. (175): 20, pl. 3, figs. C, E (*spenceri* synonymized).
Odontopsyllus dentatus (Baker). Ewing and Fox, 1943, Miscl. Publ. U. S. Dept. Agric. No. 500: 25 (*ponerus* synonymized).
Odontopsyllus dentatus (Baker). Hopkins and Rothschild, 1971, An Illustrated Catalogue... Vol. V, pp. 286-288.

Four hundred and twelve males and 182 females are reported here. Although rabbits and hares are the preferred hosts in this genus, 307 males and 108 females were collected on predators, especially *Felis rufus* and *Canis latrans*. The species occurs throughout the area, and adult populations evidently peak in February and March, falling off rapidly in April and May.

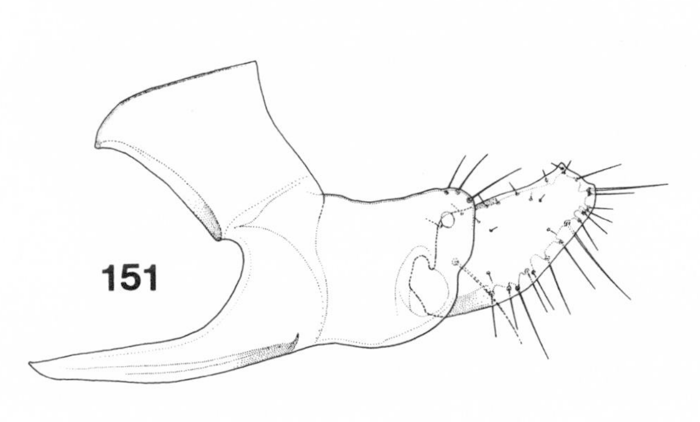

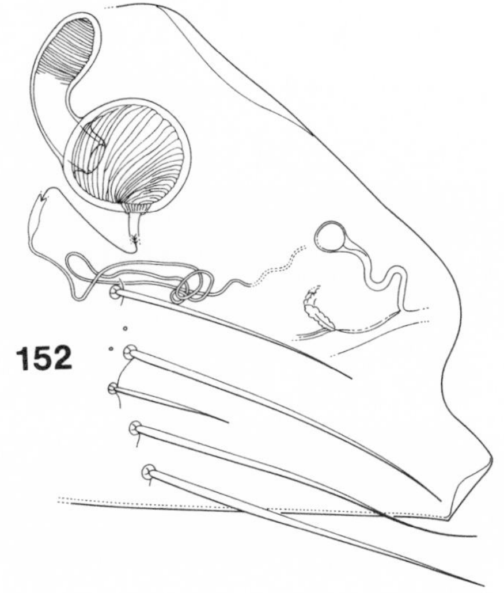

151-152 *Odontopsyllus dentatus* (Baker). **151** Male clasper. **152** Female sternum VII, spermatheca and ducts. After Holland, 1985.

GENUS **Ctenophyllus** Wagner

Ctenophyllus. Wagner, 1927, Konowia 6: 108, 112 (type species by subsequent designation: [Wagner, 1930, p. 14] *C. armatus* Wagner, 1901).

Ctenophyllus Wagner, 1927. Smit, 1975, Senckenbergiana biol. 56: 251-255 (genus restricted to *armatus* (Wagner, 1901), *rigidus* Darskaya, 1949, *subarmatus* (Wagner, 1901) and *tarasovi* Scalon, 1953).

The members of this genus, assigned to three subgenera by Hopkins and Rothschild (1971), are now relegated to five separate genera after Smit (1975). With two exceptions, all species are Palaearctic in distribution and all are true parasites of pikas belonging to the genus *Ochotona.* One species belonging to this genus occurs in the area under consideration, and the closely related *Geusibia ashcrafti* (Augustson, 1941) is known from California and Colorado, according to Hubbard (1947).

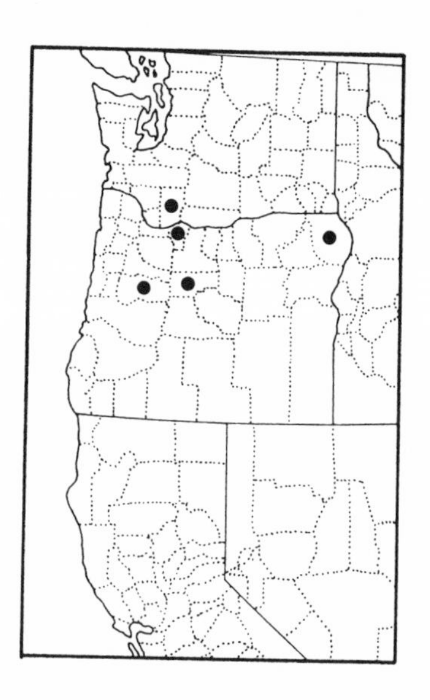

Ctenophyllus armatus (Wagner)

[90101]

(Figs. 153, 154)

Ceratophyllus armatus. Wagner, 1901, Trudy Russk. Ent. Obshch. 35: 17, pl. I, fig. 1 (Siberia, from [?] *Pteromys volans* [probably *Ochotona alpina* ssp.]).

Ceratophyllus terribilis. Rothschild, 1903, Novit. Zool. 10: 317, pl. 9, figs. 1-3 (Canadian National Park, Alberta, from *Lagomys princeps*).

Ctenophyllus armatus Wagner. Tiflov and Pavlov, 1936, Vest. Mikrobiol. Epidem. Parazit. 15: 85, figs. 2, 3.

Ctenophyllus terribilis (Rothschild). Jellison, 1941, Publ. Hlth. Rep., Wash. 56: 2343-2346, figs. 1, 2.

Ctenophyllus (Ctenophyllus) armatus (Wagner). Hopkins and Rothschild, 1971, An Illustrated Catalogue... Vol. V, pp. 297-300.

Ctenophyllus armatus (Wagner). Smit, 1975, Senckenbergiana biol. 56: 247-256 (designates *armatus* as type species).

This species has an amphiberingian type of distribution, occurring in Siberia, China, and Japan, as well as Alaska and western North America as far south as Utah and Colorado. It is a specific parasite of pikas and occurs consistently, but infrequently, on these hosts throughout its range. Nothing is known of its seasonal abundance since pikas are inaccessible during the winter months even though they do not hibernate. Hubbard (1947) speculated that either adult populations peak during the winter or that the adults are nest fleas, seldom occurring on the host when it is away from its nest.

For whatever reason, this species is seldom collected, and then only in small numbers. Experience with an allied species in Asia suggests that the adult fleas leave the host at the first sign of trauma to the host, which may contribute to their infrequent collection.

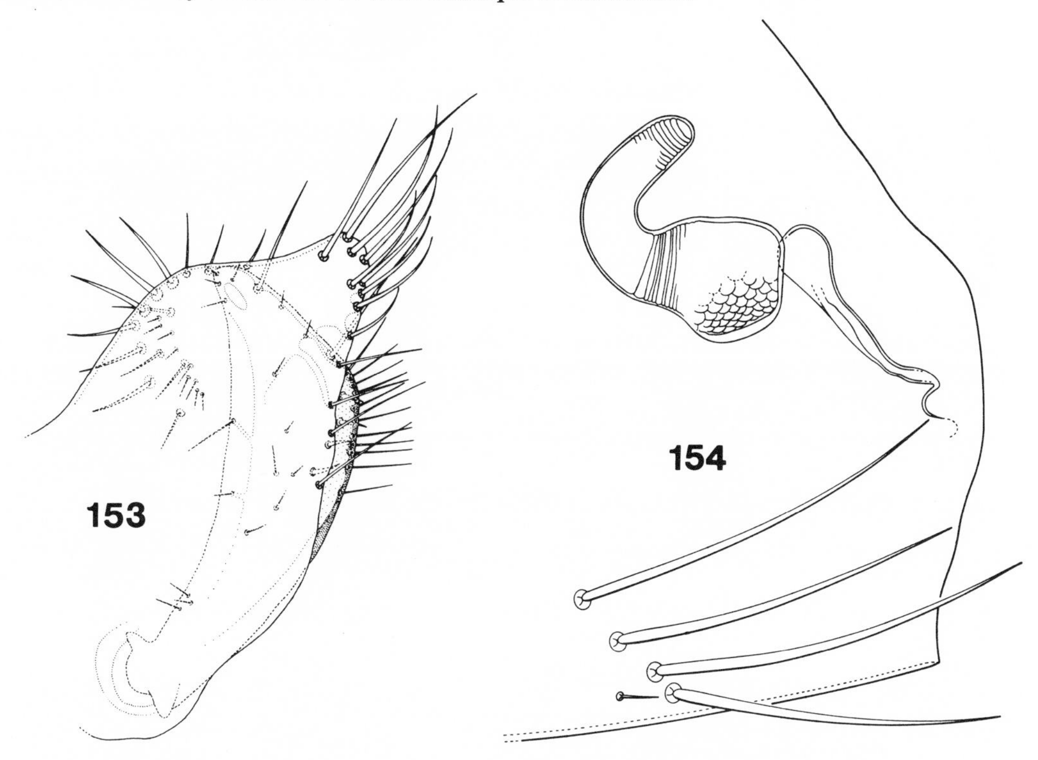

153-154 *Ctenophyllus armatus* (Wagner). **153** Male clasper. **154** Female sternum VII and spermatheca. After Holland, 1985.

GENUS **Amphipsylla** Wagner

Amphipsylla. Wagner, 1909, Izv. Kavkaz Mus. 4: 196, 201 (type by subsequent designation: [Jordan and Rothschild, 1913, Official List generic Names Zoology no. 896] *schelkovnikovi* Wagner).

This genus currently contains 52 recognized taxa, practically all of which occur in the Siberian Subregion of the Palaearctic Region. A few species are known from the Mediterranean, European and Manchurian subregions, and two are known from North America. One of these occurs in the area under consideration. All are mainly parasites of microtine rodents.

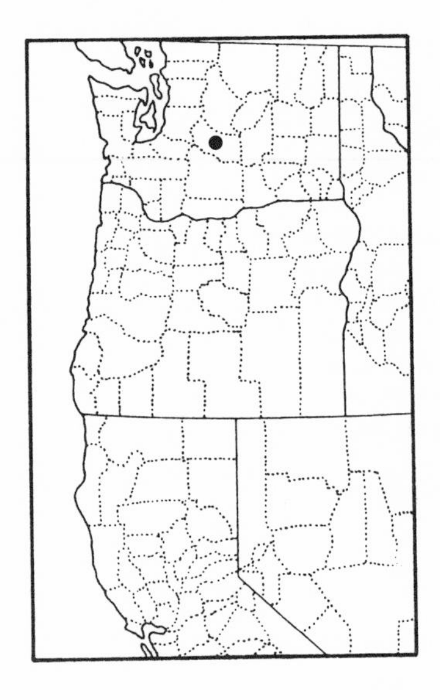

Amphipsylla washingtona Hubbard

[95418]

Amphipsylla sibirica washingtona. Hubbard, 1954, Ent. News 65: 171 (18 mi. E. of Ellensburg, Washington, from *Lagurus curtatus pauperrimus*).
Amphipsylla sibirica washingtona Hubbard, 1954. Hopkins and Rothschild, 1971, An Illustrated Catalogue... Vol. V, p. 479.
Amphipsylla washingtona Hubbard, 1954. Holland, 1985, Mem. Ent. Soc. Canada 130, p. 246.

This is a specific parasite of the pallid vole, *Lagurus curtatus,* and is known from arid, sagebrush areas of eastern Washington state. Holland (1985) also reports a pair from Youngstown, Alberta. Although it was originally described as a subspecies of *A. sibirica,* Holland suggested as early as 1958 that *A. s. washingtona* was genetically incompatible with *A. s. pollionis,* the only other North American subspecies of *A. sibirica.* In Holland (1985) it was accorded the status of full species.

FAMILY **CERATOPHYLLIDAE**

Derivation: The family name is derived from the Greek words *cerato* or *kerato* (horn) and *psylla* (a flea) and the Latin suffix *idae* and alludes to the habit of the male members of this family of erecting their antennae, thus appearing to be horned. The type genus is *Ceratophyllus* Curtis, 1832.

General Characters: Trabecula centralis present; genal ctenidium absent; anterior arm of tentorium absent (in all American species); eye usually well developed, never sinuate ventrally; upper seta in ocular row arising in front of eye; pronotal ctenidium present; pseudosetae present under mesonotal collar; inner side of hind coxae without spiniform setae; five pairs of lateral plantar setae; usually three antesensilial setae per side; two well developed acetabular setae usually present; anal stylet of female with one or two small sublateral bristles in addition to long apical bristle.

World Distribution: Overwhelmingly Holarctic, with a few genera represented in South America, Africa and Southeast Asia, and 1 in Antarctica.

Number of Northwestern Taxa: 73

General References: Traub et al. (1983), Holland (1985).

This large family currently includes 43 genera. Representatives occur on all major landmasses, but by far the best representation is that in the Northern Hemisphere. Some genera are Holarctic in distribution, while others are unique to a single continent. Australia, which has no native species, still has three species, according to Dunnet and Mardon (1974), imported on mice and domestic poultry.

Traub (1983) and Traub and Rothschild (1983) treat host preferences and evolution in this family in considerable detail. Based on available evidence, they conclude that the ancestors of this family evolved in the Nearctic Region during the early Oligocene (38-40 mybp) as parasites of squirrels, and only later did some taxa become associated with other mammalian and avian hosts.

The family has recently undergone a rigorous systematic reappraisal by Smit (1983). The new arrangement of species in the various genera, which is followed here, is likely to require some time before becoming widely accepted. Genera occurring in the area may be separated with the following key.

KEY TO THE NORTHWESTERN GENERA OF *CERATOPHYLLIDAE*

(Modified from Smit 1983)

1 Movable process of male bearing a long, caudoventral projection which is expanded apically; anal stylet of female cylindrical, its rounded apex bearing many long setae .... **Amphalius**

Not as above **2**

2 Fore femur with a small seta adjoining apical guard seta; first hind tarsal segment longer than segments II-IV combined; on tree squirrels .... **Tarsopsylla**

Fore femur with a single apical guard seta; first hind tarsal segment shorter than next three combined **3**

3 Frontal row of one or a few setae bordering upper half of antennal fossa, postantennal region of head with at most one long seta in addition to those of occipital row; dorsal setae of male meso- and metanotum and tergum I forming an erect mane; female anal stylet long and distinctly curved upwards; on flying squirrels .... **Opisodasys (Opisodasys)**

Not with this combination of characters **4**

4 Eye much reduced or virtually absent, unpigmented; ocular row of four to eight setae; squamulum of metepisternum poorly developed or absent; outer side of mid and hind femur very densely striated; on pocket gophers **5**

Eye normal, dark; ocular row usually of three setae; squamulum usually well developed; not on pocket gophers **7**

5 First and third lateral plantar setae of fifth tarsal segment displaced somewhat mesad onto the plantar surface; with distinct, minute spicules on abdominal sternites .... **Dactylopsylla**

First pair of lateral plantar setae of fifth tarsal segment not displaced laterally onto plantar surface; sternal spicules very indistinct or absent **6**

6 Male tergum VIII without spicules on inner side; distal arm of st. IX with a dense patch of five to 25 long, slender setae; base of female bursa copulatrix heavily sclerotized, large fleas .... **Spicata**

Male tergum VIII with spicules at least in the dorsal region; distal arm of st. IX without a dense patch of setae; base of female bursa copulatrix not particularly heavily sclerotized; small fleas .... **Foxella**

7 Frontal tubercle in a small pit which extends distinctly into the frontal lumen; male movable process with four to seven short, sharply pointed, spiniform setae which point upward; ventral margin of female anal sternum markedly angular; on sciurid and peromyscine rodents .... **Orchopeas**

Frontal tubercle not arising in a pit; male movable process lacking the short, sharply pointed, upward directed spiniforms; female anal sternum with rounded angle on ventral margin **8**

8 Outside of fore femur without lateral setae (rarely with one); first pair of lateral plantar setae arising between the second pair .... **Opisodasys (Oxypsylla)**

Outer side of fore femur with lateral setae; lateral plantar setae variable **9**

9 Apex of fore femur with a single stout guard seta; pronotal ctenidium of thirty to forty spines .... **Dasypsyllus (Dasypsyllus)**

Apex of fore femur with a seta adjoining the stout guard seta; pronotal ctenidium variable, usually with fewer than 25 spines **10**

10 Seta adjoining guard seta quite stout and reaching well beyond the middle of the guard seta; outer surface of mid and hind femur densely striated **11**

This seta slender and usually not reaching beyond middle of guard seta; striations of outer surface of mid and hind femur rather coarse **15**

11 Basal abdominal sternum with a patch of lateral setae often extending onto upper anterior half **12**

Basal abdominal sternum with at most one or two lateral setae on middle or lower half **13**

12 Frontal row consisting of usually one ventral and one or two dorsal setae of medium length (rarely only one); lower minute antesensilial seta absent in male; st. VIII long and narrow, extending apically in a long, filamentous vexillum, the apex of which is usually split; female bursa copulatrix with small but distinct semicircular sclerotization around opening of ductus obturatus; on ground squirrels .... **Oropsylla (Opisocrostis)**

Frontal row consisting of one very small ventral seta which may be absent; lower minute antepygidial seta present; st. VIII of medium length, with an upright, sickle-shaped vexillum with an entire apex; bursa copulatrix without a dark sclerotization; on ground squirrels and prairie dogs .... **Oropsylla (Hubbardipsylla)**

13 Male st. VIII much reduced, without setae; movable process long, slender, curved forward; female spermatheca with a ventral constriction between bulga and hilla; anal stylet usually with only one long, ventral, preapical seta, without a dorsal seta; on ground squirrels .... **Oropsylla (Diamanus)**

Male st. VIII with setae; movable process fairly straight; female spermatheca with less distinct constriction between hilla and bulga; anal stylet often with two or three lateral setae, but sometimes with only a ventral one or none at all **14**

14 Male st. VIII rectangular or triangular, when apical half narrowed, the apex without very long setae; female spermathecal bulga globular, hilla usually much longer than bulga, its apical papilla often small or absent; ground squirrels and marmots .... **Oropsylla (Thrassis)**

Male st. VIII with a narrow apex bearing two or three very long apical setae; bulga of spermatheca usually pyriform, hilla at most a little longer than bulga, with a conspicuous apical papilla; ground squirrels and marmots .... **Oropsylla (Oropsylla)**

15 Postantennal region of head with only one long seta (rarely absent) above antennal fossa in addition to the usually small bordering setae, and sometimes with a short seta in front of it; cuticular striation of outer side of mid and hind femur and tibia usually not very coarse **16**

Postantennal region of head with one or more setae above the large, central seta, and usually also at least one seta in front of that row; striation of outer side of mid and hind femur and tibia usually very coarse **17**

16 Mesepimeron without lateral setae other than some preapical ones; metepimeron with only two or three setae, none of them placed adjacent to the spiracular fossa; on wood rats and various other rodents .... **Amaradix**

Mesepimeron with one or more lateral setae; metepimeron with five to seven setae, one of which is placed in close proximity to the spiracular fossa; male st. VIII vestigial .... **Nosopsyllus**

17 Pronotal ctenidium with fewer than 24 spines **19**

With more than 24 spines **18**

18 Apophysis of male st. IX making at least one convolution; basal part of spermathecal duct not strongly dialated and wrinkled; base of ductus obturatus strongly sclerotized; bulga much longer than hilla .... **Ceratophyllus (Ceratophyllus)**

Apophysis of male st. IX making at most one convolution; basal part of spermathecal duct dialated and wrinkled; ductus obturatus weakly sclerotized throughout; bulga swollen, about as long as hilla .... **Ceratophyllus (Emmareus)**

19 Inner side of mid and hind tibia with cuticular sculpturing tending to form acute scales **20**

Inner side of mid and hind tibia with a different type of sculpturing **21**

20 Second postantennal row of two setae; males with several apical setae of first and second hind tarsal segments quite long; t. VIII without a spiculose area; st. VIII vestigial; movable process with some pointed spiniforms and anal sternum usually large; hilla of spermatheca without a papilla; on microtine and peromyscine rodents .... **Malaraeus**

Second postantennal row with three to five setae; males with none of the apical setae of first and second hind tarsal segments very long; t. VIII with a spiculose area; st. VIII long and narrow, with an apical vexillum; movable process with two or three closely set, short, obtuse, spiniform setae as well as one or more stout ones; anal sternum not very large .... **Amalaraeus**

21 Male st. VIII long, narrow and curved, with an apical vexillum which is at most only slightly spiculose basally; movable process straight, expanded apically and bearing two stout spiniform setae, below which is a fairly large, strong or subspiniform seta; spermathecal hilla with a small preapical papilla causing it to be markedly asymmetrical .... **Pleochaetis**

Apex of male st. VIII usually with a spiculose vexillum; usually at least part of the female genital ducts strongly sclerotized; basal part of spermathecal duct often dilated and modified; bulga with an apical external orifice **22**

22 First two postantennal rows with one or two and three to six setae; respectively; spiracular fossa of t.VII enlarged; the two short spiniforms of male movable process close together near posterior margin; base of spermathecal duct surrounded by a transversely striated, wide 'glandular' structure; on microtine rodents, chipmunks .... **Megabothris (Megabothris)**

First two postantennal rows with none or one and two (rarely three) setae respectively; spiracular fossa of t. VII not enlarged, except in *Megabothris (Amegabothris)* **23**

23 Movable process of male without dark spiniforms; bulga of spermatheca wider than hilla, either elongate, cylindrical or pyriform; on murine and sciurid rodents, pikas and mustelids .... **Ceratophyllus (Monopsyllus)**

Movable process of male with three spiniform setae, the upper two of which are usually blunt; bulga either long and narrower than hilla or broad and barrel-shaped, much wider than hilla **24**

24 The three spiniform setae of male movable process close together and arising well away from the dorsoposterior margin; ductus bursae partly strongly sclerotized; hilla of spermatheca wider than the long, curved, sausage-shaped bulga .... **Aetheca**

At least the upper of the three spiniforms of the male movable process placed near or on the dorsoposterior angle; ductus bursae not strongly sclerotized; hilla narrower than bulga **25**

25 Spiracular fossa of t. VIII enlarged; basal portion of male movable process with dense vertical striation; base of spermathecal duct widened but not transversely striated; bulga barrel-shaped; parasites of microtine rodents .... **Megabothris (Amegabothris)**

Spiracular fossa of t.VIII not enlarged; males with Wagner's Organ present; bulga of spermatheca either barrel-shaped or pyriform **26**

26 First or first and second tarsal segments of mid leg of male with a number of long setae which reach to or beyond apex of following segment; the two acetabular setae arising at or above the level of the acetabulum; the three spiniform setae of the movable process well separated from each other; female ductus bursae and bursa copulatrix forming one long or short, continuous, wrinkled duct; spermatheca rather thin, bulga barrel-shaped; on tree and ground squirrels....... **Eumolpianus**

None of the setae of the first two tarsal segments of the mid leg of the male very long; acetabular setae arising at mid-level of acetabulum; upper two short, blunt spiniform setae of the movable process close together on the projecting dorsoposterior angle; bulga of spermatheca globular; parasites of tree and ground squirrels............ .... **Ceratophyllus (Amonopsyllus)**

GENUS **Aetheca** Smit

Aetheca. Smit and Wright, 1978b, A catalogue of primary type-specimens..., p. 47.
Aetheca. Smit, in: Traub et al., 1983, p. 12 (type species: *"Ceratophyllus" wagneri* Baker, 1904).
Aetheca Smit, 1983. Haddow et al., in: Traub et al., 1983, p. 42.

This genus contains but two species that are restricted to North America north of Mexico, mainly west of the Great Lakes. Only one occurs in the Pacific Northwest. Preferred hosts are species of *Peromyscus,* but examples of this flea have been taken from most of the small mammal species in the area, as well as from their predators.

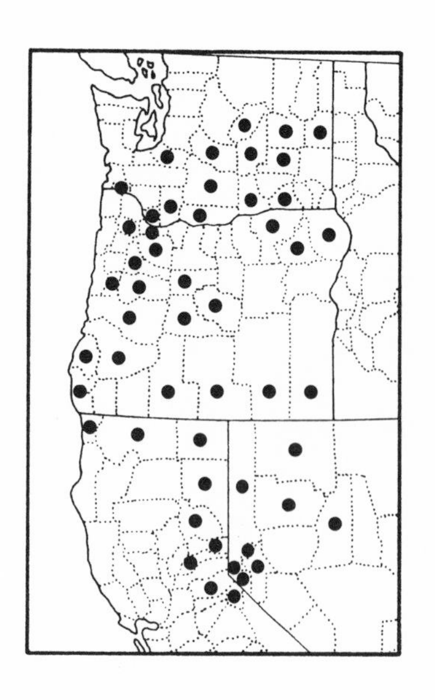

Aetheca wagneri (Baker)

[90423]

(Figs. 155, 156)

Ceratophyllus wagneri. Baker, 1904, Proc. U. S. Nat. Mus. 27: 387, 405-406, 448 (Moscow, Latah Co., Idaho, from *Peromyscus leucopus* and house mouse).
Ceratophyllus peromysci. Stewart, 1928, Canad. Ent. 60: 148 (Cortez, Colorado, from *Peromyscus* sp.).
Ceratophyllus wagneri systaltus. Jordan, 1929, Novit. Zool. 35: 35 (Blackfalls [Blackfalds], Alberta, on mouse).
Ceratophyllus wagneri ophidius. Jordan, 1929, Novit. Zool. 35: 36 (San Francisco, California, on *Putorius xanthogenys [=Mustela frenata xanthogenys]*).
Monopsyllus wagneri. Jordan, 1933, Novit. Zool. 39: 78.
Monopsyllus wagneri kylei. Hubbard, 1949, Bull. S. Calif. Acad. Sci. 48: 52 (Kyle Canyon, N.W. of Las Vegas, Clark Co., Nevada, from *Peromyscus maniculatus*).
Monopsyllus wagneri (Baker). Johnson, 1961, U. S. Dept. Agric. Tech. Bull. 1227: 31, 35 (*systaltus, ophidius,* and *kylei* synonymized).

Aetheca wagneri (Baker, 1904). Smit, in: Traub et al., 1983, p. 34.
Aetheca wagneri (Baker). Haddow et al., in: Traub et al., 1983, p. 42-43.

Because this species is so commonly collected it is also frequently mentioned in the literature. It is found in a broad range of habitats from the arid southwest to subalpine areas in the Rocky Mountains. It is most commonly collected on *Peromyscus maniculatus* and other members of that genus, but frequently is taken on other small mammals as well as their predators. Collection records show that this flea is present in the adult stage throughout the year, being most frequently taken from June through September. Our records are based on 134 males and 177 females.

Although this species is frequently mentioned in the plague literature, there seems to be little evidence that it is important in disease transmission, though it may play a minor role in maintaining the disease in the small rodent population.

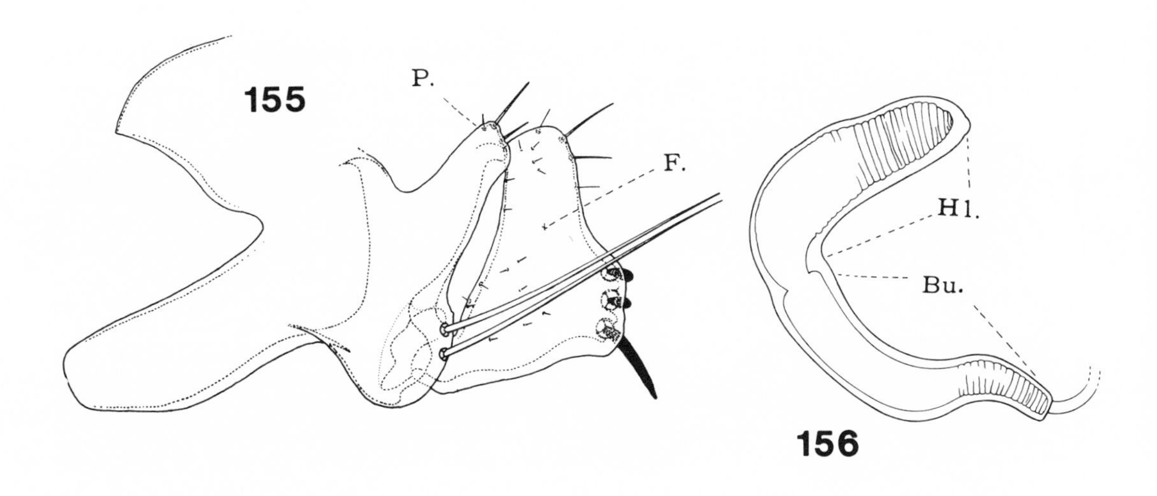

155-156 *Aetheca wagneri* (Baker). **155** Male clasper. **156** Female spermatheca. After Holland, 1985.

GENUS **Amalaraeus** Ioff, 1936

Amalaraeus. Ioff, 1936, Z. Parasitenk. 9: 98 (erected as a section of the subgenus *Malaraeus,* genus *Ceratophyllus* for *C. (M.) penicilliger* Grube, 1851).
Amalaraeus Ioff, 1936. Smit, in: Traub et al., 1983, p. 32.
Amalaraeus Ioff, 1936. Haddow et al., in: Traub et al., 1983, p. 43.

This name has had a varied history, to which the synonymy above only addresses the beginning and the present. Usually it has been treated as the name of a 'section' or a subgenus by eastern workers and ignored by western pulicologists. It is now accorded generic status to include seven species and fourteen subspecies, with *A. p. penicilliger* (Grube, 1851) as its generitype. The only representative of this taxon known from the area under consideration is the rare *A. dobbsi* (Hubbard, 1940).

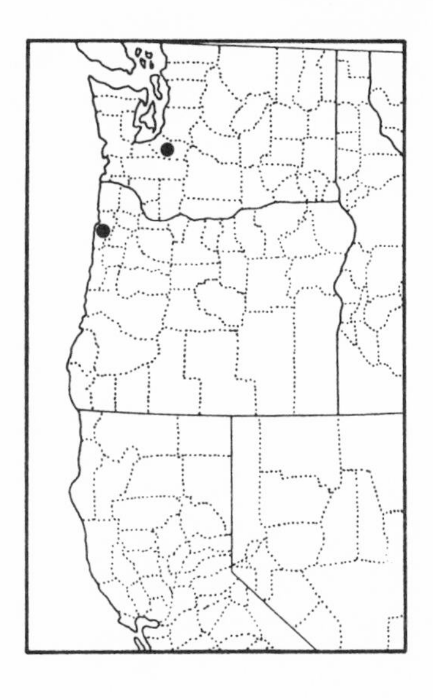

Amalaraeus dobbsi (Hubbard)

[94051]

(Figs. 157, 158)

Malaraeus dobbsi. Hubbard, 1940, Pacific Univ. Bull. 37: 3-4 (10 males, 6 females from Tillamook, Tillamook Co., Oregon, from *Microtus oregoni oregoni,* 30-VIII-1937).
Amalaraeus dobbsi (Hubbard, 1940). Smit, in: Traub et al., 1983, p. 32.
Amalaraeus dobbsi (Hubbard, 1940). Haddow et al., in: Traub et al., 1983, p. 45.

This is evidently a rare faunal element and the only known published records are those in the original description, subsequently repeated by Hubbard (1947). There is a single male in the Lewis Collection taken on *Microtus oregoni,* Pierce County, Washington (T16N, R4W), 17-VIII-1979, P. Gunther *leg.* Holland (1985) reports one male collected in Olympic National Park, Washington, from *Clethrionomys gapperi* ssp.

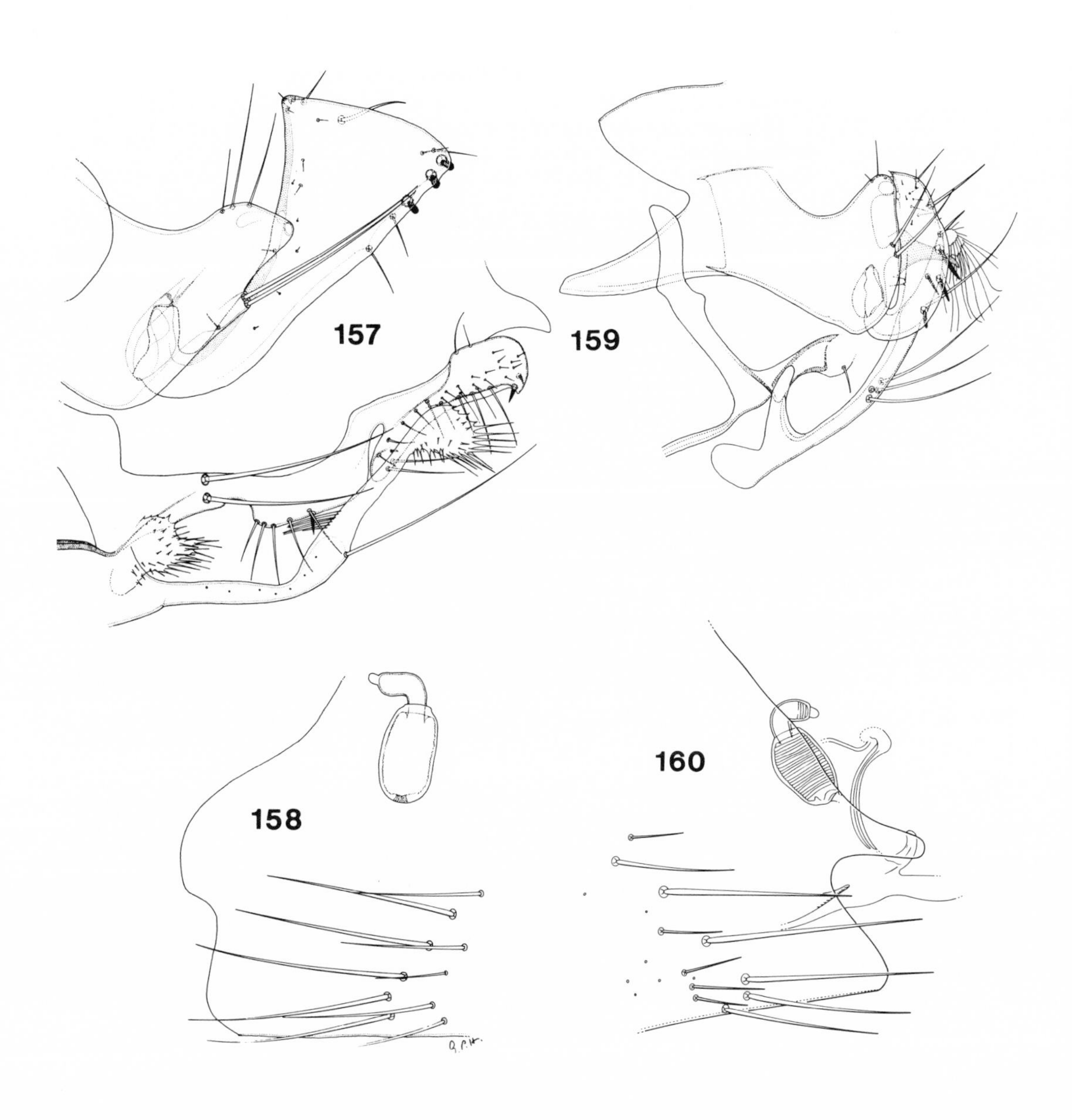

157-158 *Amalaraeus dobbsi* (Hubbard). **157** Male clasper and sternites VIII and IX. **158** Female sternum VII and spermatheca. After Holland, 1985.
159-160 *Amaradix bitterrootensis bitterrootensis* (Dunn). **159** Male clasper. **160** Female sternum VII and spermatheca. After Holland, 1985.

GENUS **Amaradix** Smit

Amaradix. Smit and Wright, 1978b, A catalogue of primary type-specimens..., p. 19. (type species: *"Ceratophyllus" bitterrootensis* Dunn, 1923).
Amaradix. Smit, in: Traub et al., 1983, p. 12-13.
Amaradix Smit, 1983. Haddow et al., in: Traub et al., 1983, p. 47.

This genus was erected for two species originally assigned to the genus *Malaraeus.* Both are restricted to western North America where they are associated primarily with *Neotoma, Ochotona,* and *Peromyscus* species.

KEY TO THE NORTHWESTERN SPECIES OF *AMARADIX*

1 Movable process of male with distinct caudoventral angle, its caudal margin bearing five long setae, the lower two of which are slightly spiniform; st. VIII with two setae apically, its basal arms extending at a right angle to the long axis of the segment; female st. VII with a broad, finger-like, caudal lobe subtended by a deep sinus; bulga of spermatheca broad at duct end, tapering gradually to junction with hilla .... **euphorbi**

Movable process of male crescentric, its caudal margin smoothly rounded and lacking a distinct caudoventral angle, and caudal margin with four marginal setae, the ventral three being shorter, spiniform and subequal in length; st. VIII with four or five setae apically, its basal arms flexed strongly apicad; female st. VII with sharp, triangular lobe subtended by a deep sinus; spermathecal bulga less strongly tapered toward junction with the hilla .... **bitterrootensis bitterrootensis**

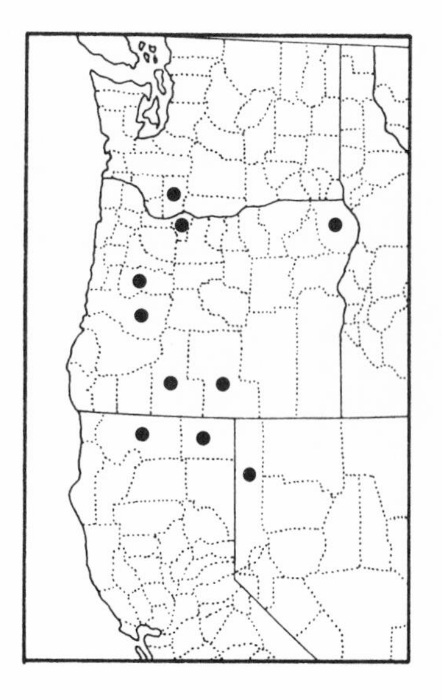

Amaradix bitterrootensis bitterrootensis (Dunn)

[92342]

(Figs. 159, 160)

Ceratophyllus bitterrootensis. Dunn, 1923, (in: Dunn and Parker), Publ. Hlth. Rep., Wash. 38: 2771-2772, 2775 (Spoon Creek, S.W. of Darby, Montana, from *Neotoma cinerea*).
Ceratophyllus isus. Jordan, 1925, Novit. Zool. 32: 110 (Red Deer River, Canadian Rocky Mountains, from *Mus*).
Ceratophyllus bitterrootensis Dunn and Parker. Jordan, 1928, Novit. Zool. 35: 36 (*isus* synonymized).
Malaraeus bitterrootensis Dunn and Parker. Jordan, 1933, Novit. Zool. 39: 76.
Malaraeus bitterrootensis (Dunn). Holland, 1949, The Siphonaptera of Canada, p. 154.
"Ceratophyllus" bitterrootensis Dunn, 1923. Smit, in: Traub et al., 1983, p. 13.
Amaradix bitterrootensis (Dunn, 1923). Haddow et al., in: Traub et al., 1983, p. 47.

Although there are no additional records of this taxon to report, the species is an established faunal element from records cited in Hubbard (1947). Haddow et al. (1983) indicate that there is some question as to the subspecific identity of material from California and Oregon. However, they list these specimens under the nominate subspecies.

Hubbard (1947) suggested that his collection records indicated that the preferred hosts, if not the true hosts, were species of *Ochotona* native to western United States, the implication being that when the species occurred on other hosts, they shared a common habitat with pikas. Perhaps our lack of new records reflects the fact that few pikas were collected during the surveys upon which most of this report is based.

Amaradix euphorbi (Rothschild)

[90509]

(Figs. 161, 162)

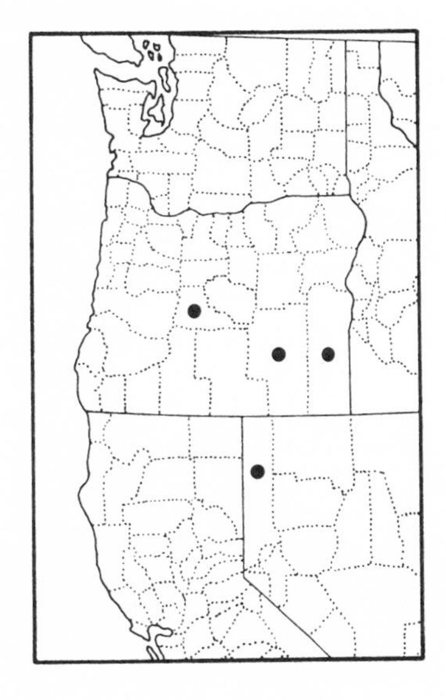

Ceratophyllus euphorbi. Rothschild, 1905, Novit. Zool. 12: 165, pl. VI, fig. 11 (Horse Creek, British Columbia, Canada, from *Peromyscus maniculatus*).

Malaraeus euphorbi Rothschild, 1905. Hubbard, 1947, Fleas of Western North America, p. 206.

"Ceratophyllus" euphorbi Rothschild, 1905. Smit, in: Traub et al., 1983, p. 13.

Amaradix euphorbi (Rothschild, 1905). Haddow et al., in: Traub et al., 1983, p. 48.

A single male of this species was collected at Bogus Lake, Malheur County, Oregon, on *Neotoma cinerea.* It is evidently a montane species associated with species of *Neotoma* and *Peromyscus,* but must not be common on these hosts.

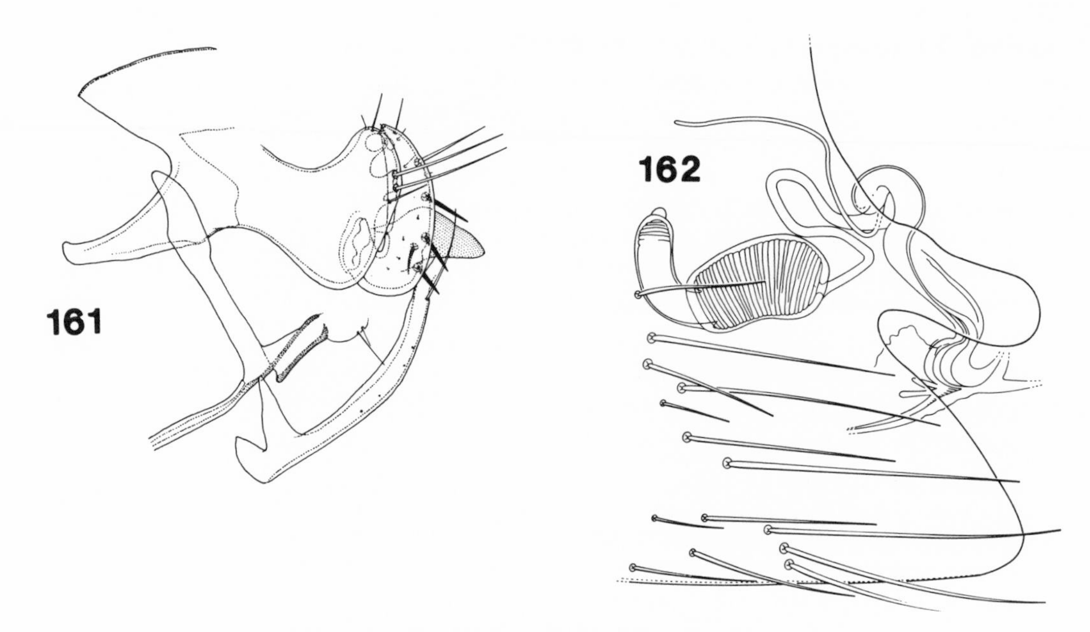

161-162 *Amaradix euphorbi* (Rothschild). **161** Male clasper. **162** Female sternum VII and spermatheca. After Holland, 1985.

GENUS **Ceratophyllus (Ceratophyllus)** Curtis

Ceratophyllus. Curtis, 1832, *British Entomology* 9: 417 (type by designation: *Ceratophyllus hirundinis* Curtis, 1826).
Ceratophyllus Curtis, 1832. Jordan, 1933, Novit. Zool. 39: 75.
Ceratophyllus (Ceratophyllus) Curtis, 1832. Smit, in: Traub et al., 1983, p. 35.
Ceratophyllus Curtis, 1832. Haddow et al., in: Traub et al., 1983, p. 56.

As currently restricted, this subgenus contains 37 species and 10 subspecies, all of which parasitize birds. Except for two forms which are Neotropical, the genus is Holarctic in distribution, showing a strong preference for birds that nest in holes in banks, cliffs, or trees or those constructing their nests of mud plastered to cliff faces or under the eaves of buildings. Also, birds that construct bulky nests on or near the ground are parasitized. More than half of the known species are associated with swallows or martins. Two species are known as pests of domestic poultry and one, *C. (C.) gallinae* (Schrank, 1803), is cosmopolitan in occurrence, though not yet known from the Pacific Northwest. Although there are only two common species reported here, the additional taxa are likely to be encountered when sufficient attention is directed to the collection of birds' nests.

KEY TO THE NORTHWESTERN SPECIES OF *CERATOPHYLLUS S. S.*

(or species likely to be encountered)

1 Males 2

Females 8

2 Movable process triangular, its apex tapering to a blunt point; apex of st. VIII ending in four to six long setae and a long, narrow, tapering vexillum .... **petrochelidoni**

Movable process and apex of st. VIII shaped otherwise 3

3 Movable process barely extending beyond apex of fixed process, clavate in outline and widest near apex ....**idius**

Movable process either much longer than fixed process, or fusiform in shape, extending little beyond apex of fixed process 4

4 Movable process fusiform, widest in the middle, with incrassation in anterior margin very near apex; mesal surface of movable process with conspicuous, longitudinal sclerotized rod; fixed process about six times as long as broad .... **scopulorum**

Movable process clavate or rectangular; fixed process much shorter 5

5 Acetabular setae arising on the level of the articulation of the movable process with the fixed process, approximately one-half of the former extending beyond apex of the latter 6

Acetabular setae arising either well above or well below dorsal articulation of movable process with fixed process, approximately one-half of the former extending beyond the latter 7

6 Apical one-half of anterior margin of movable process distinctly concave between incrassation and apex; domestic poultry and various birds .... **niger**

Apical one-half of anterior margin of movable process straight between incrassation and apex; on pelicans, gulls, etc. .... **pelecani**

7 Acetabular setae arising well below dorsal articulation of movable process with fixed process; fixed process narrow, about three times as long as broad, its posterior margin strongly concave .... **celsus celsus**

Acetabular setae arising near dorsal articulation of movable process with fixed process; fixed process much broader, its posterior margin only slightly concave .... **styx riparius**

8 Sternum VII entire 9

Sternum VII with sinus or other modification 10

9 On poultry and various wild birds .... **niger**

On cliff swallows and in their nests .... (dimorphic form of) **petrochelidoni**

10	Lower lobe of st. VII distinctly shorter than upper lobe .... (typical form of)	**petrochelidoni**
	Upper and lower lobes of st. VII otherwise	**11**
11	Sternum VII variable, its posterior margin being entire or sinuate; with a small, sclerotized structure in the bursa copulatrix; light in color, parsites of swallows	**celsus celsus**
	Sternum VII either with a blunt lobe or divided into an upper and lower lobe by a variable sinus; color and hosts variable	**12**
12	St. VII with a pronounced, blunt lobe on caudal margin subtended by a deep sinus with a sclerotized margin	**scopulorum**
	St. VII divided into dorsal and ventral lobes separated by a variable sinus	**13**
13	Upper and lower lobes of st. VII about equal in size, with smoothly rounded margins and separated by a shallow sinus	**pelecani**
	Upper and lower lobes of st. VII dissimilar	**14**
14	Upper and lower lobes of st. VII angular apically; upper lobe larger and extending beyond apex of lower lobe	**idius**
	Upper lobe of st. VII blunt, forming an angle above sinus; ventral lobe more acuminate, projecting beyond upper lobe	**styx riparius**

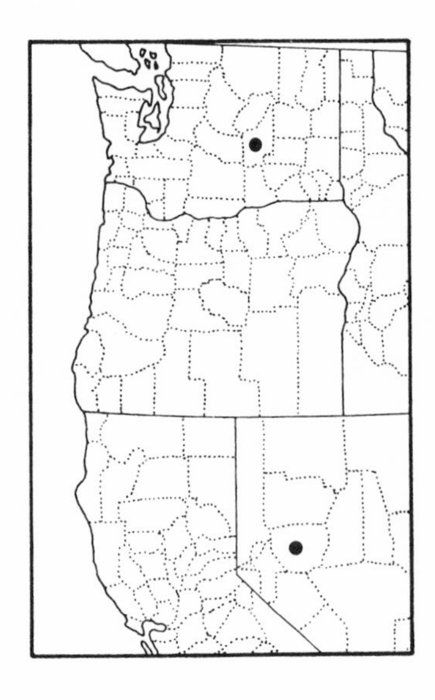

Ceratophyllus (Ceratophyllus) celsus celsus Jordan

[92624]

Ceratophyllus celsus. Jordan, 1926, Novit Zool. 33: 387-388 (Okanagan Falls, British Columbia, from *Riparia riparia*).
Ceratophyllus celsus celsus Jordan. Jordan, 1933, *op. cit.* 39: 75.

This is a widespread species associated with swallows, especially those belonging to the genera *Riparia* and *Petrochelidon.* Holland (1985) lists it from Alaska to New Brunswick. Benton (1980) reports it from Illinois, Michigan, New York, and Vermont. In the west it has been collected in Nevada, Washington, and Montana. It is most often found in the nests of its hosts and may occur in large numbers during the nesting season.

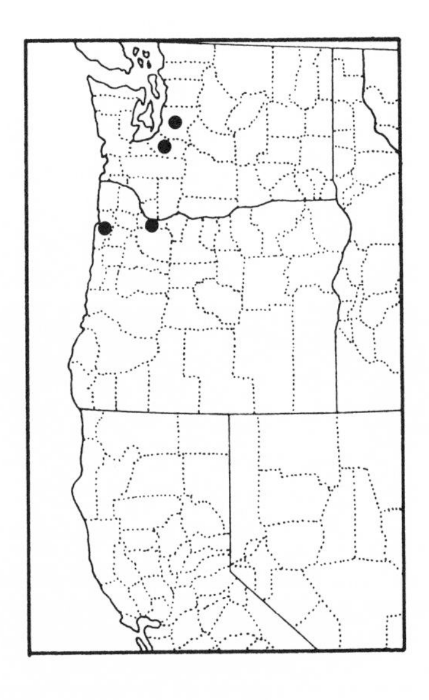

Ceratophyllus (Ceratophyllus) niger C. Fox

[90834]

(Figs. 163, 164)

Ceratophyllus gallinae (Schrank). Harvey, 1907, Ent. Soc. B. C. Bull. No. 7, p. 1.
Ceratophyllus niger. C. Fox, 1908, Ent. News 19: 434-435 (male from unknown locality from human and *Rattus norvegicus*).
Ceratophyllus niger Fox. Jordan and Rothschild, 1920, Ectoparasites, I: 70-71 (Port Essington, B. C., from 'hen', and Okanagan Landing, B. C., from *Turdus migratorius propinqus,* also from San Francisco, off human and from Tacoma, Washington, and Bridgeport, California, in a henhouse and from chicken nests).
Ceratophyllus niger inflexus. Jordan, 1929, Novit. Zool. 35: 37 (female from Custer County, Colorado, no host given).
Ceratophyllus niger C. Fox. Wagner, 1936, Canad. Ent. 68: 201 (synonymizes *C. n. inflexus* Jordan).
Ceratophyllus niger niger Fox. Jordan, 1929, Novit. Zool. 35: 89-92.
Ceratophyllus niger C. Fox. Holland, 1949, The Siphonaptera of Canada, p. 149.
Ceratophyllus (Ceratophyllus) niger C. Fox. Smit and Wright, 1978b, A catalogue of primary type-specimens.., p. 33 (selection of lectotype male).
Ceratophyllus (Ceratophyllus) niger Fox, 1908. Smit, in: Traub et al., 1983, p. 35.
Ceratophyllus (Ceratophyllus) niger C. Fox, 1908. Haddow et al., in: Traub et al., 1983, p. 63.

This species was reported by Hubbard (1947) from a long series collected in a chickenhouse at Cannon Beach, Tillamook County, Oregon, in November 1940. It is also common in purple martin nests, according to Benton (personal communication). It is known as the western chicken flea since it is a common pest of chickens in northwestern North America as far south as San Francisco, California, and northern New Mexico, east to Colorado and Montana. As Holland (1949, 1985) points out, it is a dark, very active flea which readily bites humans.

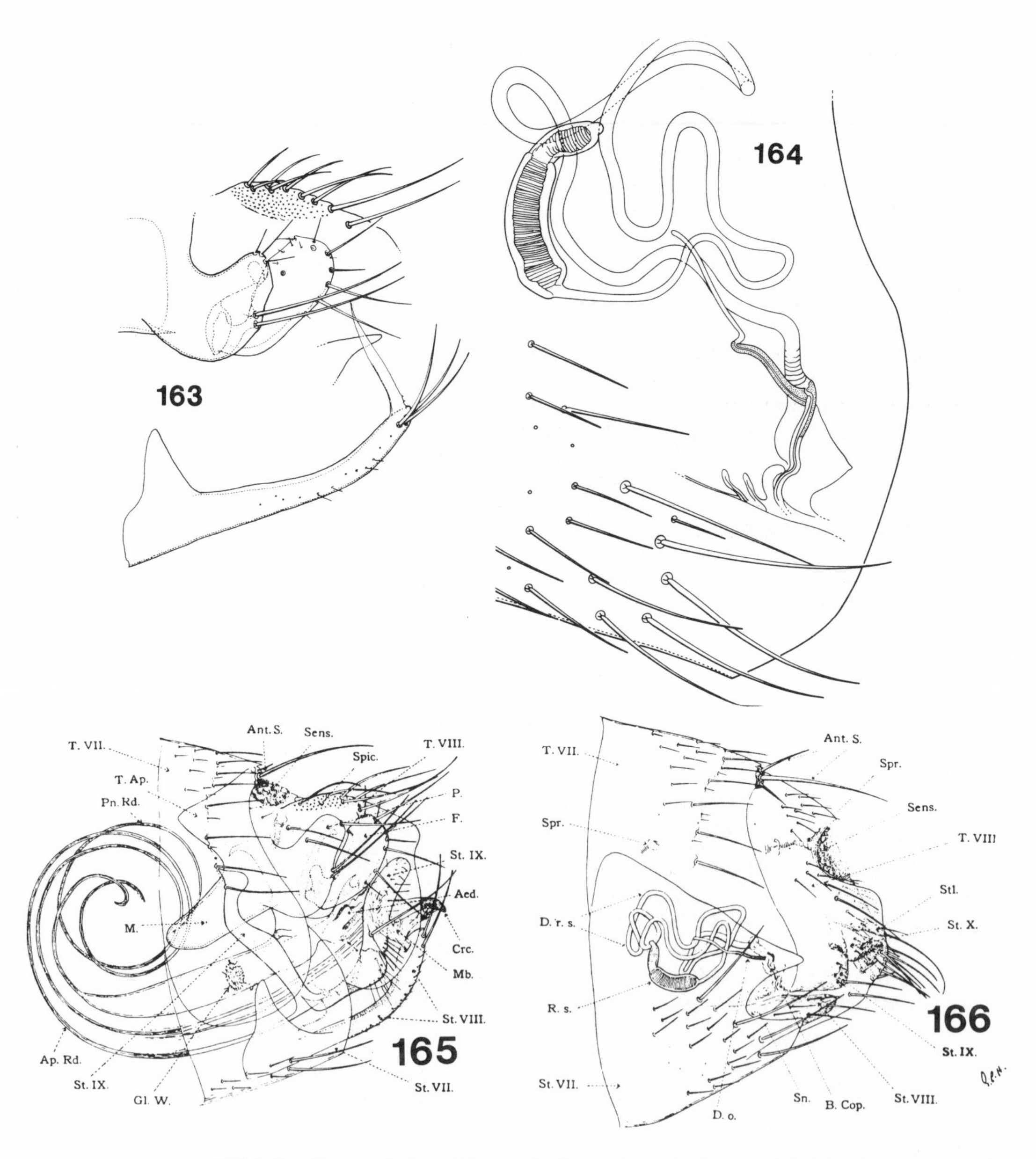

163-164 *Ceratophyllus (Ceratophyllus) niger* C. Fox. **163** Male clasper, dorsal margin of tergum VIII and sternum VIII. **164** Female sternum VII, spermatheca and ducts. After Holland, 1985.

165-166 *Ceratophyllus (Ceratophyllus) idius* J. & R. **165** Terminal abdominal segments of male. **166** Terminal abdominal segments of female. After Holland, 1985.

Ceratophyllus (Ceratophyllus) idius Jordan and Rothschild

[92004]

(Figs. 165, 166)

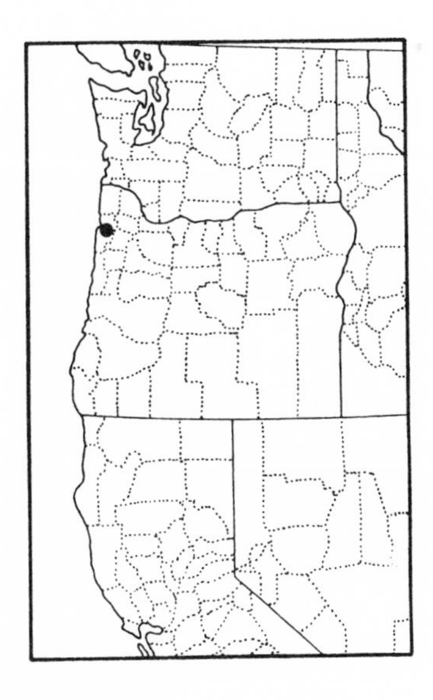

Ceratophyllus idius. Jordan and Rothschild, 1920, Ectoparasites I: 73-76 (Okanagan Landing, B. C., from *Iridoprocne bicolor*).
Ceratophyllus idius Jordan and Rothschild. Holland, 1949, The Siphonaptera of Canada, p. 148.
Ceratophyllus sternacuminatus. Brown, 1968, Canad. Ent. 100: 492-495 (Fredericton, New Brunswick, from nest of *Iridoprocne bicolor*).
Ceratophyllus idius Jordan and Rothschild, 1920. Smit and Wright, 1978a, A list of code numbers..., pp. 13, 42 (*sternacuminatus* synonymized).
Ceratophyllus (Ceratophyllus) idius Jordan and Rothschild, 1920. Smit and Wright, 1978b, A catalogue of primary type specimens..., p. 24 (selection of lectotype male).
Ceratophyllus (Ceratophyllus) idius Jordan and Rothschild, 1920. Smit, in: Traub et al., 1983, p. 35.
Ceratophyllus (Ceratophyllus) idius Jordan and Rothschild, 1920. Haddow et al., in: Traub et al., 1983, p. 62.
Ceratophyllus (Ceratophyllus) idius Jordan and Rothschild. Holland, 1985, Mem. ent. Soc. Canada No. 130, pp. 288-290.

This is another species known to occur in the area but not collected during the surveys upon which the bulk of this report is based. It is associated mainly with the nests of the tree swallow, *Tachycineta bicolor,* but is likewise common in the nests of the purple martin, *Progne subis,* the bluebird, *Sialia sialis,* and the violet-green swallow, *Tachycineta thalassina.* The species is distributed across the northern portion of North America from central Alaska south to Oregon and east in southern Canada and the northern United States.

Ceratophyllus (Ceratophyllus) pelecani Augustson

[94214]

Ceratophyllus pelecani. Augustson, 1942, Trans. San Diego Soc. Nat. Hist. 9: 437-438, figs. 1, 2 (Los Coronados Islands, Lower California, Mexico, from the brown pelican, *Pelecanus occidentalis californicus*).
Ceratophyllus pelecani Augustson, 1942. Hubbard, 1947, Fleas of Western North America, p. 251.
Ceratophyllus (Ceratophyllus) pelecani Augustson, 1942. Haddow et al., in: Traub et al., 1983, p. 63.

This species is native to the Pacific Coast of North America and, although it has not yet been reported from the area under consideration, it is known from as far north as Vancouver Island. In addition to pelicans, it has been reported from the nests of *Larus occidentalis* (western gull) and *Asio otus* (long-eared owl) and it is suspected to occur on other species of colonial seabirds.

Ceratophyllus (Ceratophyllus) petrochelidoni Wagner

[93634]

(Figs. 167, 168)

Ceratophyllus petrochelidoni. Wagner, 1936, Z. Parasitenk. 8: 655-656, fig. 2 (Chilcotin, British Columbia, Canada, from *Petrochelidon lunifrons* [= *pyrrhonota*]).
Ceratophyllus petrochelidoni Wagner, 1936. Hubbard, 1947, Fleas of Western North America, p. 255.
Ceratophyllus petrochelidoni Wagner. Holland, 1949, The Siphonaptera of Canada, p. 149.
Ceratophyllus (Ceratophyllus) petrochelidoni Wagner, 1936. Haddow et al., in: Traub et al., 1983, p. 64.

This is a specific parasite of the cliff swallow, a bird which has a broad range throughout western North America. Though seldom found on the birds themselves, older mud nests frequently yield this and other species in large numbers.

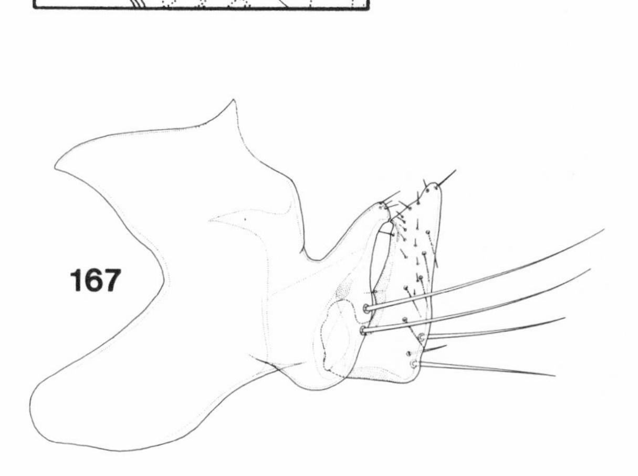

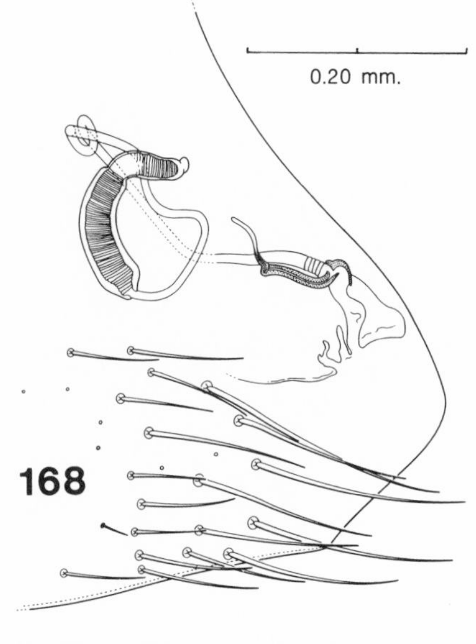

167-168 *Ceratophyllus (Ceratophyllus) petrochelidoni* Wagner. **167** Male clasper. **168** Female sternum VII and spermatheca. After Holland, 1985.

Ceratophyllus (Ceratophyllus) scopulorum Holland

[95221]

(Figs. 169, 170, 171)

Ceratophyllus scopulorum. Holland, 1952, Canad. Ent. 84: 297-299, figs. 1-7 (Rampart House, Yukon Territory, Canada, from nests of *Petrochelidon pyrrhonota albifrons,* the cliff swallow).

Ceratophyllus (Ceratophyllus) scopulorum Holland, 1952. Haddow et al., in: Traub et al., 1983, p. 65.

This species was described from the cliff swallow and its nests; it has been taken on *Hirundo rustica* (barn swallow) and occasionally on *Riparia riparia* (bank swallow). It is known from a number of records in Alaska and one record from New Brunswick. While it has not been reported from the Pacific Northwest, cliff swallow nests from higher elevations should be productive. There are specimens from North Dakota in the Lewis Collection.

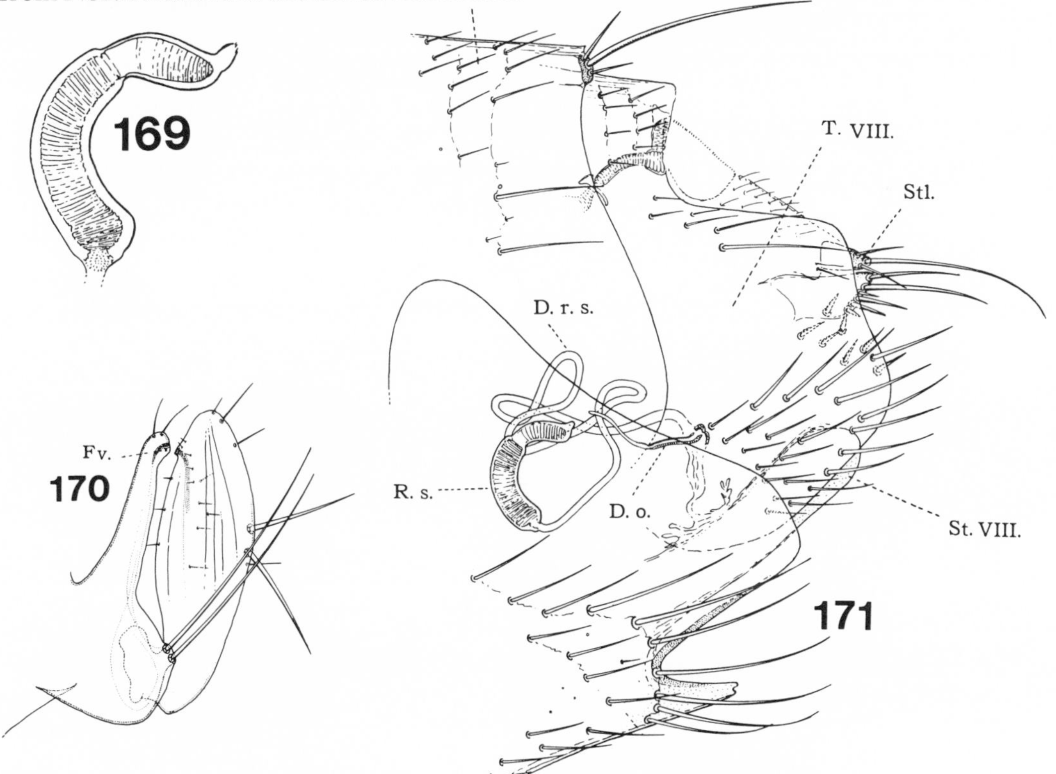

169-171 *Ceratophyllus (Ceratophyllus) scopulorum* Holland. **169** Female spermatheca. **170** Clasper of male. **171** Apical abdominal segments of female. After Holland, 1985.

Ceratophyllus (Ceratophyllus) styx riparius J. & R.

[92003]

(Figs. 172, 173)

Ceratophyllus riparius. Jordan and Rothschild, 1920, Ectoparasites I: 65-76, fig. 67 (Bay View, Milwaukee, Wisconsin, from nest of *Riparia riparia*, the bank swallow).

Ceratophyllus riparius Jordan and Rothschild, 1920. Hubbard, 1947, Fleas of Western North America, p. 251-252.

Ceratophyllus riparius Jordan and Rothschild. Holland, 1949, The Siphonaptera of Canada, p. 149-150.

Ceratophyllus styx riparius Jordan and Rothschild, 1920. Haddow et al., in: Traub et al., 1983, p. 67.

This is a Holarctic species, occuring both in North America and in eastern Asia. It is specific to the bank swallow (also known as the sand martin in Europe), but has been reported from other birds as accidental infestations. It has also been reported from the red squirrel, *Tamiasciurus hudsonicus*.

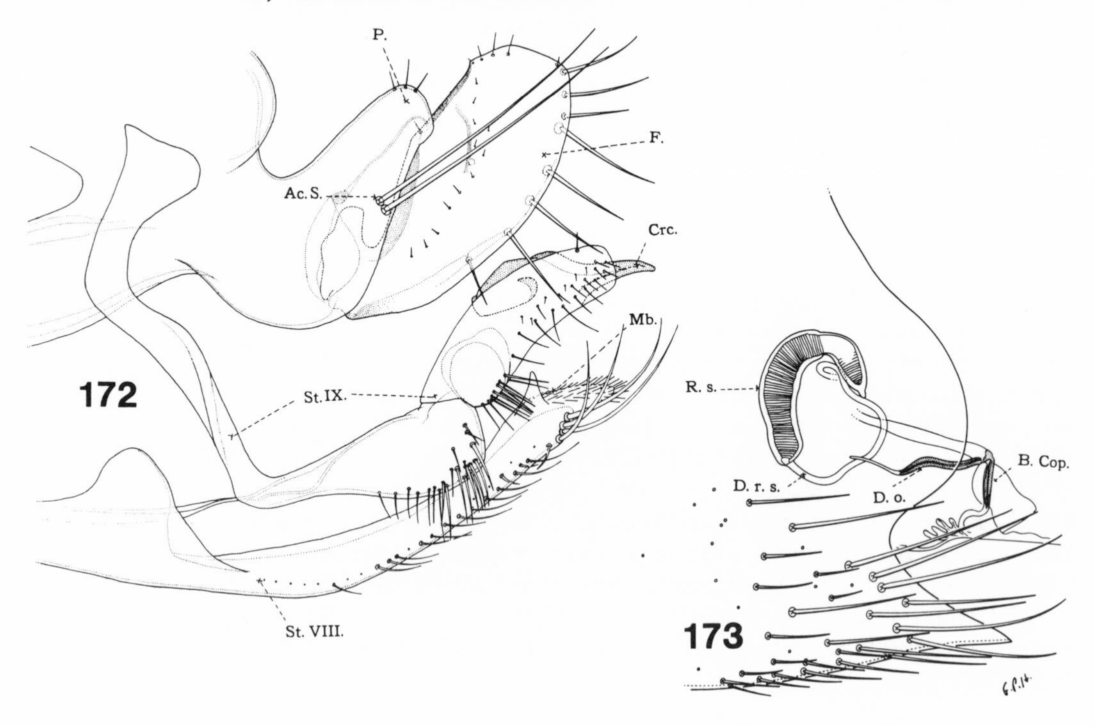

172-173 *Ceratophyllus (Ceratophyllus) styx riparius* J. & R. **172** Modified abdominal segments of male. **173** Sternum VII and spermatheca of female. After Holland, 1985.

GENUS **Ceratophyllus (Emmareus)** Smit

(=C. (Emmareus) g.) Smit and Wright, 1978b, A catalogue of primary type-specimens..., p. 20.
Ceratophyllus (Emmareus) Smit, 1983. Smit, in: Traub et al., 1983, pp. 14, 36 (type species: *"Pulex" columbae* Gervais, 1844).
Emmareus Smit, 1983. Haddow et al., in: Traub et al., 1983, p. 72.

Erected by Smit as a subgenus of *Ceratophyllus* to include ten species which infest birds nesting on or near the ground, in burrows, on rocks or on cliff faces. Eight of these are limited to the Palaearctic, one is Nearctic, and one is Holarctic.

Ceratophyllus (Emmareus) garei Rothschild

[90204]

(Figs. 174, 175, 176)

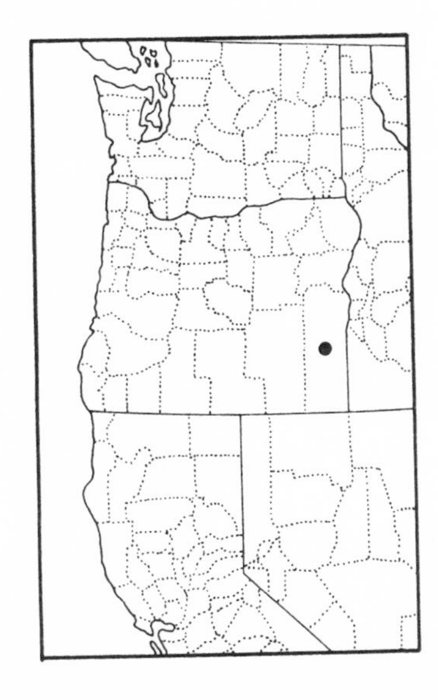

Ceratophyllus garei. Rothschild, 1902, Ent. mon. Mag. 13: 225 (Tring, England, from nest of *Gallinula chloropus*).
Ceratophyllus utahensis. Chapin, 1919, Bull. Brooklyn ent. Soc. 14: 60-62 (Bear River, Utah, from *Steganopus tricolor* or *Spatula clypeata*).
Ceratophyllus garei Rothschild. Jordan and Rothschild, 1920, Ectoparasites I: 69 (*C. utahensis* synonymized).
Ceratophyllus quebecensis. I. Fox, 1940, Proc. ent. Soc. Wash. 42: 65-66 (St. Mary's Island, Quebec, from "eider down").
Ceratophyllus garei Rothschild. Holland, 1949, The Siphonaptera of Canada, p. 148 (*C. quebecensis* synonymized).

This is a Holarctic species that is quite common in the nests of water birds and other birds nesting in moist situations. Holland (1985) lists the species from Alaska south to British Columbia and east to Labrador. While it is a true parasite of birds, it also has been reported from a number of small mammals. Haddow et al., in Traub et al. (1983) list over sixty genera of birds as hosts of this species, as well as three genera of mammals. Hubbard (1947) reports the species from Utah and Montana in addition to records from Canada and Alaska.

We have found but a single record of this species for the area under consideration, although it is certainly much more common than this suggests.

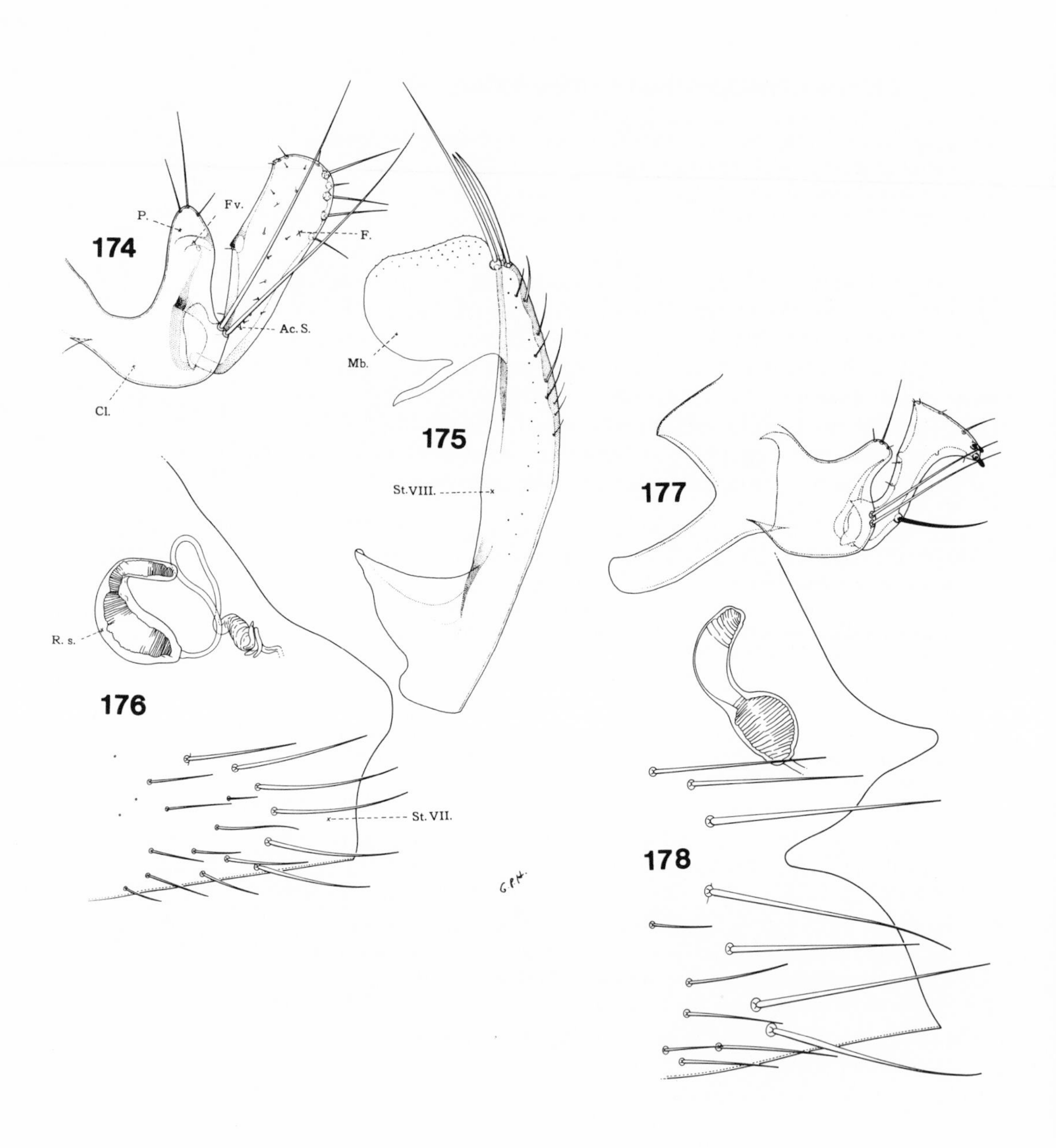

174-176 *Ceratophyllus (Emmareus) garei* Rothschild. **174** Male clasper. **175** Male sternum VIII. **176** Female sternum VII and spermatheca. After Holland, 1985.
177-178 *Ceratophyllus (Amonopsyllus) ciliatus protinus* (Jordan). **177** Male clasper. **178** Female sternum VII and spermatheca. After Holland, 1985.

GENUS **Ceratophyllus (Amonopsyllus)** Wagner

Amonopsyllus. Ioff, 1936, Z. Parasitenk. 9: 96-97.
Amonopsyllus Ioff, 1936. Wagner, 1938, Konowia 17: 9 (designates *Monopsyllus ciliatus* as generitype of *Amonopsyllus*).
Ceratophyllus (Amonopsyllus) Wagner, 1938. Smit, in: Traub et al., 1983, p. 35.
Ceratophyllus (Amonopsyllus) Wagner, 1938. Haddow et al., in: Traub et al., 1983, p. 69.

This subgenus includes the four subspecies of what previously was known as *Monopsyllus ciliatus,* parasites of chipmunks of the genus *Eutamias* and chickaree squirrels of the genus *Tamiasciurus* in northwestern North America. All of these taxa might be expected to occur in the area under consideration, although only one is a dominant faunal element and a certain amount of intergradation is known from the south and east parts of the area. Because of this, a key to the subspecies is not particularly useful to the nonspecialist.

KEY TO THE NORTHWESTERN SUBSPECIES OF *AMONOPSYLLUS*

1 Apex of movable process of male not strongly expanded; anterior margin lacking pronounced incrassate projection; upper lobe of female st. VII evenly rounded and broad but less so than lower lobe; sinus a narrow, sharp incision .... **ciliatus ciliatus**

Not as above **2**

2 Anterior margin of movable process of male with pronounced incrassate projection approximately three-fifths down from apex; upper and lower lobes of female st. VII separated by a shallow, smoothly rounded sinus .... **ciliatus mononis**

Anterior margin of movable process of male with reduced incrassate projection, or, it is absent completely; sinus in female st. VII forming an angle **3**

3 Incrassate projection on anterior margin of male movable process occurring about midway down, but much reduced; ventral margin of upper lobe of female st. VII undulate, ventral sinus ending in less than a right angle .... **ciliatus protinus**

Incrassate projection absent; sinus in female st. VII forming approximately a right angle .... **ciliatus kincaidi**

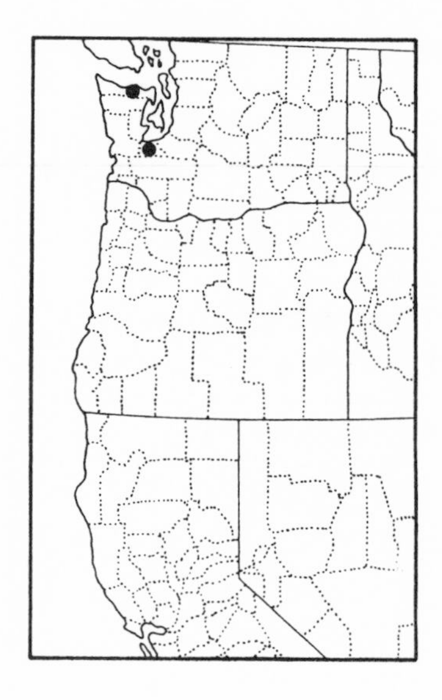

Ceratophyllus (Amonopsyllus) ciliatus ciliatus Baker

[90418]

Ceratophyllus ciliatus. Baker, 1904, Proc. U. S. Nat. Mus. 27: 397 (Mountain View, California, from "chipmunk").
Ceratophyllus ciliatus ciliatus Baker, (1904). Jordan, 1929, Novit. Zool. 35: 34.
Monopsyllus ciliatus Baker, 1904. Jordan, 1933, Novit. Zool. 39: 78.
Ceratophyllus (Monopsyllus) ciliatus B., 1904 [section *Amonopsyllus*]. Ioff, 1936, Z. Parasitenk. 9: 96.
Monopsyllus ciliatus ciliatus Bak., 1904. Wagner, 1936, Canad. Ent. 68: 200.
Trichopsylla (Trichopsylla) ciliata ciliata (Baker). Ewing and Fox, 1943, U. S. Dept. Agric. Miscl. Publ. 500, p. 59.
Monopsyllus ciliatus ciliatus (Baker). Johnson, 1961, U. S. Dept. Agric. Tech. Bull. 1227: 20-24.
Ceratophyllus (Amonopsyllus) ciliatus ciliatus Baker, 1904. Smit, in: Traub et al., 1983, p. 35.

If Traub et al. (1983) are correct, this taxon is extralimital to the area under consideration, being known only from collections south of the San Francisco Bay area. However, we note two records from the state of Washington, clearly in the midst of the range of *C. (A.) ciliatus protinus.* Very likely these are misidentifications, if the various subspecies assigned here are valid, but this cannot be proved without examining the specimens in question. Traub et al., (1983) show what appear to be intergrades along the California/Nevada border, but the subspecies are so similar that a good case could be made for synonymizing them all.

Ceratophyllus (Amonopsyllus) ciliatus protinus Jordan

[92910]

(Figs. 177, 178)

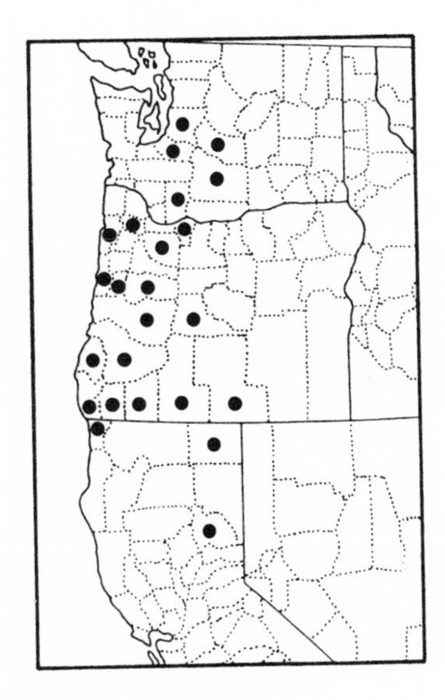

Ceratophyllus ciliatus protinus. Jordan, 1929, Novit. Zool. 35: 34 (Sumas, "British Columbia" [Washington?] from *Eutamias townsendi*).
Monopsyllus ciliatus protinus (Jordan). Wagner, 1936, Canad. Ent. 68: 200.
Monopsyllus ciliatus protinus (Jordan). Johnson, 1961, U. S. Dept. Agric. Tech. Bull. 1227: 24-25.
Ceratophyllus (Amonopsyllus) c. protinus Jordan, 1929. Smit, in: Traub et al., 1983, p. 35.
Ceratophyllus (Amonopsyllus) ciliatus protinus Jordan, 1929. Haddow et al., in: Traub et al., 1983, p. 70.

This was certainly one of the most prevalent species collected during the surveys. Its preferred host is *Eutamias townsendi* but, as shown below, it occurs on a broad range of hosts, particularly tree squirrels of the genus *Tamiasciurus*. Haddow et al. (1983) list the distribution of this subspecies as northwestern California, north to the south coast of Alaska. Our collection records show the subspecies to be limited, in Oregon, to the western half of the state.

Host associations in *Ceratophyllus (Amonopsyllus) ciliatus protinus* Jordan

Hosts	Males	Females	%
Eutamias townsendi	440	563	66.4
Eutamias amoenus	6	11	1.1
Tamiasciurus douglasi	139	162	19.9
Sciurus griseus	17	22	2.5
Spermophilus beecheyi	3	1	0.2
Glaucomys sabrinus	0	3	0.2
Peromyscus maniculatus	24	32	3.7
Predators	7	15	1.4
Accidentals	17	32	3.2
Nests	8	8	1.0
Totals	661	849	
	43.7%	56.2%	

Seasonal distribution of *Ceratophyllus (Amonopsyllus) ciliatus protinus* Jordan collections

	J	F	M	A	M	J	J	A	S	O	N	D
Males	18	4	10	11	27	15	52	214	254	14	9	31
Females	17	8	18	28	40	20	44	257	347	20	16	36

Ceratophyllus (Amonopsyllus) ciliatus kincaidi (Hubbard)

[94704]

(Fig. 179)

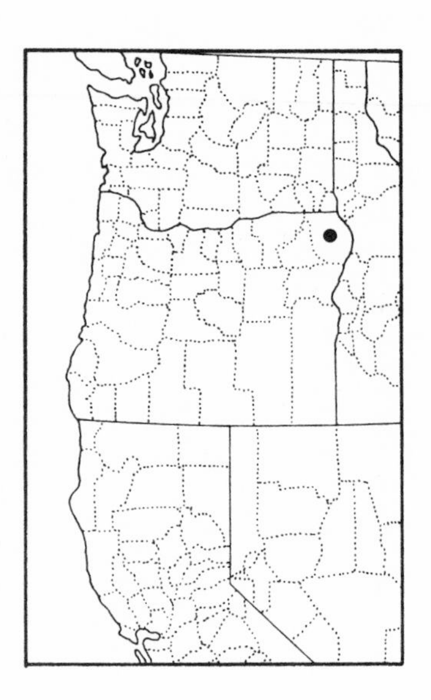

Monopsyllus ciliatus kincaidi. Hubbard, 1947, Fleas of Western North America, pp. 232-233 (Wallowa Lake, Wallowa Co., Oregon, from *Eutamias amoenus luteiventris*).

Monopsyllus ciliatus fasteni. Hubbard, 1954, Ent. News 65: 174 (10 mi. N. of Potlatch, Latah Co., Idaho, from *Eutamias*).

Monopsyllus ciliatus kincaidi Hubbard. Johnson, 1961, U. S. Dept. Agric. Tech. Bull. 1227: 26 (*M. c. fasteni* synonymized).

Ceratophyllus (Amonopsyllus) ciliatus kincaidi (Hubbard, 1947). Smit, in: Traub et al., 1983, p. 35.

Ceratophyllus (Amonopsyllus) ciliatus kincaidi (Hubbard, 1947). Haddow et al., in: Traub et al., 1983, p. 69.

Reported from the Wallowa Mountains of northeastern Oregon, extending into western Montana and central and southwestern Utah, according to Haddow et al. (1983).

Ceratophyllus (Amonopsyllus) ciliatus mononis Jordan

[92911]

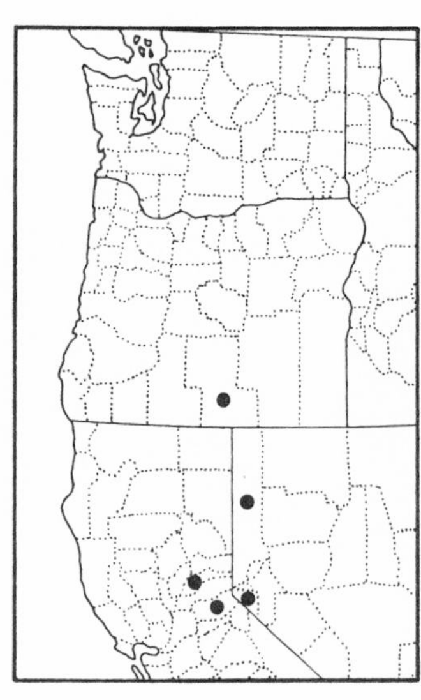

Ceratophyllus ciliatus mononis. Jordan, 1929, Novit. Zool. 35: 35, pl. I, figs. 17, 18 (California: Pine City, Mono Co., from *Mustela arizonensis* and *Eutamias frater*).

Monopsyllus ciliatus Baker, 1904. Jordan, 1933, Novit. Zool. 39: 78.

Trichopsylla (Trichopsylla) ciliata mononis (Jordan). Ewing and Fox, 1943, U. S. Dept. Agric. Miscl. Publ. 500: 59.

Monopsyllus ciliatus mononis (Jordan). Johnson, 1961, U. S. Dept. Agric. Tech. Bull. 1227: 25.

Ceratophyllus (Amonopsyllus) c. mononis Jordan, 1929. Smit, in: Traub et al., 1983, p. 35.

Ceratophyllus (Amonopsyllus) ciliatus mononis Jordan, 1929. Haddow et al., in: Traub et al., 1983, p. 69.

According to Haddow et al. (1983), this subspecies occurs in southern Oregon and eastern California, in the Sierra Nevada and San Bernardino Mountains. It is said to intergrade with *C. (A.) c. protinus* in northeastern California.

GENUS **Ceratophyllus (Monopsyllus)** Kolenati

Monopsyllus. Kolenati, 1857, Wien ent. Monatschr. 1: 65 (type species by subsequent designation: *M. sciurorum* Schrank, 1803).
Monopsyllus Kolenati, 1857. Jordan, 1933, Novit. Zool. 39: 78.
Trichopsylla Kolenati. Ewing and Fox, 1943, U. S. Dept. Agric. Miscl. Publ. 500: 55-57 (*partim*).
Ceratophyllus (Monopsyllus) Kolenati, 1857. Smit, in: Traub et al., 1983, p.34.
Ceratophyllus (Monopsyllus) Kolenati, 1857. Haddow et al., in: Traub et al., 1983, p. 75.

As restricted by Smit (1983), only one taxon belonging to this subgenus occurs in North America, where it is a parasite of red squirrels belonging to the genus *Tamiasciurus*.

Ceratophyllus (Monopsyllus) vison Baker

[90426]

(Figs 180, 181)

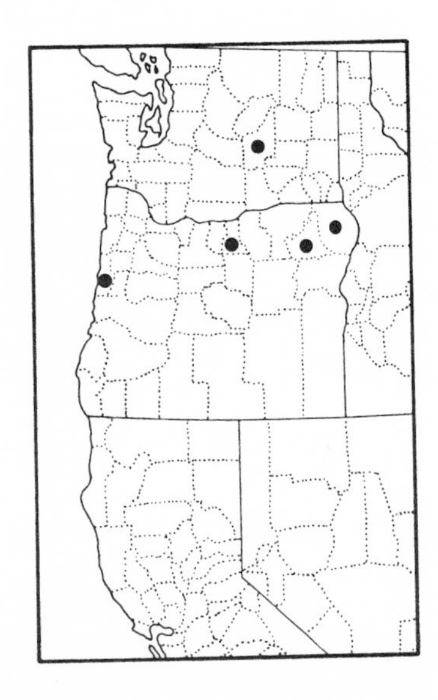

Ceratophyllus vison. Baker, 1904, Proc. U. S. Nat. Mus. 27: 388, 408 (Peterboro, New York, from *Mustela vison*).
Ceratophyllus lucidus. Baker, 1904, Proc. U. S. Nat. Mus. 27: 388, 410 (Pagosa Peak, Colorado, from spruce squirrels).
Ceratophyllus vison. Jordan, 1929, Novit Zool. 35: 35 (*lucidus* synonymized).
Monopsyllus vison Baker, 1904. Jordan, 1933, Novit. Zool. 39: 78.
Monopsyllus vison reeheri. Hubbard, 1954, Ent. News 65: 174 (Heppner, Oregon, from *Tamiasciurus hudsonicus richardsoni*).
Monopsyllus vison (Baker). Johnson, 1961, U. S. Dept. Agric. Tech. Bull. 1227: 11-12 (*reeheri* synonymized).
Ceratophyllus (Monopsyllus) vison Baker, 1904. Smit, in: Traub et al., 1983, p. 34.
Ceratophyllus (Monopsyllus) vison Baker, 1904. Haddow et al., in: Traub et al., 1983, p. 77.

The true host of this species, *Tamiasciurus hudsonicus,* is restricted to the northeastern part of Oregon and Washington. Hubbard (1947) reports it from Wallowa County, Oregon, on this host. It was not collected during this survey.

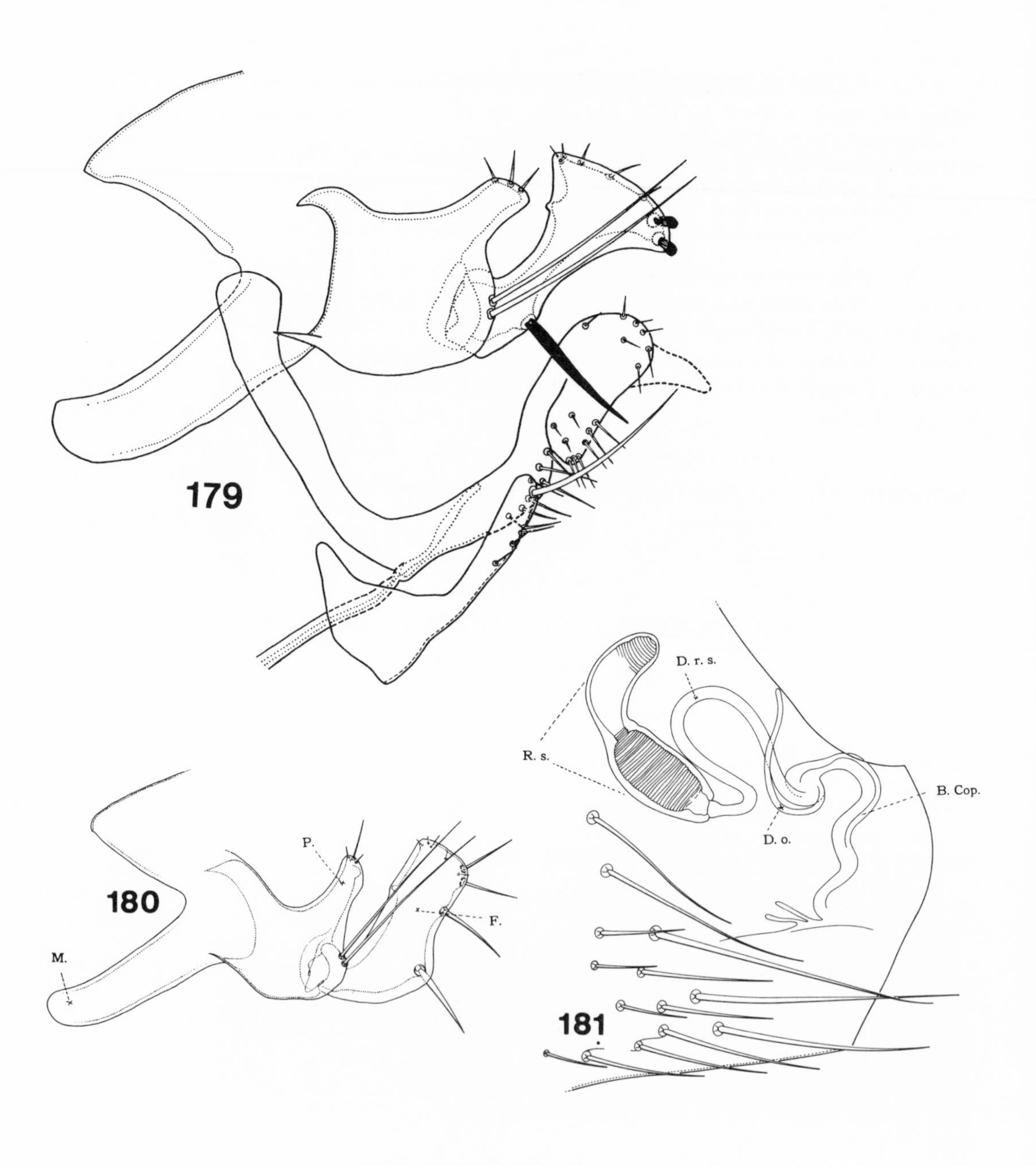

179 *Ceratophyllus (Amonopsyllus) ciliatus kincaidi* (Hubbard). Male clasper and sternites VIII and IX. Original.

180-181 *Ceratophyllus (Monopsyllus) vison* (Baker). **180** Male clasper. **181** Female sternum VII and spermatheca. After Holland, 1985.

GENUS **Dasypsyllus (Dasypsyllus)** Baker

Dasypsyllus. Baker, 1905, Proc. U. S. Nat. Mus. 29: 129, 146 (type species by monotypy: *Ceratophyllus perpinnatus* Baker, 1904).
Dasypsyllus Baker, 1905. Jordan, 1933, Novit. Zool. 39: 76 (type species by designation: *D. gallinulae* Dale, 1878).
Dasypsyllus (Dasypsyllus) Baker, 1905. Smit, in: Traub et al., 1983, p. 25.
Dasypsyllus (Dasypsyllus) Baker, 1905. Haddow et al., in: Traub et al., 1983, p. 84.

This subgenus contains a single species with three recognized subspecies. The nominate taxon is widely distributed in the Palaearctic Region in low, moist nests according to Haddow et al. (1983). It is a common parasite of a broad range of birds and is even known from a number of small mammals. One subspecies is known from western North America.

Dasypsyllus (Dasypsyllus) gallinulae perpinnatus (Baker)

[90413]

(Figs. 182, 183)

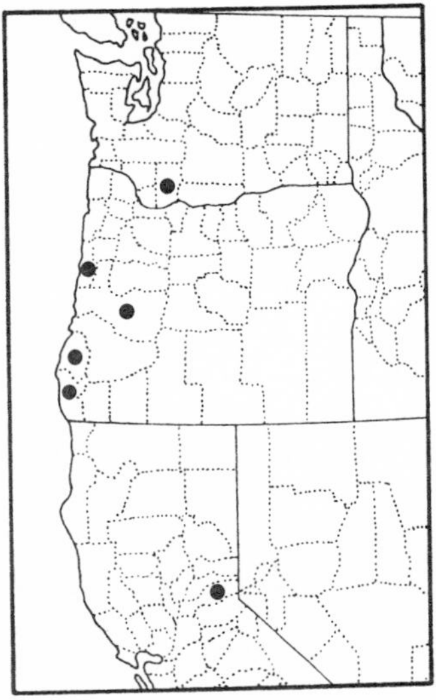

Ceratophyllus perpinnatus. Baker, 1904, Proc. U. S. Nat. Mus. 27: 386, 391-392, 445 (male from Queen Charlotte Island, B. C., no host given).
Ceratophyllus gallinulae Dale. Jordan and Rothschild, 1920, Ectoparasites I: 69.
Ceratophyllus gallinulae perpinnatus Baker. Jordan, 1926, Novit. Zool. 33: 386.
Dasypsyllus gallinulae perpinnatus (Baker). Wagner, 1936, Canad. Ent. 68: 201.
Dasypsyllus gallinulae perpinnatus (Baker). Holland, 1949, The Siphonaptera of Canada, pp. 150-151.
Dasypsyllus (Dasypsyllus) g. perpinnatus (Baker, 1904). Smit, in: Traub et al., 1983, p. 25.
Dasypsyllus (Dasypsyllus) gallinulae perpinnatus (Baker, 1904). Haddow et al., in: Traub et al., 1983, p. 85.

This subspecies is a parasite of various passerine birds that nest on or near ground level. It is distributed in western North and Central America, from Alaska south to the mountains of Panama and Venezuela. Tipton and Machado-Allison (1972) also note a record from Argentina from a bird's nest.

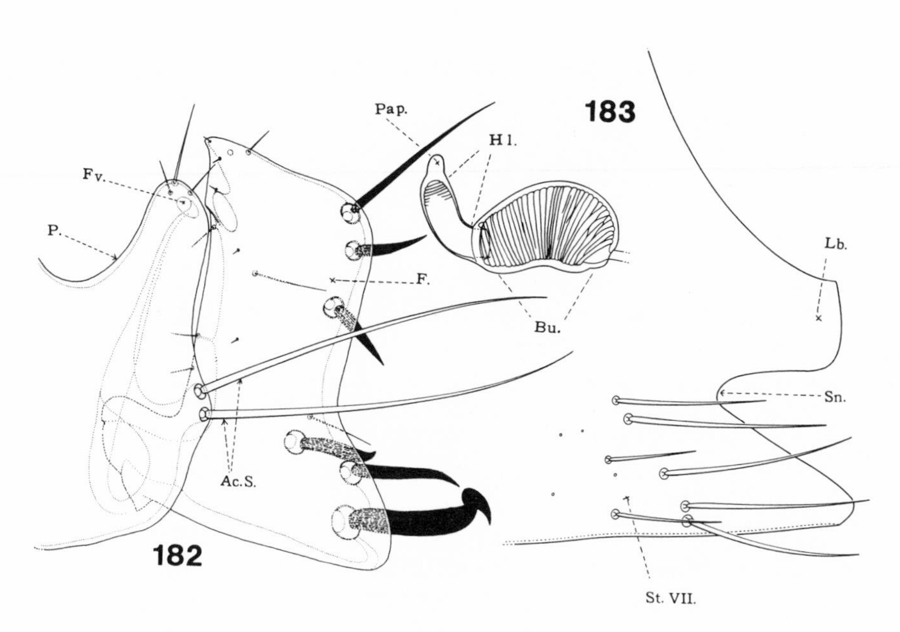

182-183 *Dasypsyllus (Dasypsyllus) gallinulae perpinnatus* (Baker). **182** Male clasper. **183** Female sternum VII and spermatheca. After Holland, 1985.

GENUS **Eumolpianus** Smit

Eumolpianus. Smit, in: Traub et al., 1983, pp. 15, 35 (type species by designation: *"Ceratophyllus" eumolpi* Rothschild, 1905).
Eumolpianus Smit, 1983. Haddow et al., in: Traub et al., 1983, p. 88.

This genus was recently erected to include a number of species originally assigned to the genus *Monopsyllus*. Although there are a number of records of accidental host associations, members of this genus are primarily parasites of chipmunks of the genus *Eutamias*. The species may be separated with the following key modified from that of Johnson (1961).

KEY TO THE NORTHWESTERN SPECIES OF *EUMOLPIANUS*

1 Males **2**

Females **5**

2 Distal arm of st. IX straight, its apex bearing a ventral subapical tooth, causing it to appear barbed .... **eumolpi eumolpi**

Distal arm of st. IX bent, its apex forming a right angle to the remainder of the segment **3**

3 Apical portion of movable process above notch in anterior margin wider than high .... **wallowensis**

Apical portion of movable process not wider than high **4**

4 Median dorsal lobe of aedeagus angulate along margin; armature of inner tube not crested; apical portion of distal arm of st. IX with posteroventral corner drawn out and acute .... **orarius**

Median dorsal lobe of aedeagus smoothly convex; armature of inner tube crested; apical portion of distal arm of st. IX with posteroventral corner blunter and more rounded .... **cyrturus**

5 Bursa copulatrix long and coiled **6**

Bursa copulatrix otherwise **7**

6 Ventral anal lobe strongly angled posteroventrally, almost forming a right angle; st. VII with broad lobe .... **cyrturus**

Ventral anal lobe not so strongly angled posteroventrally, angle greater than 90 degrees; lobe of st. VII broad and usually squared .... **eumolpi eumolpi**

7 St. VII lacking a sinus or dorsal projection .... **orarius**

St. VII with dorsal projection subtended by a shallow sinus .... **wallowensis**

Eumolpianus cyrturus (Jordan)

[92909]

(Figs. 184, 185)

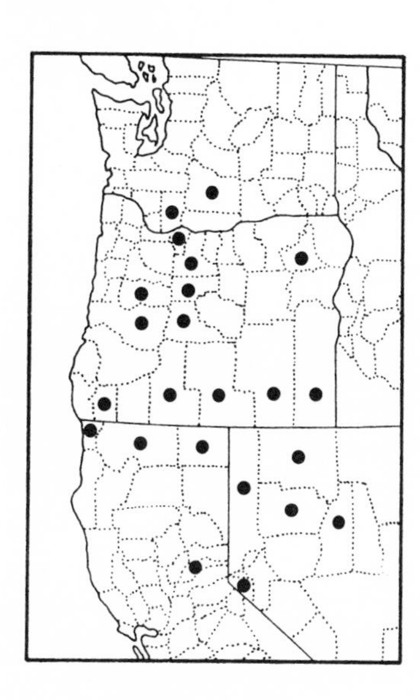

Ceratophyllus eumolpi cyrturus. Jordan, 1929, Novit. Zool. 35: 34 (Paradise, Cochise Co., Arizona, from *Mephitis* sp.).

Monopsyllus eumolpi charlestonensis. Hubbard, 1950, Ent. News 60: 254 (Charleston Peak, 25 mi. N.W. Las Vegas, Clark Co., Nevada, from *Eutamias palmeri*).

Monopsyllus cyrturus (Jordan). Johnson, 1961, U. S. Dept. Agric. Tech. Bull. 1227: 50-53 (*charlestonensis* synonymized).

Eumolpianus cyrturus Jordan, 1929a. Smit and Wright, 1978b, A catalogue of primary type-specimens..., p. 15.

Eumolpianus cyrturus (Jordan, 1929). Smit, in: Traub et al., 1983, p. 35.

Eumolpianus cyrturus (Jordan, 1929). Haddow et al., in: Traub et al., 1983, p. 88.

As indicated for the genus, this species is mainly a parasite of *Eutamias* species and the bulk of our collections came from *E. minimus, amoenus,* and *townsendi.* An additional few were taken on *Sciurus griseus, Glaucomys sabrinus, Spermophilus* species,

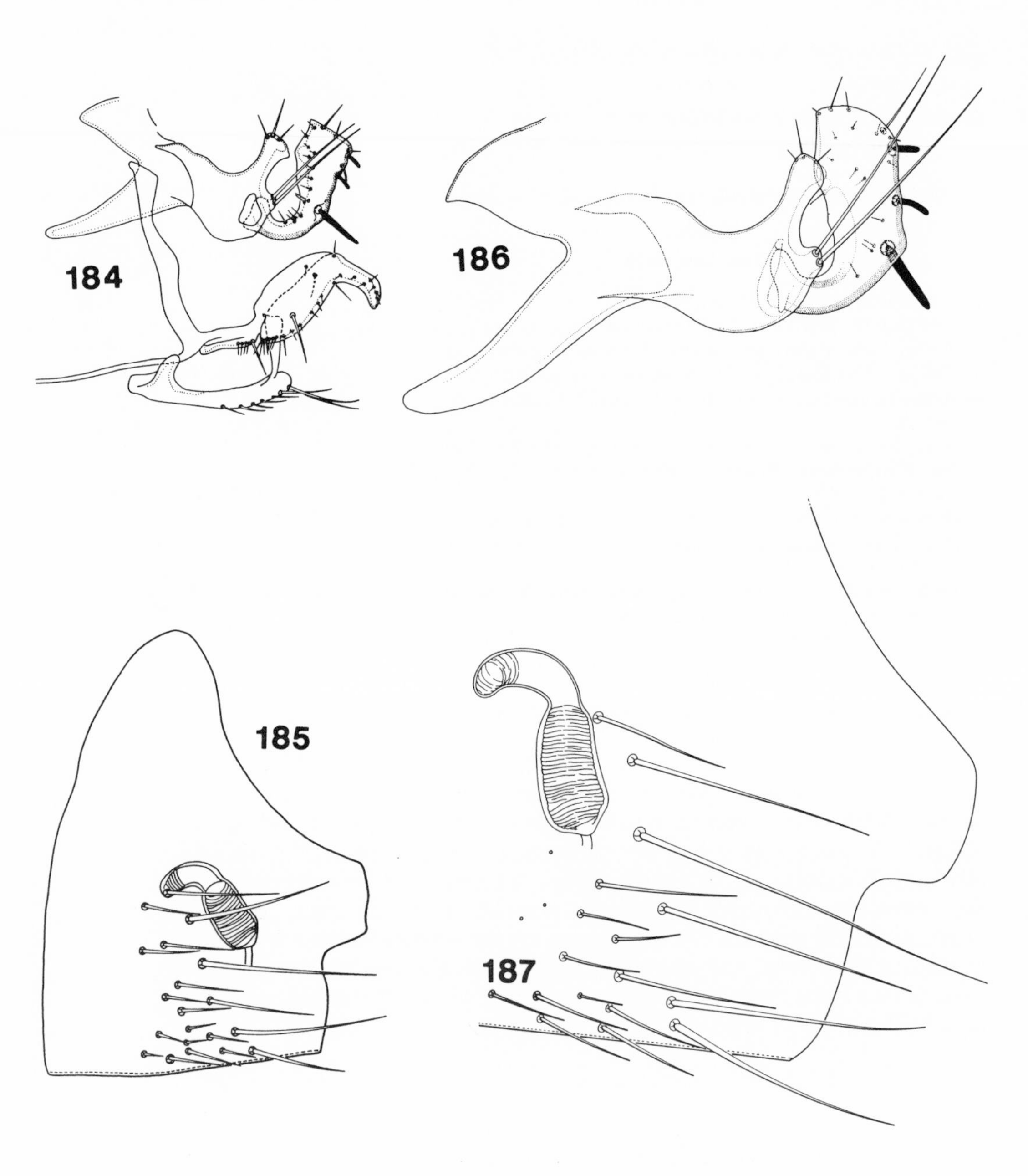

184-185 *Eumolpianus cyrturus* (Jordan). **184** Male clasper and sternites VIII and IX. **185** Female sternum VII and spermatheca. Original.
186-187 *Eumolpianus eumolpi eumolpi* (Rothschild). **186** Male clasper. **187** Female sternum VII and spermatheca. After Holland, 1985.

and small mice. Collections ranged from March to September with peak collections during July to September. As the distribution map shows, the species is widespread.

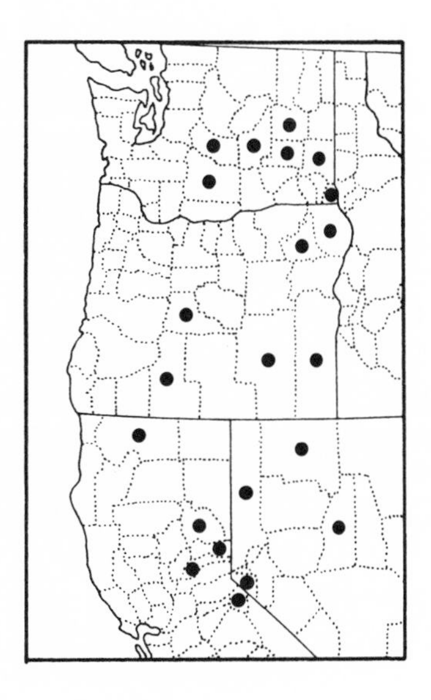

Eumolpianus eumolpi eumolpi (Rothschild)

[90506]

(Figs. 186, 187)

Ceratophyllus eumolpi. Rothschild, 1905, Novit. Zool. 12: 161 (Banff, Red Deer, and Canadian National Park, Alberta, from *Eutamias minimus borealis* and Okanagan, British Columbia, from *Eutamias amoenus affinis*).

Monopsyllus eumolpi Roths., 1905. Jordan, 1933, Novit. Zool. 39: 78.

Monopsyllus eumolpi canadensis. Hubbard, 1950, Ent. News 60: 259 (Malachi, Kenora District, Ontario, from "chipmunk").

Monopsyllus eumolpi eumolpi (Rothschild). Johnson, 1961, U.S. Dept. Agric. Tech. Bull. 1227: 42-49 (*canadensis* synonymized).

Eumolpianus eumolpi Rothschild, 1905a. Smit and Wright, 1978b, A catalogue of primary type-specimens..., p. 19 (selection of lectotype male).

Eumolpianus e. eumolpi (Rothschild, 1905). Smit, in: Traub et al., 1983, p. 35.

Eumolpianus eumolpi eumolpi (Rothschild, 1905). Haddow et al., in: Traub et al., 1983, p. 88.

This is a common parasite of squirrels in the western United States. While its preferred hosts are evidently chipmunks, other squirrels are commonly infested and there are also records from other small rodents. In Colorado, Utah, Arizona, and New Mexico, the nominate subspecies intergrades with *E. e. americanus,* and Johnson (1961) shows a few collections of the nominate form from Michigan, Wisconsin, and southern Canada. She also reports aberrant populations from Nevada, South and North Dakota, Wyoming, and eastern Washington state.

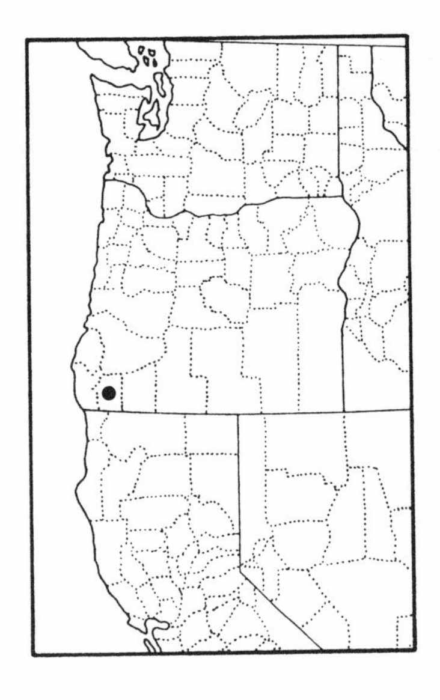

Eumolpianus orarius (Johnson)

[96105]

Monopsyllus orarius. Johnson, 1961, U. S. Dept. Agric. Tech. Bull. 1227: 55-57, pl. XIX, figs. 119, 123, pl. XX, figs. 126-128 (Bolan Lake, Josephine Co., Oregon, from *Eutamias townsendi siskiyou*).
Eumolpianus orarius (Johnson, 1961). Smit, in: Traub et al., 1983, p. 35.
Eumolpianus orarius (Johnson, 1961). Haddow et al., in: Traub et al., 1983, p. 89.

No new records of this species are reported here and as far as is known this taxon is known only from the type series that was collected in extreme southwestern Oregon.

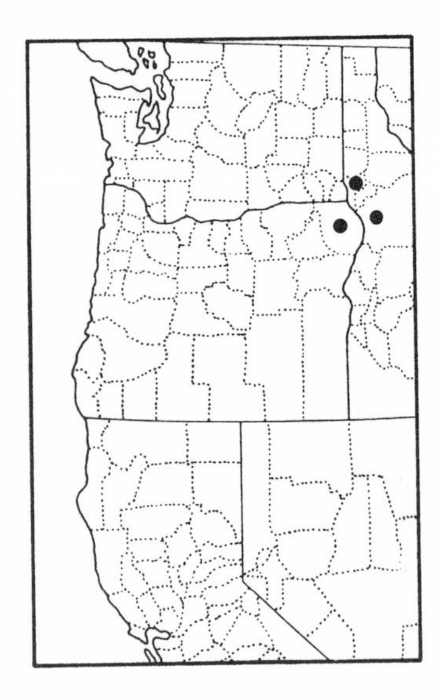

Eumolpianus wallowensis (Hubbard)

[95004]

Monopsyllus eumolpi wallowensis. Hubbard, 1950, Ent. News 60: 255, fig. 1 (Wallowa Lake, Wallowa Co., Oregon, from *Eutamias amoenus albiventris*).
Monopsyllus wallowensis Hubbard. Johnson, 1961, U. S. Dept. Agric. Tech. Bull. 1227: 53-55, pl. XIX, figs. 120, 121, pl. XX, figs. 124, 125, 129.
Eumolpianus wallowensis (Hubbard, 1950). Smit, in: Traub et al., 1983, p. 35.
Eumolpianus wallowensis (Hubbard, 1950). Haddow et al., in: Traub et al., 1983, p. 89.

As indicated by Johnson (1961), this species evidently has a limited distribution in extreme northeastern Oregon and adjacent Idaho. No additional records are reported here.

GENUS **Malaraeus** Jordan

Malaraeus. Jordan, 1933, Novit. Zool. 39: 76 (type species by designation: *Ceratophyllus telchinus* Rothschild, 1905).
Trichopsylla Kolenati. Ewing and Fox, 1943, U. S. Dept. Agric. Miscl. Publ. 500: 55 (*partim*).
Malaraeus Jordan, 1933. Smit, in: Traub et al., 1983, p. 31.
Malaraeus Jordan, 1933. Haddow et al., in: Traub et al., 1983, p. 107.

This genus has recently been restricted to three species, all of which are mainly parasites of *Peromyscus* species in western North America. Two of these are known from the area, one commonly and one less so.

KEY TO THE NORTHWESTERN SPECIES OF *MALARAEUS*

1 Movable process of male rectangular, its caudal margin bearing three (rarely two) short, dark spiniforms; caudal margin of female st. VII with a smoothly rounded convexity .... **telchinus**

Movable process of male triangular, its ventrocaudal angle bearing a stout, long seta; caudal margin of female st. VII with much more pronounced, pointed lobe .... **sinomus**

Malaraeus telchinus (Rothschild)

[90501]

(Figs. 188, 189)

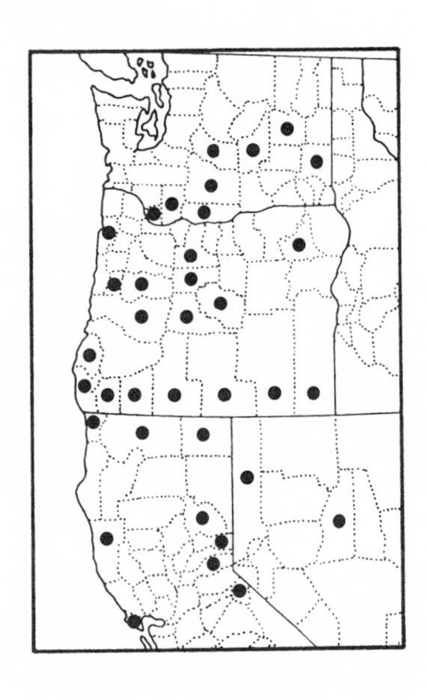

Ceratophyllus telchinum. Rothschild, 1905, Novit. Zool. 12: 153-155 (males from Kicking Horse Canyon, British Columbia, from *Clethrionomys gapperi saturatus*).
Malaraeus telchinum (Rothschild). Jordan, 1933, Novit. Zool. 39: 76.
Malaraeus telchinum (Rothschild). Hubbard, 1940, Pacific Univ. Bull. 37: 1-2 (redescription. Both sexes figured).
Malaraeus telchinum (Rothschild). Holland, 1949, The Siphonaptera of Canada, pp. 155-156.
Malaraeus telchinus. Smit and Wright, 1978b, A catalogue of primary type-specimens..., p. 46 (selection of lectotype male).
Malaraeus telchinus (Rothschild, 1905). Smit, in: Traub et al., 1983, p. 31.
Malaraeus telchinus (Rothschild, 1905). Haddow et al., in: Traub et al., 1983, p. 108.

Of the two local species belonging to this genus, this is by far the commonest. We have collections from seventeen counties in Oregon and it is safe to assume that the species occurs in all parts of the state, as it does in the adjacent states where suitable habitat exists. It shows a preference for much more mesic niches in arid areas and is thus much less restricted in distribution than the following taxon. Although stated by various authors to be mainly a parasite of *Peromyscus* species, our collections suggest a preference for microtines as shown in the following table.

Host associations of *Malaraeus telchinus* in Oregon

	Males	Females
Peromyscus maniculatus	16	50
Peromyscus crinitus	0	5
Clethrionomys californicus	44	84
Microtus montanus	28	34
Microtus oregoni	1	1
Microtus longicaudus	1	2
Lagurus curtatus	16	24
Accidental associations	3	4
Totals	109	204

The species is present as adults throughout the year and shows no particular population peaks, though somewhat more common during the cooler months, as shown in the following table.

Seasonal distribution of *Malaraeus telchinus*

	J	F	M	A	M	J	J	A	S	O	N	D
Males	3	2	4	7	3	4	7	11	26	3	4	34
Females	22	18	1	15	7	3	9	13	46	13	6	48

(Differences in totals between the two tables are caused by four specimens that lacked the date of collection and were thus not included in the second table.)

This species is known to become infected with plague organisms under experimental conditions. However, according to Eskey and Haas (1940), it is a poor vector. Its true role in the maintenance of sylvatic plague has not been demonstrated.

Malaraeus sinomus (Jordan)

[92535]

(Figs. 190, 191)

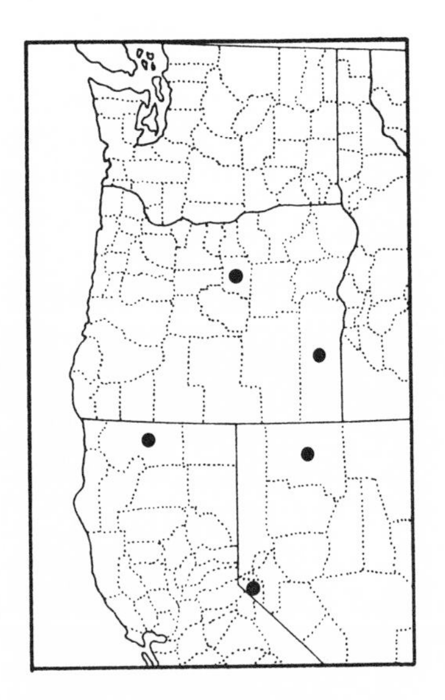

Ceratophyllus sinomus. Jordan, 1925, Novit. Zool. 32: 110 (Paradise, Cochise Co., Arizona, from *Mus* sp.).
Malaraeus sinomus (Jordan). Jordan, 1933, Novit. Zool. 39: 76.
Malaraeus sinomus Jordan, 1925. Hubbard, 1947, Fleas of Western North America, pp. 201-202, fig. 99.
Malaraeus sinomus (Jordan, 1925). Smit, in: Traub et al., 1983, p. 31.
Malaraeus sinomus (Jordan, 1925). Haddow et al., in: Traub et al., 1983, p. 107.

This species has a broad distribution in the arid parts of western North America where it is mainly a parasite of *Peromyscus* species. According to Haddow et al. (1983), it occurs at elevations between 700 and 2000 meters, but mainly at the higher elevations. The few Oregon collections reported here came from Malheur County from *Peromyscus crinitus.*

GENUS **Megabothris** Jordan

Megabothris. Jordan, 1933, Novit. Zool. 39: 77 (type species by designation: *Ceratophyllus walkeri* Rothschild, 1902).
Megabothris (Megabothris) Jordan, 1933. Smit, in: Traub et al., 1983, p. 33.
Megabothris (Megabothris) Jordan, 1933. Haddow et al., in: Traub et al., 1983, p. 109.

Smit (1983) divided this genus into four subgenera, two of which contain members of the Northwestern fauna. The subgenera may be separated by the following short key.

KEY TO THE NORTHWESTERN SUBGENERA OF *MEGABOTHRIS*

1 First two postantennal setal rows of one or two and three to six setae, respectively .... **(Megabothris)**

First two postantennal setal rows of none or one and two (rarely three) setae, respectively .... **(Amegabothris)**

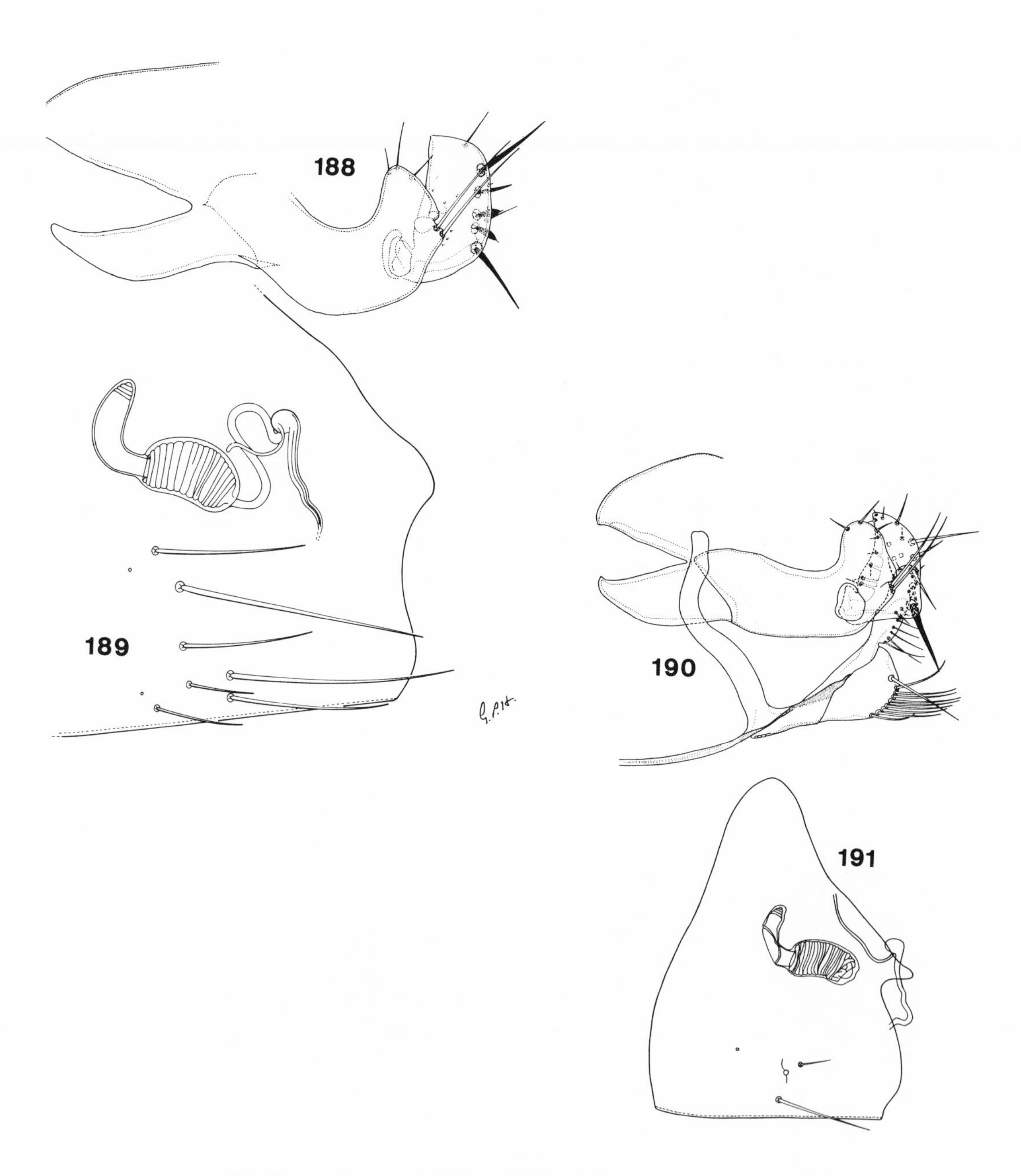

188-189 *Malaraeus telchinus* (Rothschild). **188** Male clasper. **189** Female sternum VII and spermatheca. After Holland, 1985.
190-191 *Malaraeus sinomus* (Jordan). **190** Male clasper. **191** Female sternum VII and spermatheca. Original.

Megabothris (Megabothris) asio megacolpus (Jordan)

[92907]

(Figs. 192, 193)

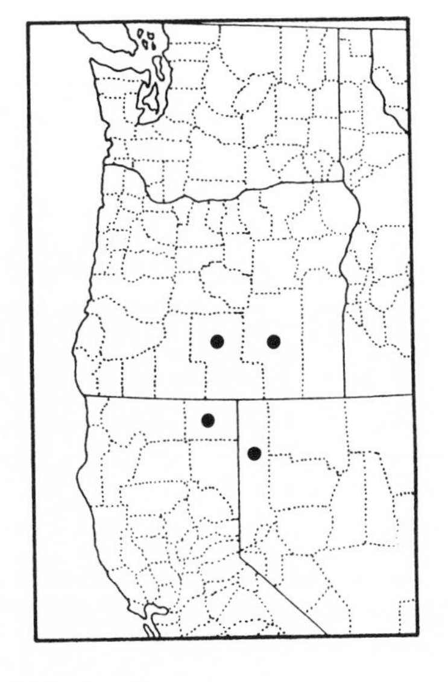

Ceratophyllus megacolpus. Jordan, 1929, Novit. Zool. 35: 33 (female from Okanagan Landing, British Columbia, from *Microtus pennsylvanicus drummondi*).

Megabothris megacolpus (Jordan). Jordan, 1933, Novit. Zool. 39: 77.

Megabothris asio (Baker). Wagner, 1936, Canad. Ent. 68: 201-202.

Megabothris asio orectus. Jordan, 1938, Novit. Zool. 41: 122.

Megabothris asio megacolpus (Jordan). Holland, 1949, The Siphonaptera of Canada, p. 160.

Megabothris asio megacolpus (Jordan). Holland, 1950, Canad. Ent. 82: 126-132 (*orectus* synonymized).

Megabothris (Megabothris) a. megacolpus (Jordan, 1929). Smit, in: Traub et al., 1983, p. 33.

Megabothris (Megabothris) asio megacolpus (Jordan, 1929). Haddow et al., in: Traub et al., 1983, p. 109.

This species was not collected during the surveys reported here, but Hubbard (1947) reported a single male collected in Lake County, Oregon, on *Microtus longicaudus*. Obviously it is not a common species.

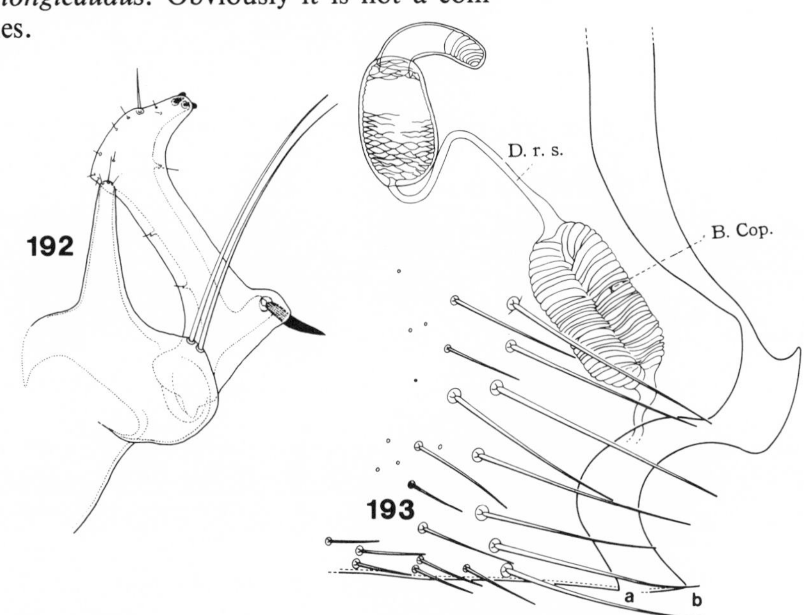

192-193 *Megabothris (Megabothris) asio megacolpus* (Jordan). **192** Male clasper. **193** Female sternum VII and spermatheca. After Holland, 1985.

GENUS **Megabothris (Amegabothris)** Smit

Megabothris. Jordan, 1933, Novit. Zool. 39: 77 (type species by designation: *Ceratophyllus walkeri* Rothschild, 1902).

Trichopsylla Kolenati. Ewing and Fox, 1943, U. S. Dept. Agric. Miscl. Publ. 500: 55-56 (*partim*).

Megabothris (Amegabothris). Smit, in: Traub et al., 1983, pp. 17, 34 (type species: *"Ceratophyllus" groenlandicus* Wahlgren, 1903).

Megabothris (Amegabothris) Smit, 1983. Haddow et al., in: Traub et al., 1983, p. 111.

Three species belonging to this subgenus are reported here. One of these has not been collected during the surveys and is listed here solely on the basis of its inclusion in Hubbard (1947). They may be separated with the following key.

KEY TO THE SPECIES OF THE SUBGENUS *AMEGABOTHRIS*

1 Males **2**

Females **4**

2 Posterior margin of movable process with sharp setae above the long spiniform setae arising on caudoventral projection; apex of process projecting caudad .... **quirini**

With short, blunt spiniforms in the above locations **3**

3 Movable process widest in apical half .... **abantis**

Movable process widest in proximal half .... **clantoni**

4 Duct of spermatheca sclerotized for part of its length; st. VII usually with a lobe but never a sinus .... **quirini**

Duct of spermatheca not sclerotized **5**

5 Tergum VIII with two long setae below sensilium; st. VII with blunt marginal lobe subtended by a shallow sinus .... **abantis**

Tergum VIII with three long setae below sensilium; st. VII with blunt marginal lobe but no sinus .... **clantoni**

Megabothris (Amegabothris) quirini (Rothschild)

[90507]

(Figs. 194, 195)

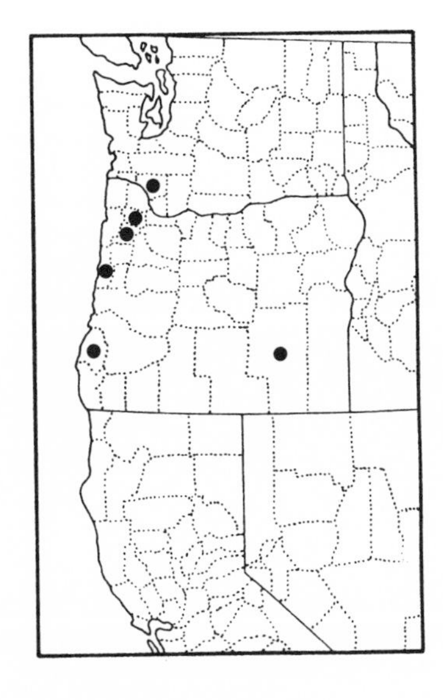

Ceratophyllus quirini. Rothschild, 1905, Novit. Zool. 12: 163-164 (males from Red Deer, Alberta, from *Clethrionomys gapperi*).

Megabothris quirini (Rothschild). Jordan, 1933, Novit. Zool. 39: 77.

Megabothris quirini (Rothschild). I. Fox, 1940, Fleas of Eastern United States, p. 70 (female described).

Megabothris quirini (Rothschild). Holland, 1949, The Siphonaptera of Canada, pp. 163-164.

Megabothris (Amegabothris) quirini (Rothschild, 1905). Smit, in: Traub et al., 1983, p. 34.

Megabothris (Amegabothris) quirini (Rothschild, 1905). Haddow et al., in: Traub et al., 1983, p. 112.

This species is included in the fauna on the strength of its listing in Hubbard (1947), as no specimens were taken during the surveys. Holland (1985) shows it occurring throughout Canada, and Haddow et al. (1983) show U. S. records from New York, Wisconsin, and Utah. While the species has been taken on a broad range of hosts, mainly rodents, Holland (1985) points out that it seems more prevalent on *Zapus,* and perhaps members of the family Zapodidae are the original hosts. He goes on to state that the species is distributed beyond areas where *Zapus* occurs and speculates that the species probably has specific habitat preferences and parasitizes any species of mouse that exists in that habitat. Among other genera of hosts, the species shows a strong preference for microtine rodents.

Megabothris (Amegabothris) abantis (Rothschild)

[90508]

(Figs. 196, 197)

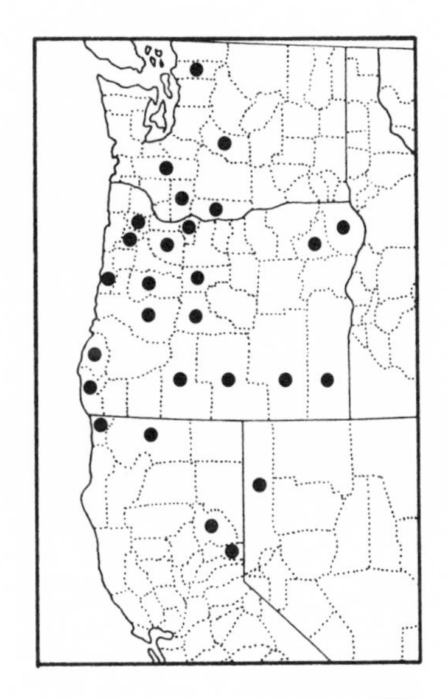

Ceratophyllus abantis. Rothschild, 1905, Novit. Zool. 12: 164-165 (males from Canadian National Park, Alberta, from *Mustela frenata* and from Horse Creek, British Columbia, from *Microtus pennsylvanicus*).

Megabothris abantis (Rothschild). Jordan, 1933, Novit. Zool. 39: 77.

Megabothris adversus. Wagner, 1936, Z. Parasitenk. 8: 654, 656-657 (female from Vancouver, British Columbia, from *Peromyscus maniculatus*).
Megabothris abantis (Rothschild). Holland, 1942, Canad. Ent. 74: 158 (*M. adversus* synonymized).
Megabothris abantis (Rothschild). Good, 1942, J. Kans. ent. Soc. 15: 7-9 (female described).
Megabothris abantis (Rothschild). Holland, 1949, The Siphonaptera of Canada, pp. 158-159.
Megabothris (Amegabothris) abantis (Rothschild, 1905). Smit, in: Traub et al., 1983, p. 34.
Megabothris (Amegabothris) abantis (Rothschild, 1905). Haddow et al., in: Traub et al., 1983, p. 111.

This is a western species known from all of the western states including Montana south to New Mexico. Holland (1985) shows it concentrated in British Columbia and southwestern Alberta, with a cluster of collection localities in the Alaskan peninsula. The species has been taken from a wide range of hosts, but shows a distinct preference for microtines. In addition, there are the usual accidental associations with ground squirrels, shrews, and carnivores.

Collection records from Oregon include seventeen counties and there is no reason to doubt that the species is distributed throughout the state. Host and seasonal distribution are shown in the following tables. The sex ratio was roughly two females to one male.

Host associations of *Megabothris abantis* in Oregon

	Males	Females
Microtus species and microtines	19	41
Peromyscus maniculatus	7	6
Zapus trinotatus	5	18
Accidental hosts	2	4
Predators	1	6

Seasonal distribution of *M. abantis* in Oregon

	J	F	M	A	M	J	J	A	S	O	N	D
Males	0	1	0	1	0	2	4	6	11	4	5	0
Females	0	0	3	10	0	9	11	11	17	7	5	2

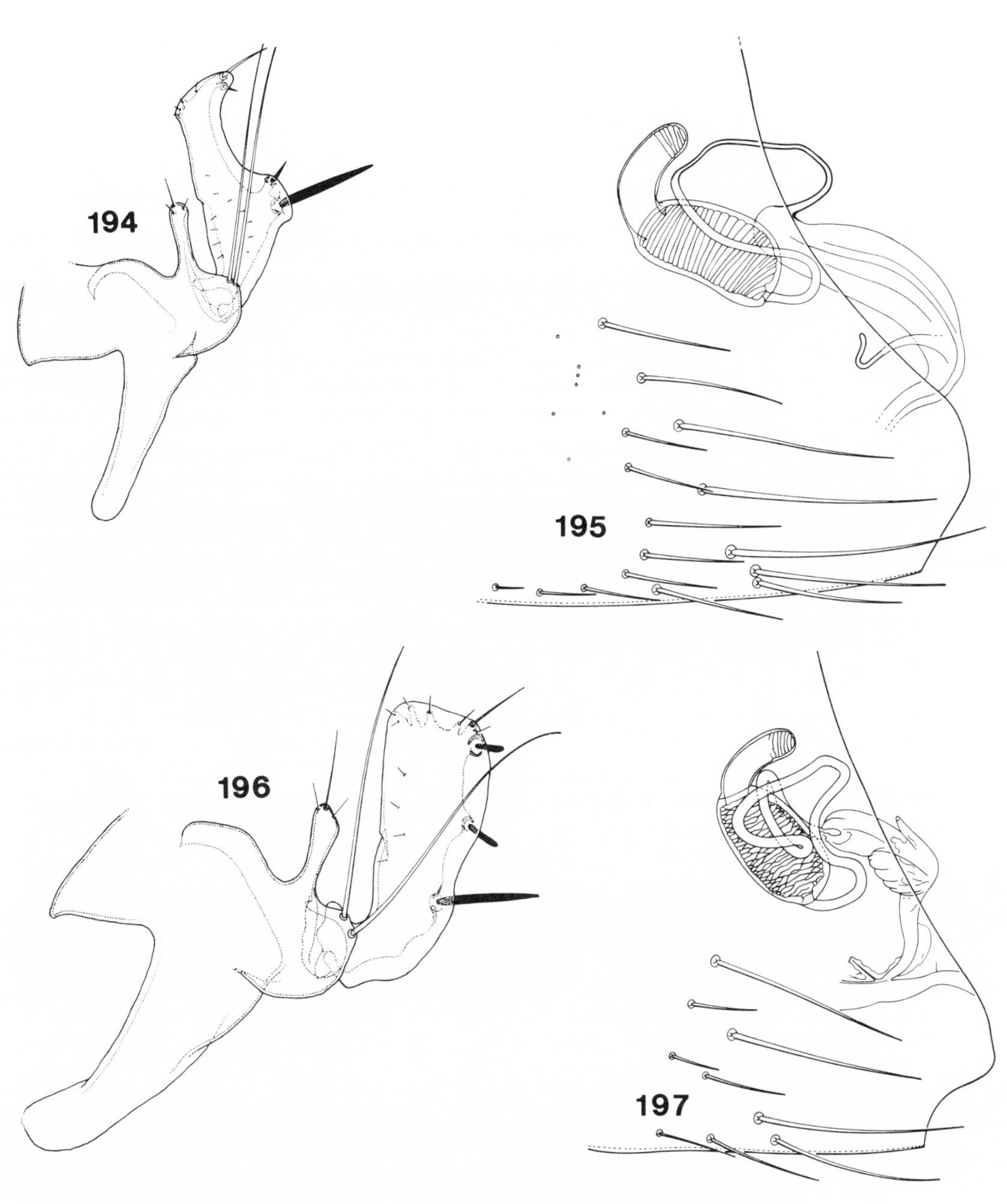

194-195 *Megabothris (Amegabothris) quirini* (Rothschild). **194** Male clasper. **195** Female sternum VII and spermatheca. After Holland, 1985.
196-197 *Megabothris (Amegabothris) abantis* (Rothschild). **196** Male clasper. **197** Female sternum VII and spermatheca. After Holland, 1985.

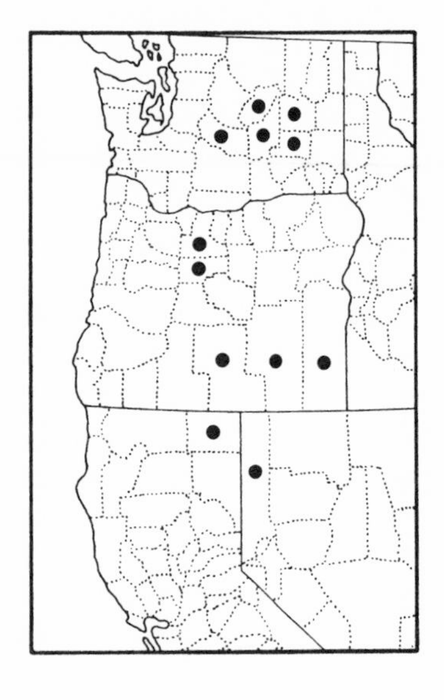

Megabothris (Amegabothris) clantoni Hubbard

[94920]

(Figs. 198, 199)

Megabothris clantoni. Hubbard, 1949, Ent. News 60: 141-144 (Davenport, Lincoln Co., Washington, from *Lagurus curtatus pauperrimus*).

Megabothris clantoni johnsoni. Hubbard, 1949, Ent. News 60: 171-172 (Ellensburg, Kittitas Co., Washington, from *Lagurus curtatus pauperrimus*). **New Synonymy**

Megabothris princei. Hubbard, 1949, Ent. News 60: 170-171 (49 Ranch House, 4 mi. W. of Vya, Washoe Co., Nevada, from *Lagurus curtatus intermedius*). **New Synonymy**

Megabothris obscurus. Holland, 1949, The Siphonaptera of Canada, pp. 162-163 (Yakima Co., Washington, from burrow of *Speotyto* sp.).

Megabothris clantoni Hubbard. Holland, 1952, Canad. Ent. 84: 70 (*obscurus* synonymized).

Megabothris (Amegabothris) clantoni princei Hubbard, 1949. Haddow et al., in: Traub et al., 1983, p. 111.

Megabothris clantoni clantoni Hubbard. Holland, 1985, Mem. Ent. Soc. Canada No. 130, pp. 348-349.

This species was originally described from material collected at Davenport, Lincoln County, Washington. Later that same year, Hubbard (1949) described two new taxa based on collections from localities in Oregon. Various authors since then have voiced doubts about the validity of the subspecies in view of the fact that they are all limited to the same host species and the area of distribution is so limited. The degree of variation in the male st. IX of our samples is ample to justify synonymizing the subspecies.

These are specific parasites of the sage vole, *Lagurus curtatus.* Holland (1985) reports what was the nominate subspecies from British Columbia, but notes that the host species also occurs in southern Alberta and Saskatchewan.

One hundred and fifty-five specimens, 64 males and 91 females, are reported here from three counties in Oregon. All were taken from *L. curtatus* or their nests and all were taken in association with *Oropsylla (Thrassis) bacchi johnsoni* Hubbard, 1949.

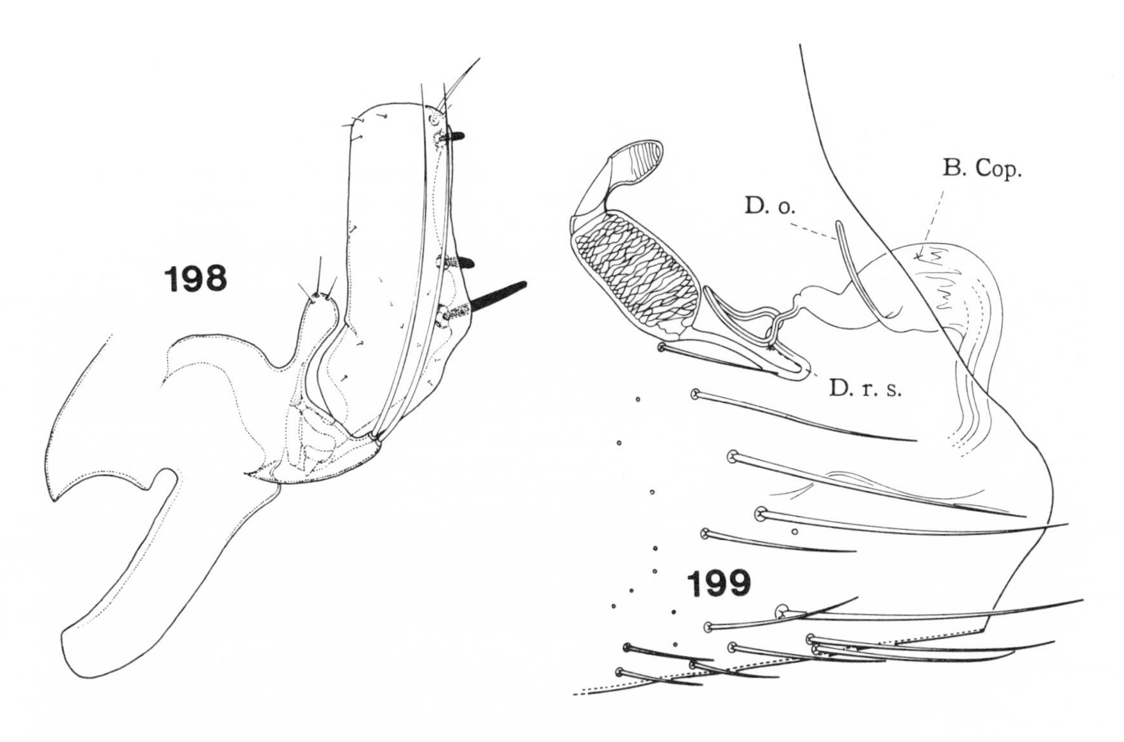

198-199 *Megabothris (Amegabothris) clantoni* Hubbard. **198** Male clasper. **199** Female sternum VII and spermatheca. After Holland, 1985.

GENUS **Nosopsyllus (Nosopsyllus)** Jordan

Nosopsyllus. Jordan, 1933, Novit. Zool. 39: 76-77 (type by designation: *Pulex fasciatus* Bosc, 1800).

This large genus is not native to the Western Hemisphere, being restricted to the Palaearctic and Indian Regions, with one taxon indigenous to Africa south of the Sahara. Two species are relatively worldwide in distribution, especially in port cities, due to their association with domestic rats. One of these, *N. londiniensis,* is not known to occur in Washington or Oregon, although it has been reported from California by Hubbard (1947). However, the following species is known from a few localities within the area.

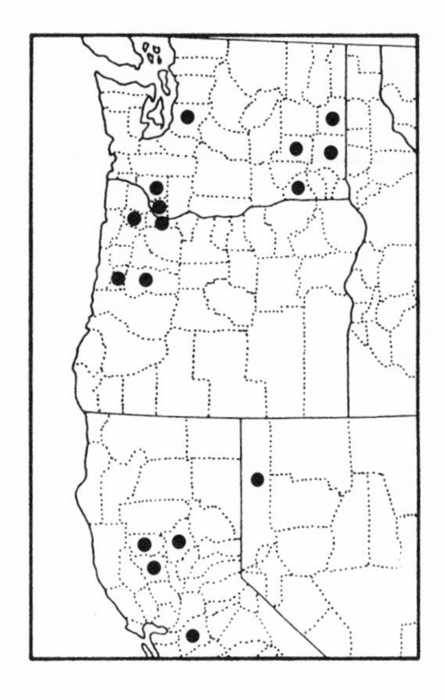

Nosopsyllus (Nosopsyllus) fasciatus (Bosc)

[80001]

(Figs. 200, 201)

Pulex fasciatus. Bosc, 1800, Bull. sci. Soc. Philom., Paris 2: 156 (France, from "un lirot" [= dormouse]).
Nosopsyllus fasciatus Bosc. Jordan, 1933, Novit. Zool. 39: 76, 77.
N. (N.) fasciatus (Bosc, 1800). Lewis, 1967, J. Med. Ent. 4: 126-129.
Nosopsyllus (Nosopsyllus) fasciatus (Bosc, 1800). Smit, in: Traub et al., 1983, p. 29.
Nosopsyllus (Nosopsyllus) fasciatus (Bosc, 1800). Haddow et al., in: Traub et al., 1983, p. 118.

For a more detailed synonymy of this species the reader is directed to Lewis (1967).

This species, commonly known as the northern rat flea, is included here on the strength of various published reports. None were collected during the surveys, as we were concerned with the parasite fauna of endemic birds and mammals and no effort was made to collect domestic rats and mice.

GENUS Opisodasys Jordan

Opisodasys. Jordan, 1933, Novit. Zool. 39: 72 (type species by designation: *Ceratophyllus vesperalis* Jordan, 1929).
Opisodasys (Opisodasys) Jordan, 1933. Smit, in: Traub et al. 1983, p. 21.
Opisodasys (Opisodasys) Jordan, 1933. Haddow et al., in: Traub et al., 1983, p. 129.

Smit (1983) divided this genus into three subgenera, two of which contain species occurring in the area. Both species in the nominate subgenus are parasites of squirrels (which is also true of the extralimital species of *Sciuropsylla*), while the two species placed in *(Oxypsylla)* are associated with peromyscine rodents. The two subgenera may be separated with the following key.

KEY TO THE NORTHWESTERN SUBGENERA OF *OPISODASYS*

1 Pleural arch of metathorax absent, preantennal portion of head with but one row of setae .... **(Opisodasys)**

Pleural arch of metathorax present, preantennal portion of head with two rows of setae .... **(Oxypsylla)**

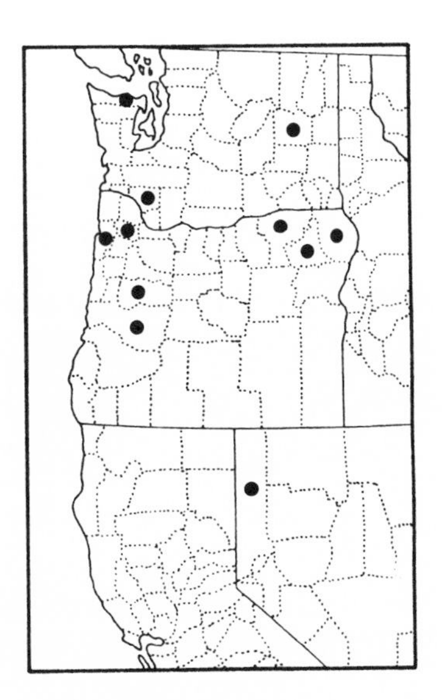

Opisodasys (Opisodasys) vesperalis (Jordan)

[92901]

(Figs. 202, 203)

Ceratophyllus vesperalis. Jordan, 1929, Novit. Zool. 35: 28 (Okanagan and Okanagan Landing, British Columbia, from *Glaucomys sabrinus columbiensis*).
Opisodasys vesperalis (Jordan). Jordan, 1933, Novit. Zool. 39: 72.
Opisodasys pseudarctomys (Baker). Wagner, 1940, Z. Parasitenk. 11: 463.
Opisodasys jellisoni. I. Fox. 1941, Ent. News 52: 45-47 (Boise, Idaho, from *Glaucomys sabrinus bangsii* [female = *vesperalis*]).
Opisodasys vesperalis (Jordan). Holland, 1949, The Siphonaptera of Canada, pp. 134-135 (*O. jellisoni* synonymized. Female = *O. vesperalis,* male = *Tarsopsylla octodecimdentata coloradensis*).
Opisodasys (Opisodasys) vesperalis (Jordan, 1929). Smit, in: Traub et al., 1983, p. 21.
Opisodasys (Opisodasys) vesperalis (Jordan, 1929). Haddow et al., in: Traub et al., 1983, p. 130.

Both this and the other species assigned to this subgenus are true parasites of *Glaucomys volans* and *G. sabrinus,* the two North American flying squirrels. This species is western in its distribution, being known from California north to British Columbia and east to Idaho and Montana. All Oregon collections came from *G. sabrinus* with no strays to any other host.

Seven hundred and eighteen specimens were collected during the surveys. Of these 322 were males and 396 were females, giving a sex ratio of 1:1.2. Seasonal abundance is shown in the following table.

Seasonal abundance of *O. (O.) vesperalis* in Oregon

	J	F	M	A	M	J	J	A	S	O	N	D
Males	0	1	42	4	180	65	0	12	9	9	0	0
Females	0	5	37	1	168	81	2	34	39	28	0	1

GENUS **Opisodasys (Oxypsylla)** Smit

Opisodasys (Oxypsylla). Smit, in: Traub et al., 1983, p. 18 (type species: *Pulex keeni* Baker, 1896).
Opisodasys (Oxypsylla) Smit, 1983. Haddow et al., in: Traub et al., 1983, p. 130.

The two species assigned to this subgenus are primarily parasites of *Peromyscus maniculatus* and, secondarily, many other small rodents. Both are restricted to the western United States.

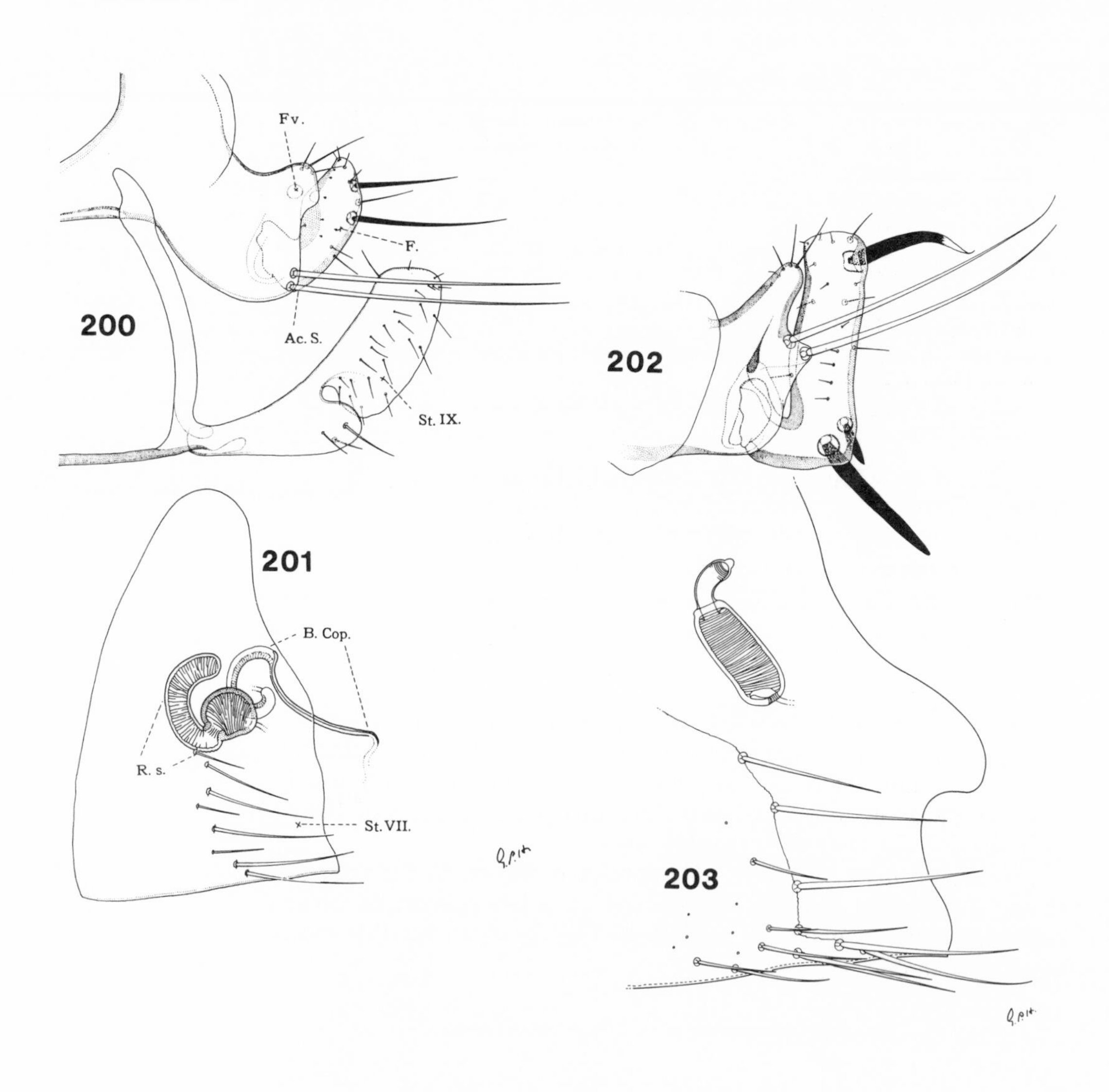

200-201 *Nosopsyllus (Nosopsyllus) fasciatus* (Bosc). **200** Male clasper and sternum IX. **201** Female sternum VII and spermatheca. After Holland, 1985.
202-203 *Opisodasys (Opisodasys) vesperalis* (Jordan). **202** Male clasper. **203** Female sternum VII and spermatheca. After Holland, 1985.

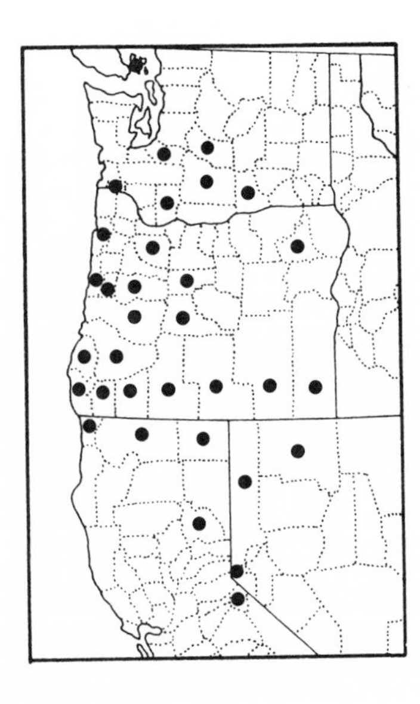

Opisodasys (Oxypsylla) keeni (Baker)

[89603]

(Figs. 204, 205)

Pulex keeni. Baker, 1896, Canad. Ent. 28: 234 (Massett, Queen Charlotte Islands, British Columbia, from *Peromyscus maniculatus keeni*).

Ceratophyllus keeni (Baker). Baker, 1904, Proc. U. S. Nat. Mus. 27: 400, 444.

Opisodasys keeni (Baker). Jordan, 1933, Novit. Zool. 39: 72.

Opisodasys keeni (Baker). Jellison, 1939, J. Parasitol. 25: 416.

Opisodasys keeni (Baker). Holland, 1949, The Siphonaptera of Canada, pp. 132-133.

Opisodasys (Oxypsylla) keeni (Baker, 1896). Smit, in: Traub et al., 1983, p. 25.

Opisodasys (Oxypsylla) keeni (Baker, 1896). Haddow et al., in: Traub et al., 1983, p. 130.

This is an ubiquitous ectoparasite of the deer mouse, *Peromyscus maniculatus,* in northwestern United States and southwestern Canada. It seems to be ecologically quite tolerant and occurs in a variety of habitats. Haddow et al. (1983) list its distribution area as California north to British Columbia, and east to Nevada, Utah, and Montana.

During the surveys reported here, 204 males and 342 females of this species were collected. Of these, 183 males and 286 females, or 85.8 percent, came from *Peromyscus maniculatus.* The remaining specimens came from 23 different species of hosts including a bat and a wren. None of these hosts bore sufficient numbers of fleas to even suggest that they were regular hosts.

Seasonal distribution of this species is shown in the following table. Differences in totals are caused by a few specimens without collection dates which were excluded. The sex ratio for this species was 1:1.68 males to females.

Seasonal distribution of *O. (O.) keeni* in Oregon

	J	F	M	A	M	J	J	A	S	O	N	D
Males	7	6	9	11	9	7	16	65	14	2	22	33
Females	8	6	15	20	10	4	48	88	45	6	40	43

GENUS **Orchopeas** Jordan

Bakerella. Wagner, 1930, Mag. Parasitol. 1: 101, 119 (no type species designated. Subsequent designation by Smit and Wright, 1965, Mitt. Hamburg Zool. Mus. Inst. 62: 6 as *Pulex wickhami* Baker, 1895 [=*howardi* Baker, 1895]. *nomen preocc.).*
Orchopeas. Jordan, 1933, Novit. Zool. 39: 71-72 (type species by designation: *O. wickhami* Baker, 1895 [=*howardi* Baker, 1895]).
Orchopeas Jordan. Smit, in: Traub et al., 1983, p. 24.
Orchopeas Jordan. Haddow et al., in: Traub et al., 1983, p. 132.

This difficult genus is well represented in the area with six species as part of the fauna. Two of these are represented by two or more subspecies if one follows the current classification. Most members of the genus are parasites of tree squirrels but two species are found on cricetid rodents.

From a taxonomic point of view this genus is sorely in need of revision. Many of the taxa need to be redefined, particularly at the subspecies level. Although the following key is not wholly satisfactory, it will separate normal individuals of the species occurring in the area under consideration.

KEY TO THE NORTHWESTERN SPECIES OF *ORCHOPEAS*

1 Ocular and frontal rows of setae complete **2**

Only ocular setal row complete, frontal row usually consisting of one long seta **3**

2 Labial palpus not extending to apex of fore coxa, usually on mice .... **leucopus**

Labial paplus as long as fore coxa, on wood rats .... **sexdentatus** ssp.

3 Movable process of male rectangular, its anterior and posterior margins almost parallel, fixed process broad and almost as long as movable process; female st. VII deeply incised, with upper lobe longer than lower .... **nepos**

Not as above **4**

4 Movable and fixed processes of male approximately equal in length; sinus in female st. VII small, deep and low down .... **howardi**

Movable process of male longer than fixed process; female st. VII variable but not as above .... **caedens** ssp.

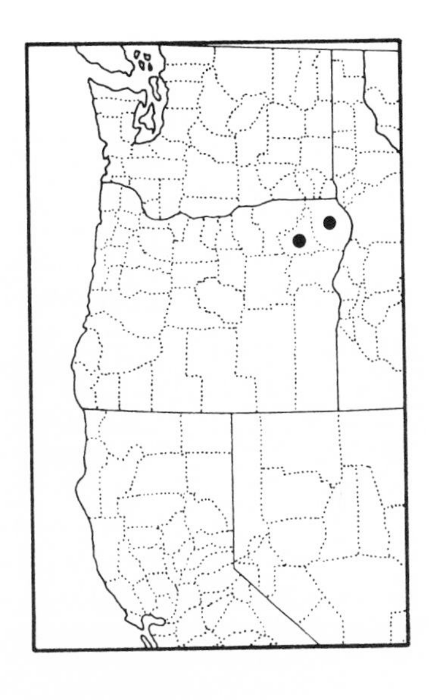

Orchopeas caedens caedens (Jordan)

[92521]

(Figs. 206, 207)

Ceratophyllus caedens. Jordan, 1925, Novit. Zool. 32: 104-105 (Banff, Alberta, from *Martes americana*).
Ceratophyllus caedens caedens Jordan. Jordan, 1929, Novit. Zool. 35: 29-30.
Orchopeas caedens (Jordan). Jordan, 1933, Novit. Zool. 39: 71-72.
Orchopeas caedens caedens Jordan, 1925. Hubbard, 1947, Fleas of Western North America, pp. 102-105.
Orchopeas caedens caedens (Jordan). Holland, 1949, The Siphonaptera of Canada, pp. 136-138.
Orchopeas caedens caedens (Jordan, 1925). Smit, in: Traub et al., 1983, p. 24.
Orchopeas caedens caedens (Jordan, 1925). Haddow et al., in: Traub et al., 1983, p. 133.

and

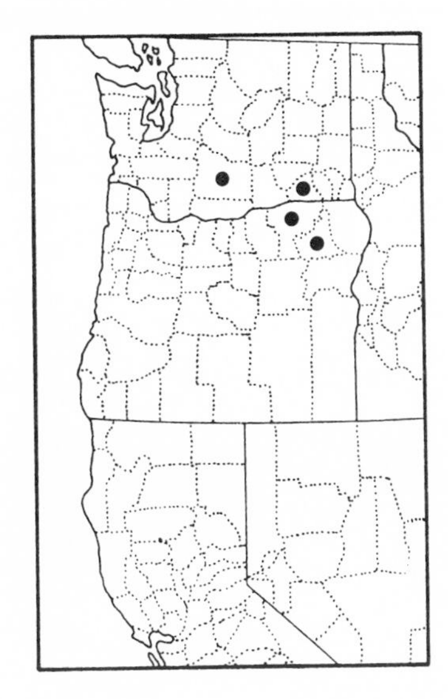

Orchopeas caedens durus (Jordan)

[92902]

(Figs. 208, 209)

Ceratophyllus caedens durus. Jordan, 1929, Novit. Zool. 35: 29-30 (female type from Okanagan, British Columbia, from *Mustela frenata* ssp., and both sexes from Blucher Hall, British Columbia, from *Tamiasciurus hudsonicus streatori*).
Ceratophyllus caedens durus Jordan. Jordan, 1932, Novit. Zool. 38: 253.
Orchopeas (Bakerella) caedens durus (Jordan). Spencer, 1936, Proc. ent. Soc. B. C. 32: 13.
Orchopeas caedens durus Jordan, 1929. Hubbard, 1947, Fleas of Western North America, p. 105.
Orchopeas caedens durus (Jordan). Holland, 1949, The Siphonaptera of Canada, pp. 138-139.
Orchopeas caedens durus (Jordan, 1929). Smit, in: Traub et al., 1983, p. 24.
Orchopeas caedens durus (Jordan, 1929). Haddow et al., in: Traub et al., 1983, p. 133.

The history of these two taxa is characteristic of the confusion generated by many species in this genus. Both Haddow et al. (1983) and Holland (1985) retain both subspecies, and the latter makes a case for the validity of both based on a character that is only visible in the female.

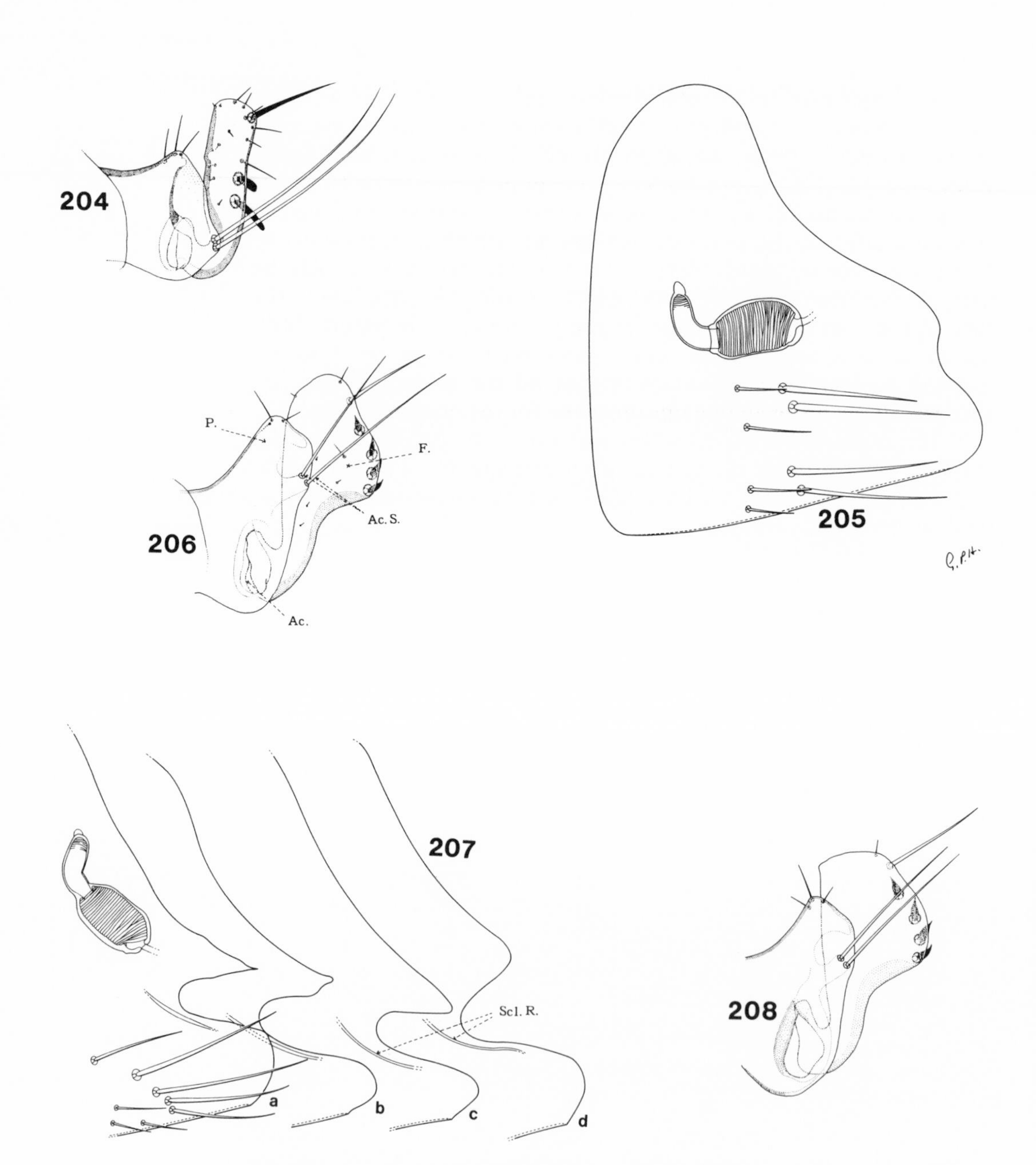

204-205 *Opisodasys (Oxypsylla) keeni* (Baker). **204** Male clasper. **205** Female sternum VII and spermatheca. After Holland, 1985.

206-208 *Orchopeas caedens* subspecies. **206** *O. caedens caedens* male clasper. **207** Female sternite VII and spermatheca of same subspecies. **208** *O. c. durus* (Jordan) male clasper. After Holland, 1985.

A comparison of the distribution maps in these two works is even more confusing. Holland (1985) shows the British Columbia records to all be *O. c. durus,* as are all his records from Alaska. Haddow et al. (1983) show both of these populations to be mixed. If one combines the records from both sources, one is left to conclude that over most of the range of this species, which includes most of North America north of Mexico, the two subspecies occur side by side. As Haddow et al. (1983) are quick to point out, this flies in the face of the concept that subspecies are geographical in nature. Still, the evidence is too bold to ignore, and even though the taxa are difficult to identify, it is unlikely that all the determinations are suspect. So the problem remains unsolved for the present.

The nominate subspecies was not taken in Oregon during the surveys. *Orchopeas c. durus* was taken infrequently and only seven males and four females are reported here from *Glaucomys sabrinus,* the Northern flying squirrel.

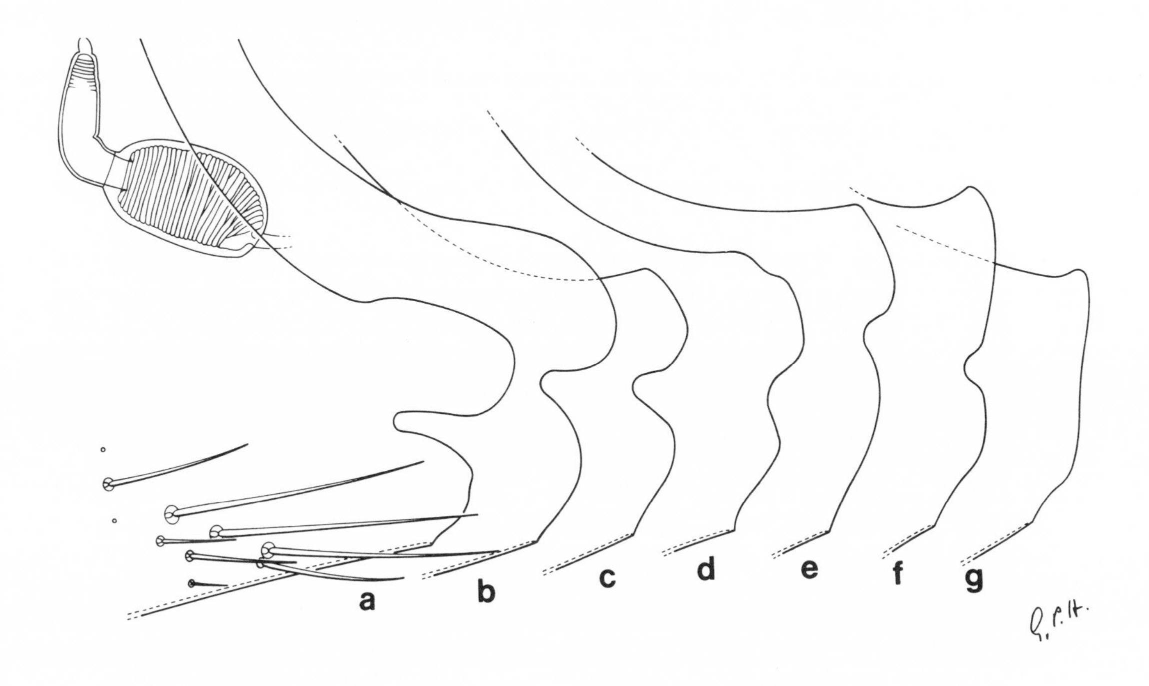

209 *Orchopeas caedens durus* (Jordan). Female sternite VII and spermatheca. After Holland, 1985.

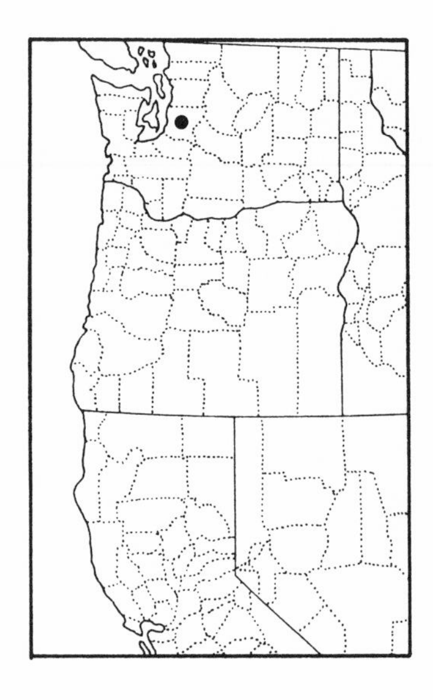

Orchopeas howardi howardi (Baker)

[89504]

Pulex wickhami. Baker, 1895, Canad. Ent. 27: 109, 111 (Iowa City, Iowa, from *Glaucomys volans*).
Pulex gillettei. Baker, 1895, Canad. Ent. 27: 109, 111 (Portland, Michigan, from *Sciurus canadensis* [= ? *Tamiasciurus hudsonicus*]).
Pulex howardii. Baker, 1895, Canad. Ent. 27: 110, 112 (Ithaca, New York, from a red squirrel; Tallula Falls, Georgia, from a squirrel; Lincoln, Nebraska, from gray or fox squirrel and field mouse nest; Ames, Iowa, no host given).
Pulex howardii Baker. Baker, 1899, Ent. News 10: 37 (selects *howardii* as proper name, reducing *wickhami* and *gillettei* to junior synonyms).
Ceratophyllus wickhami Baker. Baker, 1904, Proc. U. S. Nat. Mus. 27: 387, 403, 448 (designates *wickhami* as proper name).
Orchopeas wickhami Baker, 1895. Jordan, 1933, Novit. Zool. 39: 71 (erects the genus, with *wickhami* as generitype).
Orchopeas howardii (Baker). Ewing and Fox, 1943, U. S. Dept. Agric. Miscl. Publ. 500, p. 33 (elevate *howardii* on the basis of Baker's 1899 action).
Orchopeas howardii (Baker). Holland, 1949, The Siphonaptera of Canada, 142-143 (discussion of synonymy).
Orchopeas h. howardi (Baker, 1895). Hopkins, 1954, Entomologist 87: 197-198 (discusses synonymy and selects lectotypes for *howardi* and *wickhami*).
Orchopeas howardi howardi (Baker). Holland, 1985, Mem. Ent. Soc. Canada No. 130, pp. 417-419.

This is a true parasite of the gray squirrel and its distribution seems limited to that of the host, although the flea has been taken on other hosts within its range. The natural distribution of *Sciurus carolinensis* is all east of the 100th parallel. However, the squirrel has been introduced into various parts of the west, and its fleas along with it. Similarly, both flea and host have become naturalized elements of the fauna of the British Isles.

Orchopeas leucopus (Baker)

[90420]

(Figs. 210, 211)

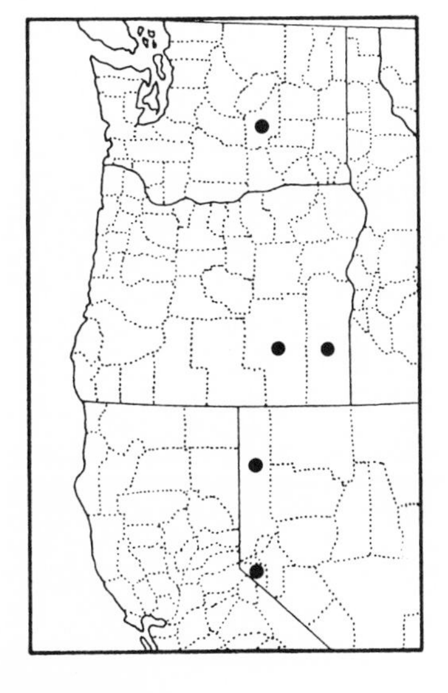

Ceratophyllus leucopus. Baker, 1904, Proc. U. S. Nat. Mus. 27: 387, 401 (Peterboro, New York, from *Peromyscus leucopus*).

Ceratophyllus aeger. Rothschild, 1905, Novit. Zool. 12: 166-167, pl. VI, figs. 5, 7, 9 (Red Deer, Alberta, from *Peromyscus maniculatus* and *Clethrionomys gapperi*).

Ceratophyllus leucopus Baker, 1904. Jordan, 1929, Novit. Zool. 35: 28-29 (*aeger* synonymized).

Orchopeas leucopus Baker, 1904. Jordan, 1933, Novit. Zool. 39: 72.

Orchopeas leucopus (Baker). Holland, 1949, The Siphonaptera of Canada, p. 140.

Orchopeas leucopus (Baker, 1904). Smit, in: Traub et al., 1983, p. 24.

Orchopeas leucopus (Baker, 1904). Haddow et al., in: Traub et al., 1983, p. 135.

This is a widespread species occurring on various species of *Peromyscus,* particularly *P. leucopus* and *P. maniculatus.* The northernmost records are from Alaska, North West Territories, and Labrador. The southernmost record is Chiapas, Mexico. As pointed out in Haddow et al. (1983), the western distribution of this species tends to be patchy, and this is another taxon that was not taken during the surveys. We know of no published records of this flea from Oregon, although it has been reported from Washington and California. Harold Egoscue (personal communication) has collected this species in Harney and Malheur counties, and he reports that here it is mainly taken on *Reithrodontomys* rather than on *Peromyscus* species.

Orchopeas latens (Jordan)

[92522]

(Figs. 212, 213)

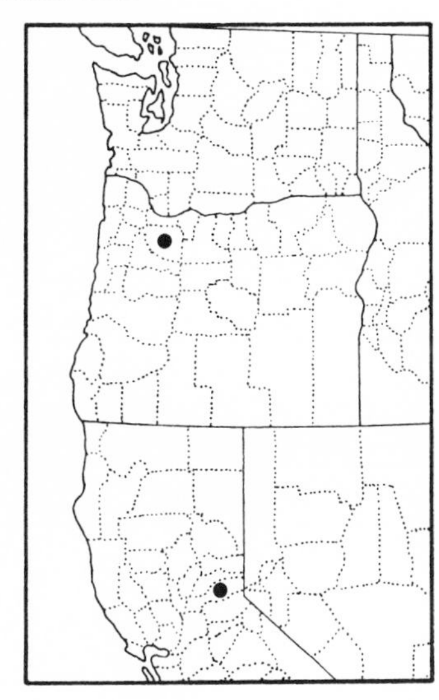

Ceratophyllus latens. Jordan, 1925, Novit. Zool. 32: 105 (Santa Cruz County, California, on gray squirrel).

Orchopeas latens (Jordan). Jordan, 1933, Novit. Zool. 39: 71.

Orchopeas latens Jordan, 1925. Hubbard, 1947, Fleas of Western North America, p. 101.

Orchopeas latens (Jordan, 1925). Smit, in: Traub et al., 1983, p. 24.

Orchopeas latens (Jordan, 1925). Haddow et al., in: Traub et al., 1983, p. 135.

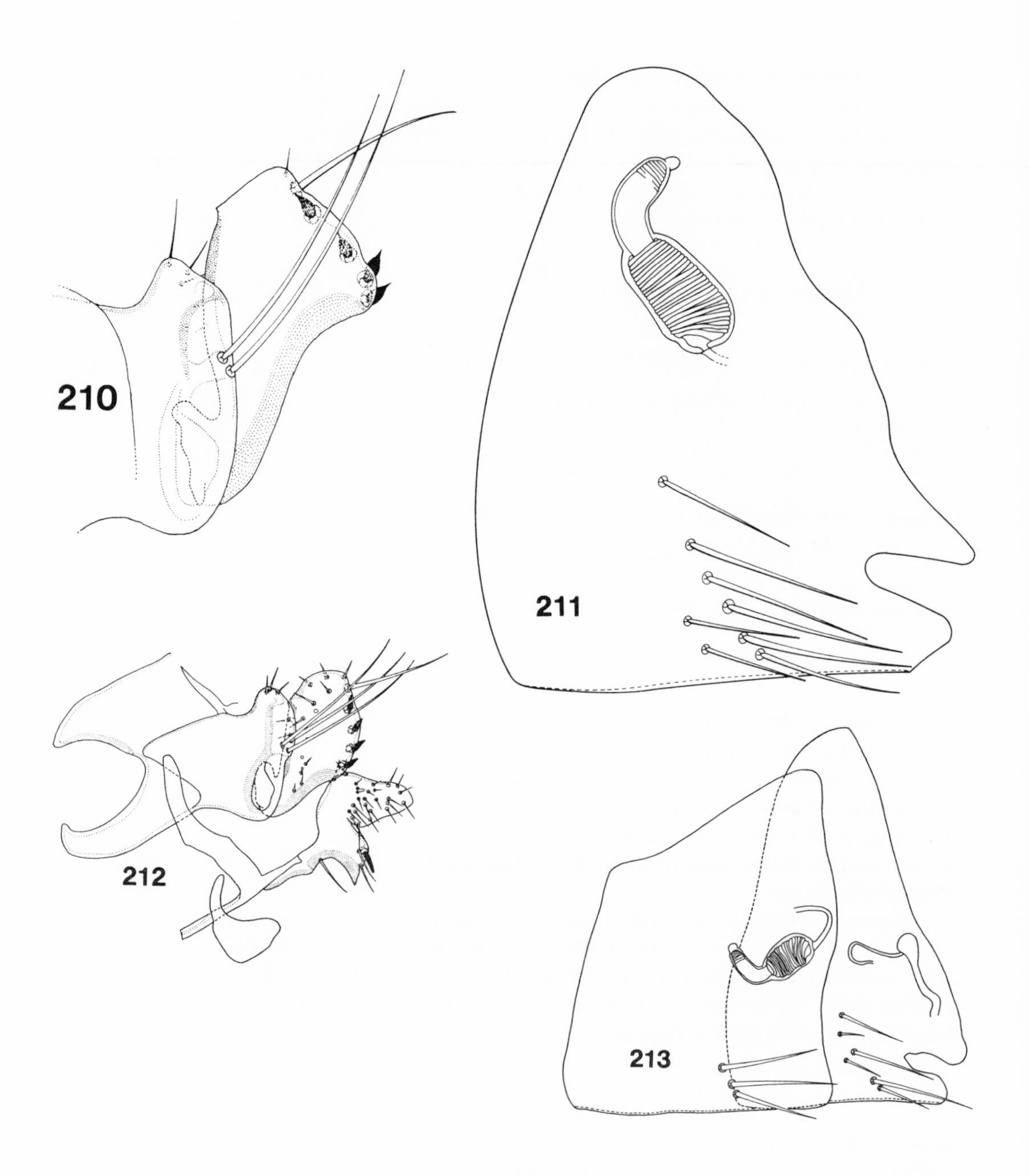

210-211 *Orchopeas leucopus* (Baker). **210** Male clasper. **211** Female sternum VII and spermatheca. After Holland, 1985.
212-213 *Orchopeas latens* (Jordan). **212** Male genitalia. **213** Modified abdominal segments of female. Original.

Haddow et al. (1983) show a single record of this species in northwestern Oregon and two records from California. The species is sufficiently similar to some of the other western members of the genus that it might be confused with them. Although it is a parasite of tree squirrels, Hubbard (1947) cites the record of Augustson from the ground squirrel, *Spermophilus beecheyi.*

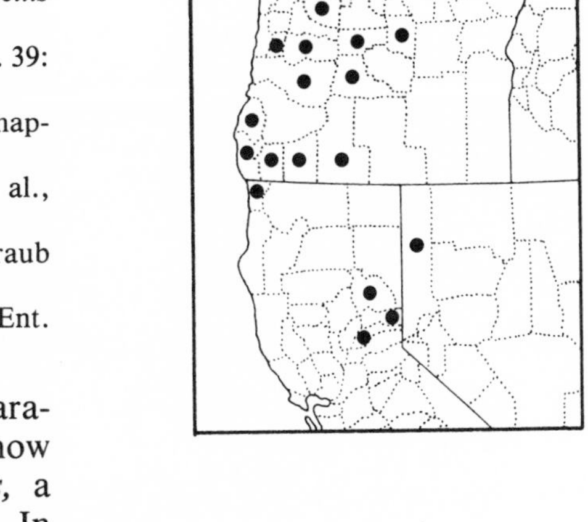

Orchopeas nepos (Rothschild)

[90512]

(Figs. 214, 215)

Ceratophyllus nepos. Rothschild, 1905, Novit. Zool. 12: 168 (Chilliwack, British Columbia, from *Spilogale gracilis olympica*).
Orchopeas nepos (Rothschild). Jordan, 1933, Novit. Zool. 39: 72.
Orchopeas nepos (Rothschild). Holland, 1949, The Siphonaptera of Canada, p. 141.
Orchopeas nepos (Rothschild, 1905). Smit, in: Traub et al., 1983, p. 24.
Orchopeas nepos (Rothschild, 1905). Haddow et al., in: Traub et al., 1983, p. 135.
Orchopeas nepos (Rothschild). Holland, 1985, Mem. Ent. Soc. Canada 130, pp. 419-420.

Holland (1985) states that this is a true parasite of *Tamiasciurus douglasi* but our records show it to be equally common on *Sciurus griseus,* a species that evidently is not common in Canada. In any case, it is a common element of the Oregon fauna, at least in the western part of the area under consideration.

Eighty-four males and 142 females of this species are reported here. Except for one specimen from a coyote and seven specimens from wood rats, all were collected on squirrels: 37 males and 72 females from *S. griseus* and 42 males and 67 females from *T. douglasi.* Seasonal distribution is shown below.

Seasonal distribution of *O. nepos* in Oregon

	J	F	M	A	M	J	J	A	S	O	N	D
Males	6	7	1	0	7	19	5	1	22	8	4	4
Females	20	5	7	1	11	23	14	0	36	9	7	9

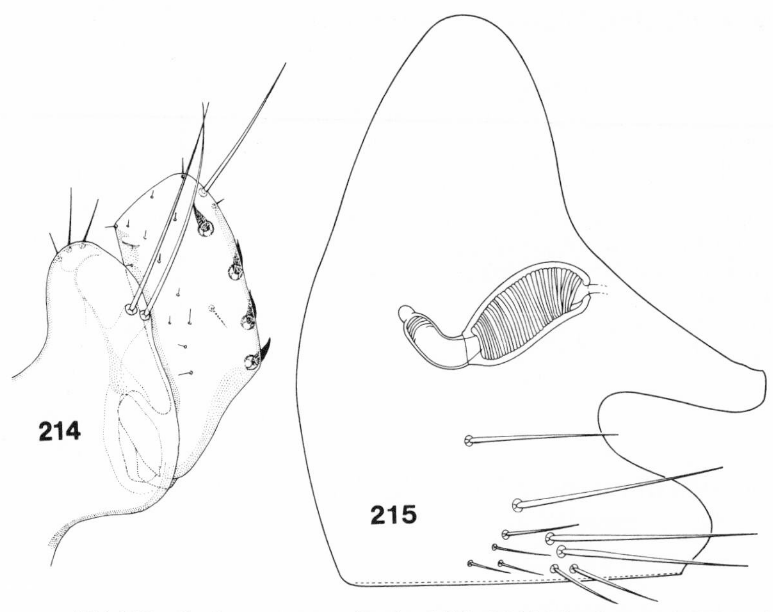

214-215 *Orchopeas nepos* (Rothschild). **214** Male clasper. **215** Female sternum VII and spermatheca. After Holland, 1985.

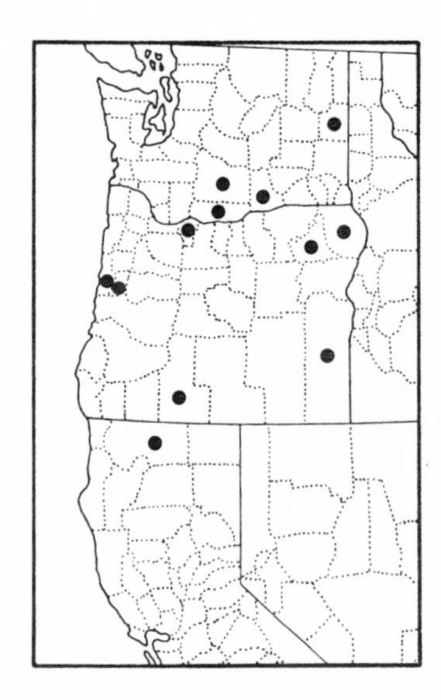

Orchopeas sexdentatus agilis (Rothschild)

[90511]

(Figs. 216, 217)

Ceratophyllus agilis. Rothschild, 1905, Novit. Zool. 12: 167-168 (Banff, Alberta, and Carpenter's Mountain, Cariboo District, British Columbia, from *Neotoma cinerea*).

Ceratophyllus sexdentatus agilis Rothschild. Jordan, 1929, Novit. Zool. 35: 30.

Orchopeas sexdentatus Baker, 1904. Jordan, 1933, Novit. Zool. 39: 72.

Orchopeas sexdentatus agilis (Rothschild). Wagner, 1936, Canad. Ent. 68: 199.

Orchopeas sexdentatus agilis (Rothschild). Holland, 1949, The Siphonaptera of Canada, pp. 141-142.

Orchopeas sexdentatus agilis (Rothschild, 1905). Smit, in: Traub et al., 1983, p. 24.

Orchopeas sexdentatus agilis (Rothschild, 1905). Haddow et al., in: Traub et al., 1983, p. 136.

Orchopeas sexdentatus agilis (Rothschild). Holland, 1985, Mem. Ent. Soc. Canada 130, pp. 426-427.

and

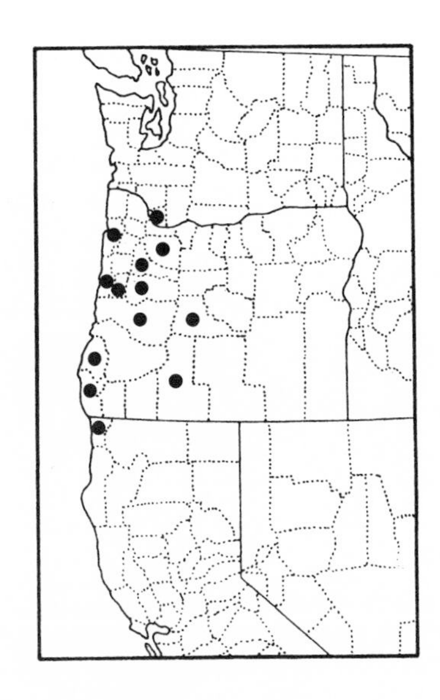

Orchopeas sexdentatus cascadensis Jordan

[93926]

(Figs. 218, 219)

Orchopeas sexdentatus cascadensis. Jordan, 1939, Novit. Zool. 41: 317 (Odell Lake, Oregon, from *Neotoma fuscipes fuscipes*).

Orchopeas sexdentatus cascadensis Jordan, 1939. Hubbard, 1947, Fleas of Western North America, pp. 96-97.

Orchopeas sexdentatus cascadensis Jordan, 1939. Smit, in: Traub et al., 1983, p. 24.

Orchopeas sexdentatus cascadensis Jordan, 1939. Haddow et al., in: Traub et al., 1983, p. 136.

and

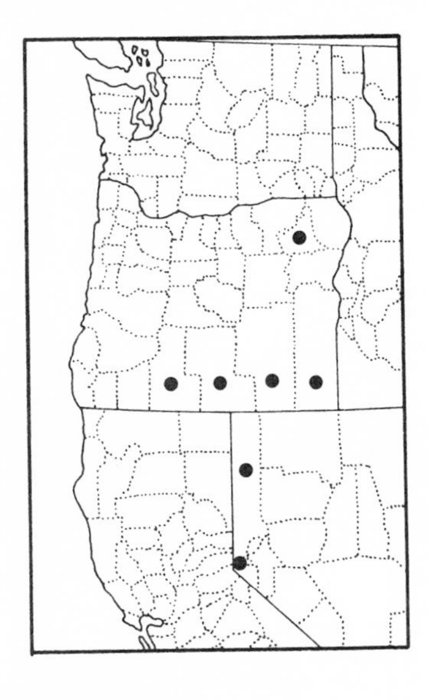

Orchopeas sexdentatus nevadensis (Jordan)

[92903]

(Figs. 220, 221)

Ceratophyllus sexdentatus nevadensis. Jordan, 1929, Novit. Zool. 35: 30 (Pine City, Mono Co., California, from *Mustela arizonensis*).

Orchopeas sexdentatus Baker, 1904. Jordan, 1933, Novit. Zool. 39: 72.

Orchopeas sexdentatus nevadensis Jordan, 1929. Hubbard, 1947, Fleas of Western North America, pp. 95-96.

Orchopeas sexdentatus nevadensis (Jordan, 1929). Smit, in: Traub et al., 1983, p. 24.

Orchopeas sexdentatus nevadensis (Jordan, 1929). Haddow et al., in: Traub et al., 1983, p. 137.

In total, 1,049 specimens belonging to this species were taken during the various surveys. This includes 364 males and 514 females of *O. s. cascadensis;* 54 males and 85 females of *O. s. nevadensis;* and twelve males and twenty females of *O. s. agilis.* Of these, 734 specimens came from various species of *Neotoma.* Another 279 specimens came from predators, with 252 of these from *Spilogale putorius,* the spotted skunk. The remaining 36 specimens represent accidental associations.

The maps show the counties in which the various collections were made, but do little to clarify the peculiar sympatric distribution of the subspecies. In truth, the assignment of the specimens to the various subspecies approached the arbitrary in some instances. As was true for *O. caedens,* the most useful character for separating the subspecies occurs in the females, not the males, and involves the development of st. VII. This is a variable character at best and many females in the collections were intermediate between the various subspecies. In general it can be said that *O. s. cascadensis* occurs in the western third of the region; *O. s. nevadensis* is restricted to the more arid parts of the south and east; and *O. s. agilis* is scattered throughout the region with many intergrades with the other two subspecies.

Combined records for the three subspecies show the following seasonal distribution.

Seasonal distribution of *O. sexdentatus* ssp. in Oregon

	J	F	M	A	M	J	J	A	S	O	N	D
Males	0	1	9	33	17	2	61	111	19	26	34	117
Females	11	0	15	61	23	1	92	138	26	45	56	151

GENUS **Oropsylla** Wagner and Ioff

Oropsylla. Wagner and Ioff, 1926, Rev. Microbiol. Epidemiol. 5: 86 (type species by designation: *Ceratophyllus silantiewi* Wagner, 1898).

Aetheopsylla. Stewart and Holland, 1940, Canad. Ent. 72: 41 (type species by designation: *A. septentrionalis* Stewart and Holland, 1940).

Oropsylla Wagner and Ioff, 1926. Jellison, 1945, J. Parasitol. 31: 88 (*Aetheopsylla* synonymized).

Oropsylla (Oropsylla) Wagner and Ioff, 1926. Smit, in: Traub et al., 1983, p. 28.

Oropsylla (Oropsylla) Wagner and Ioff, 1926. Haddow et al., in: Traub et al., 1983, p. 138.

Until recently this genus contained seven species that were associated with marmots (*Marmota*) and ground squirrels (*Spermophilus*), both ground-dwelling sciurids. However, Smit (1983) expanded the generic concept to include an additional 38 taxa originally included in four other genera. All of these taxa but one or two are also associated with the same hosts, and most are restricted to western North America. Following is a modification of the Smit (1983) key to the subgenera.

KEY TO THE SUBGENERA OF *OROPSYLLA*

1 Basal abdominal sternum with a patch of lateral setae often extending onto upper anterior half 2

Basal abdominal sternum with at most one or two lateral setae in lower half or middle of sternum 3

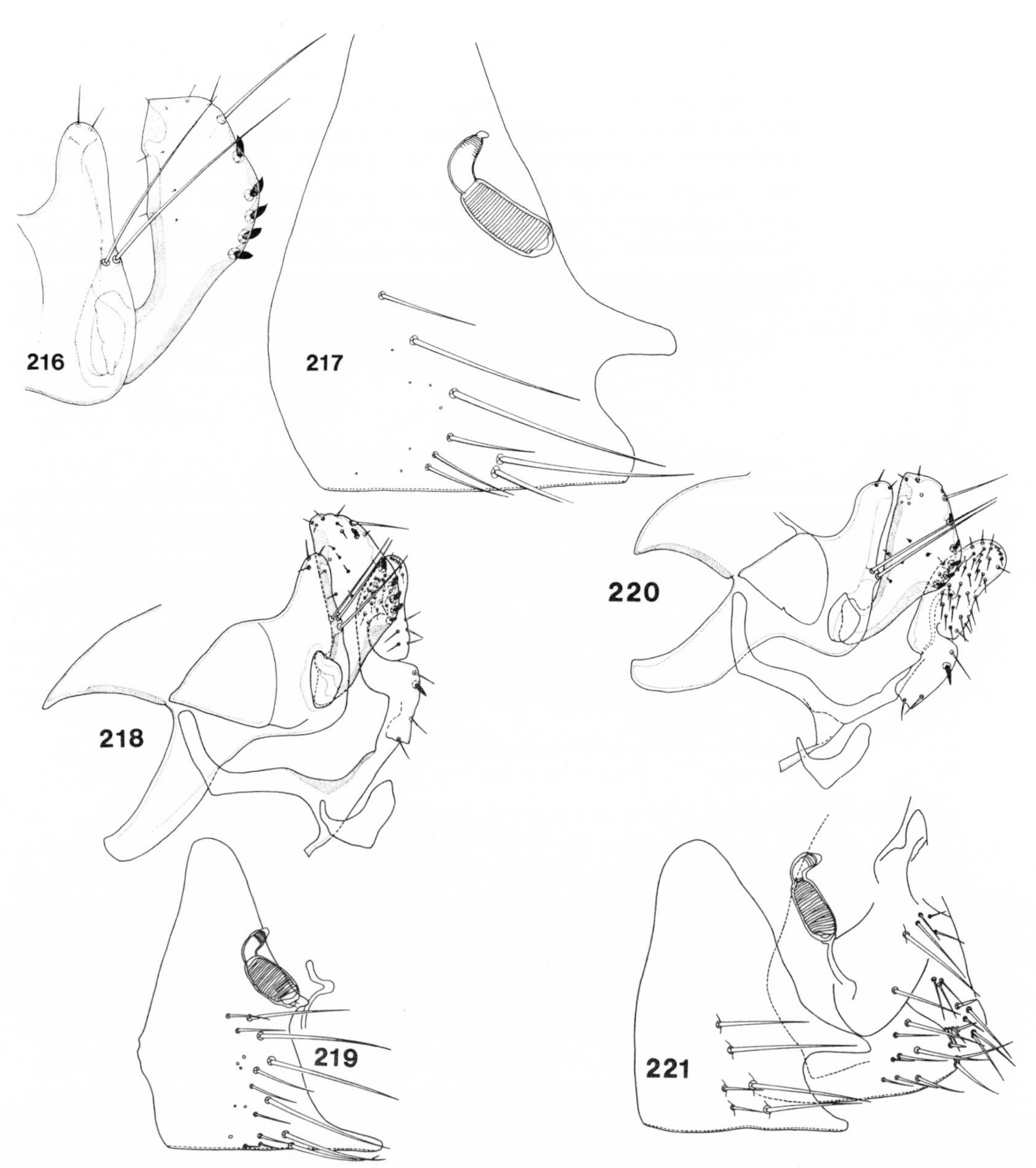

216-217 *Orchopeas sexdentatus agilis* (Rothschild). **216** Male clasper. **217** Female sternum VII and spermatheca. After Holland, 1985.
218-219 *Orchopeas sexdentatus cascadensis* Jordan. **218** Male genitalia. **219** Female sternum VII and spermatheca. Original.
220-221 *Orchopeas sexdentatus nevadensis* (Jordan). **220** Male genitalia. **221** Modified abdominal segments of female. Original.

2 Frontal row usually consisting of one ventral and one or two dorsal setae of medium length. Male: Lower minute antesensilial seta absent; st.VIII long and narrow with long, filamentous vexillum with split apex. Female: Ductus bursae fairly long, with a thin wall; bursa copulatrix with a small but distinct semicircular sclerotization around opening of ductus obturatus .... **Oropsylla (Opisocrostis)**

Frontal row consisting of only one small ventral seta which may be absent. Male: Lower minute antesensilial seta present; st. VIII of medium length with an upright, sickle-shaped vexillum with an entire apex. Female: Ductus bursae rather short, with a relatively thick wall; bursa copulatrix without sclerotization .... **Oropsylla (Hubbardipsylla)**

3 Male: Sternum VIII much reduced, without setae; movable process long and slender, curved forward. Female: Spermatheca with a ventral constriction between bulga and hilla; anal stylet usually with only one long ventral preapical seta, dorsal seta lacking .... **Oropsylla (Diamanus)**

Male: Sternum VIII with setae; movable process fairly straight. Female: Constriction between bulga and hilla less distinct or absent; anal stylet often with two or three lateral setae, sometimes with only a ventral one or none at all **4**

4 Sternum VIII of male rectangular or triangular (when apical half narrowed, the apex without very long setae). Female: Bulga of spermatheca more or less globular, hilla usually much longer than bulga, its papilla often small or absent .... **Oropsylla (Thrassis)**

Sternum VIII of male with a narrow apical half bearing two or three very long apical setae. Female: Bulga of spermatheca usually pyriform, hilla at most a little longer than bulga with a large papilla ....
.... **Oropsylla (Oropsylla)**

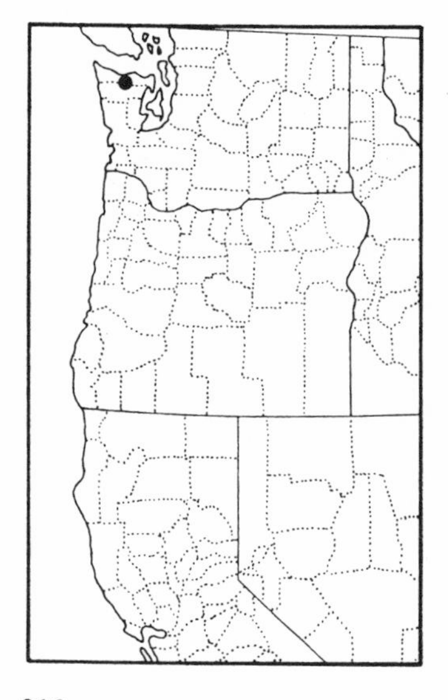

Oropsylla (Oropsylla) eatoni Hubbard

[95419]

Oropsylla arctomys eatoni. Hubbard, 1954, Ent. News 65: 173 (Olympic National Park, Washington, from *Marmota olympus*).

Oropsylla (Oropsylla) eatoni Hubbard, 1954. Smit, in: Traub et al., 1983, p. 28 (treated as a full species).

The distribution of this species of flea is evidently closely tied to the range of the Olympic Mountain marmot, which is confined to the Olympic Peninsula of Washington. Little else is known about it. Some authors consider this marmot, and the subspecies of *M. caligata,* to be only subspecifically differentiated from the Old World *Marmota marmota.* If this is true, then this flea is most likely allied to *Oropsylla (Oropsylla) silantiewi* (Wagner, 1898).

At this point it is not possible to write a key that will separate this from the following species since specimens are not available and the description does not provide diagnostic characters.

Oropsylla (Oropsylla) idahoensis (Baker)

[90430]

(Figs. 222, 223)

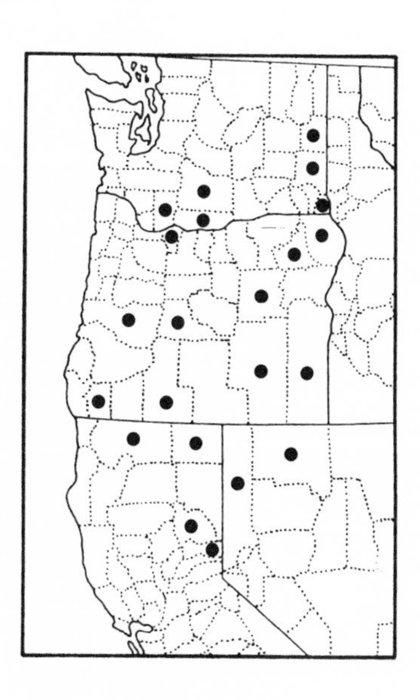

Ceratophyllus idahoensis. Baker, 1904, Proc. U. S. Nat. Mus. 27: 388, 413-415, 443 (Moscow, Idaho, from *Citellus* [= *Spermophilus*] *columbianus*).
Ceratophyllus poeantis. Rothschild, 1905, Novit. Zool. 12: 155-156 (Alberta, British Columbia, and Arizona from *Tamias, Sciurus, Spermophilus, Putorius* and *Marmota*).
Ceratophyllus bertholfi. C. Fox, 1927, Trans. Amer. ent. Soc. 53: 211 (Nagai Island, Alaska, from *Citellus* [= *Spermophilus*] [*parryi*] *nebulicola*).
Ceratophyllus idahoensis Baker. Jordan, 1929, Novit. Zool. 35: 32 (*poeantis* synonymized).
Oropsylla idahoensis Baker. Jordan, 1933, Novit. Zool. 39: 74 (*bertholfi* synonymized).
Oropsylla idahoensis Baker. Hubbard, 1947, Fleas of Western North America, pp. 163-166.
Oropsylla idahoensis (Baker). Holland, 1949, The Siphonaptera of Canada, pp. 115-116.
Oropsylla (Oropsylla) idahoensis (Baker, 1904). Smit, in: Traub et al., 1983, p. 28.
Oropsylla (Oropsylla) idahoensis (Baker, 1904). Haddow et al., in: Traub et al., 1983, p. 138.
Oropsylla idahoensis (Baker). Holland, 1985, Mem. Ent. Soc. Canada 130, pp. 436-441.

This is mainly a parasite of various species of *Spermophilus*, and 21 of our 23 specimens came from *S. lateralis, beldingi* or *S. columbianus*. The known range of the species extends from Alaska south to California and the Rocky Mountain states to northern New Mexico. This may be one of the species that functions in the maintenance of sylvatic plague in ground squirrel populations.

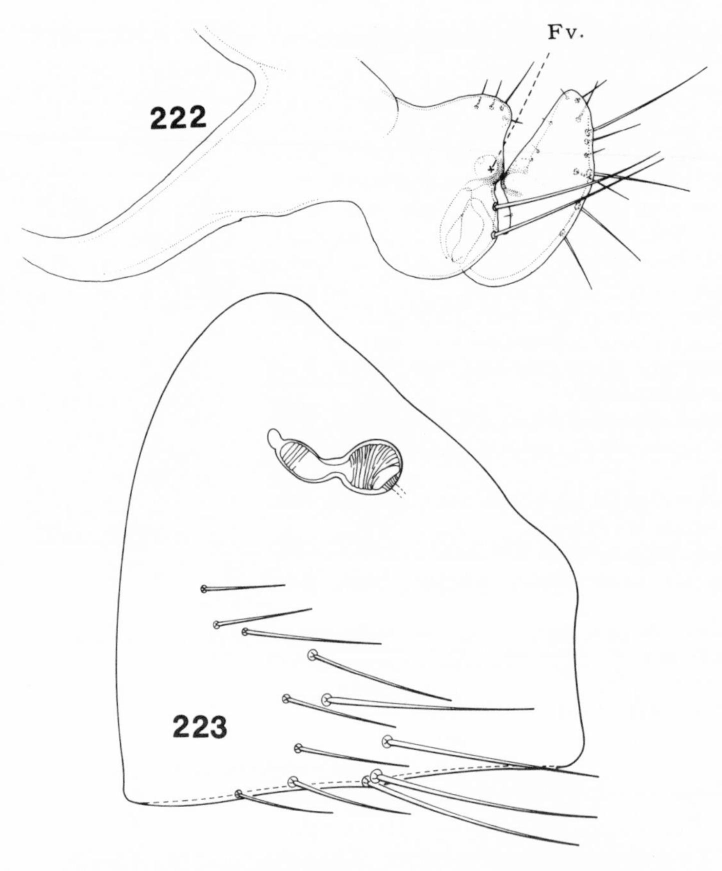

222-223 *Oropsylla (Oropsylla) idahoensis* (Baker). **222** Male clasper. **223** Female sternum VII and spermatheca. After Holland, 1985.

GENUS **Oropsylla (Diamanus)** Jordan

Diamanus. Jordan, 1933, Novit. Zool. 39: 73 (type species by designation: *Pulex montanus* Baker, 1895).
Oropsylla (Diamanus) Jordan, 1933. Smit, in: Traub et al., 1983, p. 27.
Oropsylla (Diamanus) Jordan, 1933. Haddow et al., in: Traub et al., 1983, p. 139.

Until recently this subgenus was maintained as a full genus for *D. montanus,* a widespread species in western United States. These fleas are parasites of ground squirrels, and occur only accidentally on other hosts. The following taxon is a common faunal element in the region.

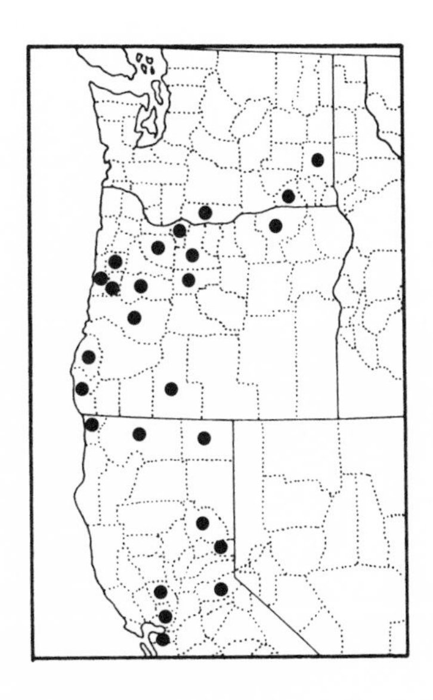

Oropsylla (Diamanus) montana (Baker)

[89509]

(Figs. 224, 225)

Pulex montanus. Baker, 1895, Canad. Ent. 25: 131-132 (Fort Collins, Colorado, from "mountain grey squirrel").

Ceratophyllus montanus (Baker). Baker, 1904, Proc. U. S. Nat. Mus. 27: 388, 411, 445 (Colorado and Santa Rita Mountains, Arizona, from "rock squirrel" and *Sciurus aberti*).

Ceratophyllus acutus. Baker, 1904, Invert. Pacifica 1: 40 (Stanford University, California, from *Spermophilus*).

Ceratophyllus montanus Baker. Jordan, 1929, Novit. Zool. 35: 31 (*acutus* synonymized).

Diamanus montanus (Baker). Jordan, 1933, Novit. Zool. 39: 73 (*Diamanus* erected for *C. montanus* and *C. mandarinus* J. & R., 1911).

Diamanus montanus (Baker). Hubbard, 1947, Fleas of Western North America, pp. 147-150.

Diamanus hopkinsi. Vargas, 1955, Rev. Inst. Salubr. enferm. trop. Mexico 15: 15-16 (Parlay Canyon, Utah, from "squirrel").

Diamanus montanus (Baker). Stark, 1958, The Siphonaptera of Utah, U.S.D.H.E.W. Publ. Hlth. Ser., CDC, pp. 141 (*hopkinsi* synonymized).

Diamanus montanus (Baker). Smit and Wright, 1978a, A list of code numbers..., p. 5 (*mandarinus* and *hopkinsi* synonymized).

Oropsylla (Diamanus) montana (Baker, 1895). Smit, in: Traub et al., 1983, p. 27.

Oropsylla (Diamanus) montana montana (Baker, 1895). Haddow et al., in: Traub et al., 1983, p. 140.

Of the 614 specimens reported here, 274 males and 315 females were taken from *Spermophilus beecheyi,* which Hubbard (1947) states to be the preferred host. The remaining specimens are all accidental host associations with other small rodents and a carnivore. According to Olsen (1981), *S. beecheyi* was the first native rodent to be found plague-positive in the United States. This discovery was made in 1908 in Contra Costa County, California. In 1934 a case of plague in a human was reported from Oregon, and today the disease is known from fifteen western states, as far east as western Kansas, Oklahoma, and Texas.

That this flea functions in the reservoir of sylvatic plague is incontestable. However, field and laboratory studies show that it is not a particularly effective vector of the disease. This deficiency is almost certainly compensated for by the fact that the species builds to huge populations in the nest and burrow systems of the host, and it is not uncommon to retrieve over fifty fleas from a single host animal.

According to Maser et al. (1981), *S. beecheyi* hibernates from November through February with occasional individuals being seen during warm spells in the winter. Following is the seasonal distribution of our Oregon collections.

Seasonal distribution of *O. (D.) montana* in Oregon

	J	F	M	A	M	J	J	A	S	O	N	D
Males	0	1	9	71	0	11	28	32	130	1	0	1
Females	0	0	6	98	0	11	41	29	136	5	0	4

GENUS **Oropsylla (Hubbardipsylla)** Smit

Oropsylla (Hubbardipsylla). Smit, in: Traub et al., 1983, p. 18.
Oropsylla (Hubbardipsylla) Smit, 1983. Haddow et al., in: Traub et al., 1983, p. 140.

A few years ago a student at Iowa State University performed a computerized phenetic analysis of the species assigned to the genus *Opisocrostis* (Perdue, 1980). Using a program described by McCammon and Wenninger (1970), dendrograms for both sexes were generated. These indicated a relationship among the species that varies from that described by Jellison (1947). Jellison described three distinct species groups in the genus: 1) *O. hirsutus* and *bruneri,* 2) *O. tuberculatus* ssp., *washingtonensis* and *oregonensis,* and 3) *O. labis.*

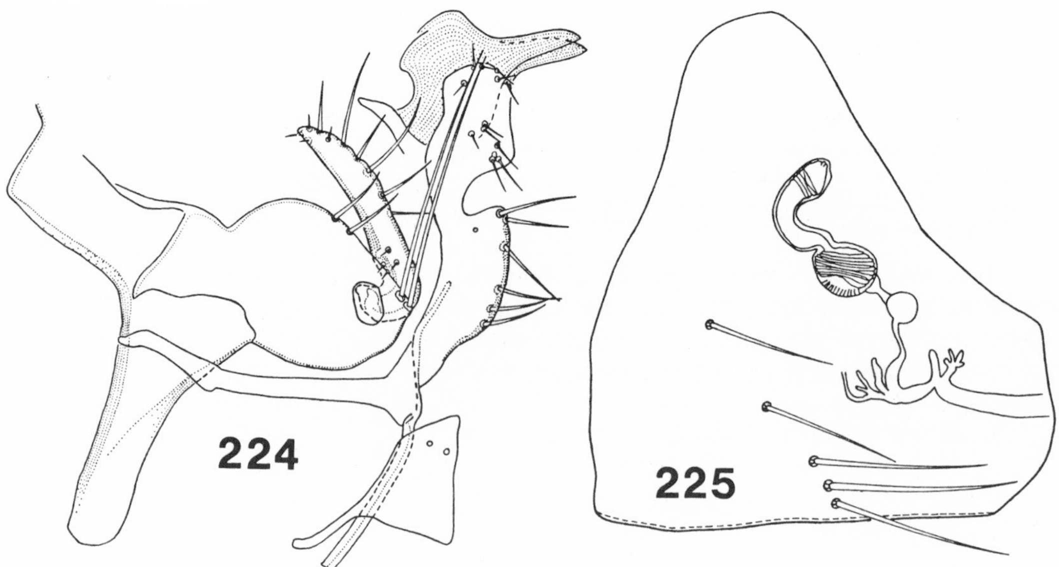

224-225 *Oropsylla (Diamanus) montana* (Baker). **224** Male genitalia. **225** Female sternum VII and spermatheca. Original.

The study by Perdue confirmed the relationship between *O. washingtonensis* and *oregonensis,* but found no relationship between these species and *O. tuberculatus* ssp. In addition, *O. labis* and *bruneri* showed a close relationship and *O. hirsutus* had no particularly close relatives. Her study did not address the relationship between these species and those assigned to the genus *Oropsylla.* Smit (1983) noted the close relationship between *O. washingtonensis* and *oregonensis* and placed them in their own subgenus in his revision of the family.

The two species assigned to this subgenus may be separated with the following key.

KEY TO THE SPECIES OF *O. (HUBBARDIPSYLLA)*

1 Male st.VIII bearing subapical bristles; female st. VII with shallow sinus, apex of upper and lower lobes rounded with the sinus appearing as a notch .... **washingtonensis**

Male st.VIII lacking subapical bristles; female st. VII with deep sinus, apex of upper lobe acute or rounded .... **oregonensis**

Oropsylla (Hubbardipsylla) oregonensis (Good and Prince)

[93922]

(Figs. 226, 227)

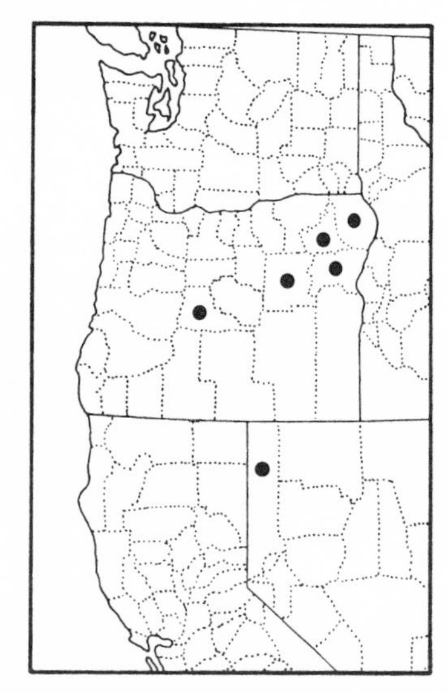

Opisocrostis oregonensis. Good and Prince, 1939, Publ. Hlth. Rep., Wash. 54: 1687-1691 (8 mi. S. Baker, Baker Co., Oregon, from *Spermophilus oregonus*).

Opisocrostis oregonensis Good and Prince. Hubbard, 1947, Fleas of Western North America, pp. 160-161.

Oropsylla (Hubbardipsylla) oregonensis (Good and Prince, 1939). Smit, in: Traub et al., 1983, p. 27.

Oropsylla (Hubbardipsylla) oregonensis (Good and Prince, 1939). Haddow et al., in: Traub et al., 1983, p. 140.

This and the following species are typically parasites of ground squirrels. This species was described from a subspecies of *Spermophilus beldingi,* but is also known from *S. columbianus, townsendi, and washingtoni.* According to Haddow et al. (1983), the species occurs in fields, meadows, and other grassy areas or in rocky terrain. It has been reported from southern Washington, Idaho, Nevada, California, and Oregon.

Oropsylla (Hubbardipsylla) washingtonensis (Good and Prince)

[93923]

(Figs. 228, 229)

Opisocrostis washingtonensis. Good and Prince, 1939, Publ. Hlth. Rep., Wash. 54: 1691-1693 (2 mi. E. Lind, Adams Co., Washington, from *Spermophilus townsendi*).

Opisocrostis washingtonensis (Good and Prince, 1939). Hubbard, 1947, Fleas of Western North America, pp. 161-162.

Oropsylla (Hubbardipsylla) washingtonensis (Good and Prince, 1939). Smit, in: Traub et al., 1983, p. 27.

Oropsylla (Hubbardipsylla) washingtonensis (Good and Prince, 1939). Haddow et al., in: Traub et al., p. 140.

This species is evidently much more restricted in its range than the preceding taxon and has been collected only in Oregon and Washington.

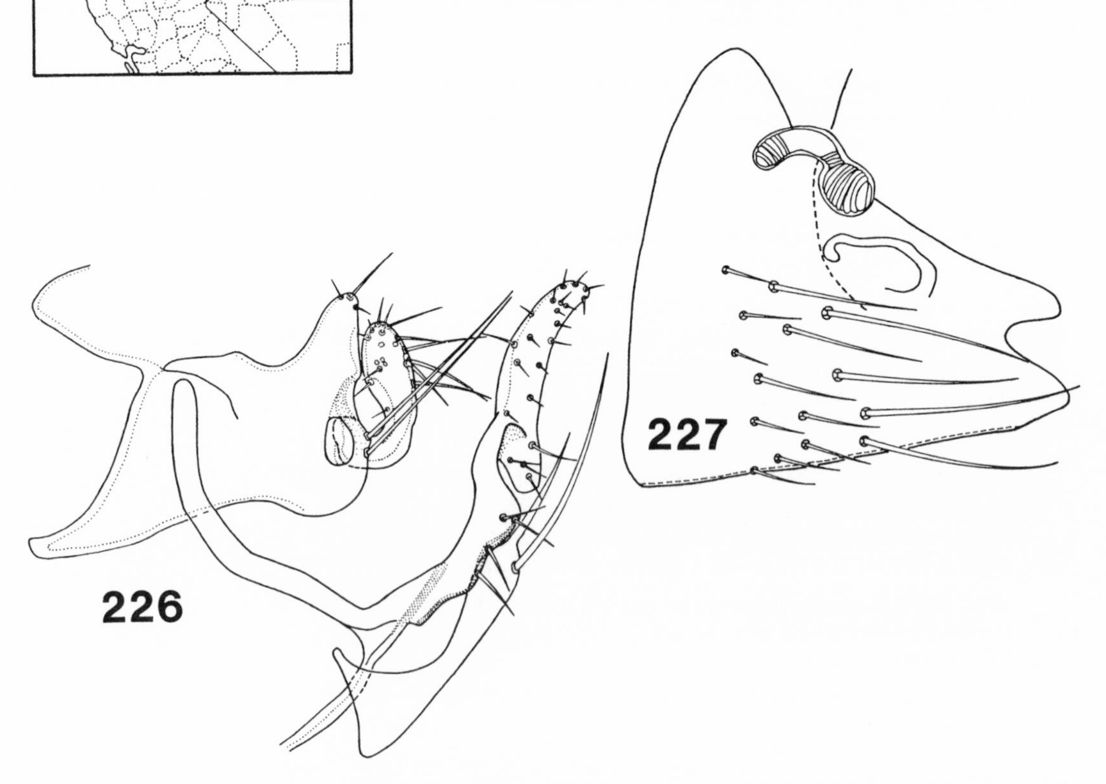

226-227 *Oropsylla (Hubbardipsylla) oregonensis* (Good & Prince). **226** Male genitalia. **227** Female sternum VII and spermatheca. Original.

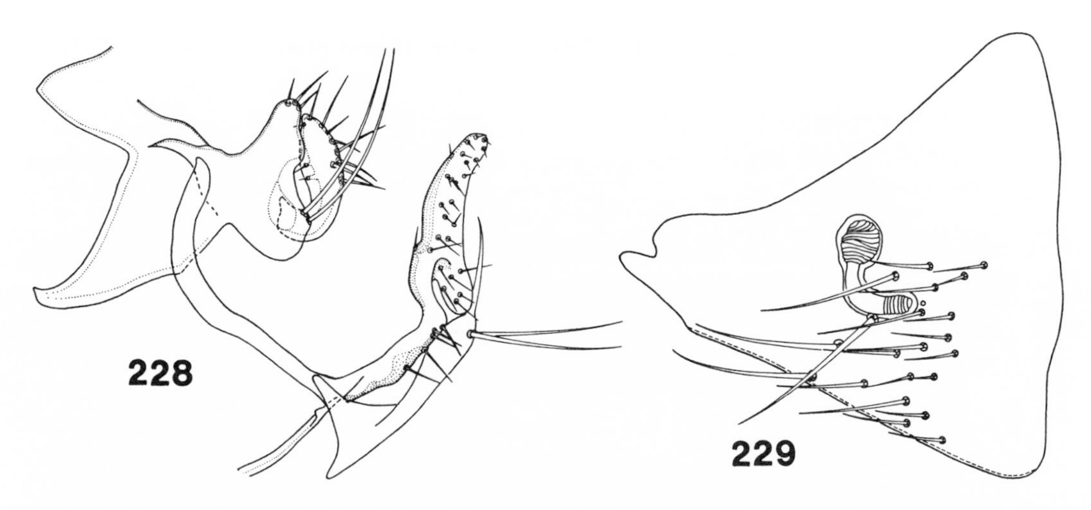

228-229 *Oropsylla (Hubbardipsylla) washingtonensis* (Good & Prince). **228** Male genitalia. **229** Female sternum VII and spermatheca. Original.

GENUS **Oropsylla (Opisocrostis)** Jordan

Opisocrostis. Jordan, 1933, Novit. Zool. 39: 73 (type species by designation: *Pulex hirsutus* Baker, 1895).
Oropsylla (Opisocrostis) Jordan, 1933. Smit, in: Traub et al., 1983, p. 27.
Oropsylla (Opisocrostis) Jordan, 1933. Haddow et al., in: Traub et al., 1983, p. 140.

As currently recognized this subgenus contains four species, one of which has two subspecies. It is exclusively Nearctic in occurrence and its species are mainly associated with species of *Spermophilus.* However, one species is specific to prairie dogs, as is one subspecies of *O. tuberculata.* The two species known from the Northwest may be separated with the following key.

KEY TO THE NORTHWESTERN SPECIES OF *O. (OPISOCROSTIS)*

1 Male movable process long, slender and flexed posteriorly at apex; female st. VII with a small marginal lobe subtended by a shallow sinus .... **labis**

Male movable process crescentric; female st. VII with pronounced marginal lobe subtended by a deep sinus .... **tuberculata tuberculata**

Oropsylla (Opisocrostis) labis (Jordan and Rothschild)

[92227]

(Figs. 230, 231)

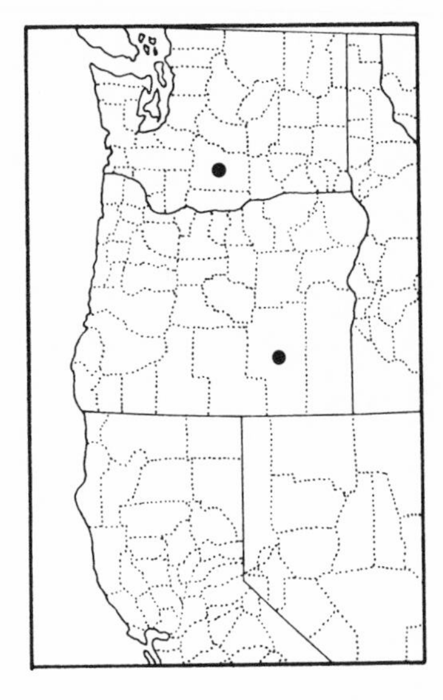

Ceratophyllus labis. Jordan and Rothschild, 1922, Ectoparasites I: 275 (Calgary, Alberta, from *Mustela frenata longicauda;* female was *Oropsylla rupestris*).

Ceratophyllus labis Jordan and Rothschild. Jordan, 1929, Novit. Zool. 35: 32 (female described).

Opisocrostis labis J. & R., 1922. Jordan, 1933, Novit. Zool. 39: 73.

Opisocrostis labis (Jordan and Rothschild). Holland, 1949, The Siphonaptera of Canada, pp. 128-129.

Oropsylla (Opisocrostis) labis Jordan and Rothschild, 1922. Smit and Wright, 1978b, A catalogue of primary type-specimens..., p. 27 (lectotype male selected).

Oropsylla (Opisocrostis) labis (Jordan and Rothschild, 1922). Smit, in: Traub et al., 1983, p. 27.

Oropsylla (Opisocrostis) labis (Jordan and Rothschild, 1922). Haddow et al., in: Traub et al., 1983, p. 141.

Opisocrostis labis (Jordan and Rothschild). Holland, 1985, Mem. Ent. Soc. Canada 130, pp. 459-462.

This is a species that has been collected only rarely in the area under consideration but very likely occurs commonly in the eastern part of the area. Hubbard (1947) reports three specimens from a grasshopper mouse, although most of his other records are from species of *Spermophilus* and *Cynomys*. Holland (1985) lists the species as an exclusive parasite of *S. richardsoni* and its predators in Canada. While the species was not collected during our surveys, it is reported from Harney County, Oregon, by Hubbard (1947).

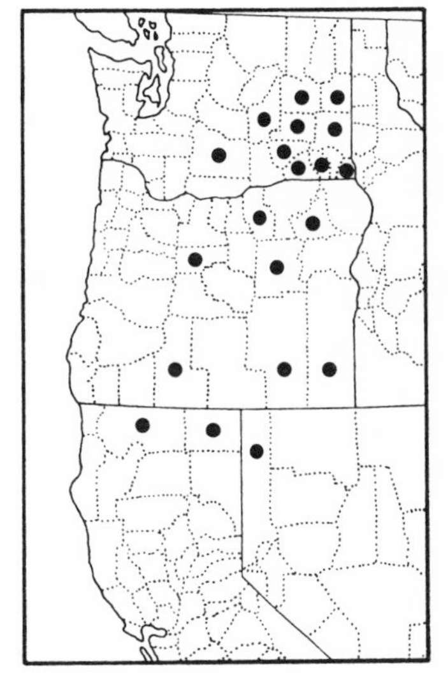

Oropsylla (Opisocrostis) tuberculata tuberculata (Baker) [90414]

(Figs. 232, 233)

Ceratophyllus tuberculatus. Baker, 1904, Proc. U. S. Nat. Mus. 27: 387, 393, 394, 447 (Moscow, Idaho, from *Spermophilus columbianus*).

Opisocrostis tuberculatus (Baker). Jordan, 1933, Novit. Zool. 39: 73.

Opisocrostis tuberculatus tuberculatus (Baker). Jellison, 1939, Publ. Hlth. Rep., Wash. 54: 843.

Oropsylla (Opisocrostis) tuberculata tuberculata (Baker). Brown, 1944, Ann. ent. Soc. Amer. 37: 210.

Opisocrostis tuberculatus tuberculatus (Baker). Holland, 1949, The Siphonaptera of Canada, pp. 130-131.

Oropsylla (Opisocrostis) t. tuberculata (Baker, 1904). Smit, in: Traub et al., 1983, p. 27.
Oropsylla (Opisocrostis) tuberculata tuberculata (Baker, 1904). Haddow et al., in: Traub et al., 1983, p. 141.
Opisocrostis tuberculatus tuberculatus (Baker). Holland, 1985, Mem. Ent. Soc. Canada 130, pp. 462-465.

Holland (1949, 1985) states that the true host for this species in Canada is probably *Spermophilus richardsoni, S. columbianus* being a more recently acquired host. All of our specimens, with the exception of a few strays, were taken from *S. beldingi.*

This is one of the common faunal elements in the Northwest and over two hundred specimens were collected, mainly from the eastern counties of Oregon. A subspecies of this flea occurs in the mountain states from Montana to New Mexico, as a parasite of prairie dogs. Jellison (1947) speculated that members of Opisocrostis originally parasitized species of *Cynomys,* all of which are restricted to North America, and later adapted to and evolved upon various species of *Spermophilus,* a Holarctic genus.

GENUS **Oropsylla (Thrassis)** Jordan

Oropsylla. Wagner, 1929, Konowia 8: 314-315 (*partim*).
Thrassis. Jordan, 1933, Novit. Zool. 39: 72-73 (type species by designation: *Ceratophyllus acamantis* Rothschild, 1905).
Thrassoides. Hubbard, 1947, Fleas of Western North America, pp. 144-145 (type species by designation: *Thrassis aridis* Prince, 1944).
Thrassis subgenus *Pandoropsylla.* Stark, 1970, Univ. Calif. Publ. Ent. 53: 82-83 (type species by designation: *T. pandorae* Jellison, 1937).
Thrassis subgenus *Nomadopsylla.* Stark, 1970, Univ. Calif. Publ. Ent. 53: 95-99 (type species by designation: *Ceratophyllus bacchi* Rothschild, 1905).
Thrassis subgenus *Thrassoides* Hubbard (**New Status**). Stark, 1970, Univ. Calif. Publ. Ent. 53: 122-123 (type species as above).
Oropsylla (Thrassis) Jordan, 1933. Smit, in: Traub et al., 1983, p. 27.
Oropsylla (Thrassis) Jordan, 1933. Haddow et al., in: Traub et al., 1983, p. 142.
Thrassis Jordan. Holland, 1985, Mem Ent. Soc. Canada 130, pp. 446-447.

The latest taxonomic revision of this difficult group of species was published by Stark in 1970. At that time *Thrassis* was still considered to be a full genus. There are eleven currently recognized species belonging to this taxon, and they are distributed throughout the more arid portions of western North America from southern Canada to northern Mexico. One species is known from as far east as Wisconsin, but most of the known taxa occur much farther west.

Members of this subgenus are mainly parasites of species of *Spermophilus* and *Marmota.* However, one species is associated with *Dipodomys* and a subspecies of another is found exclusively on *Lagurus curtatus.* There are 29 recognized species and subspecies

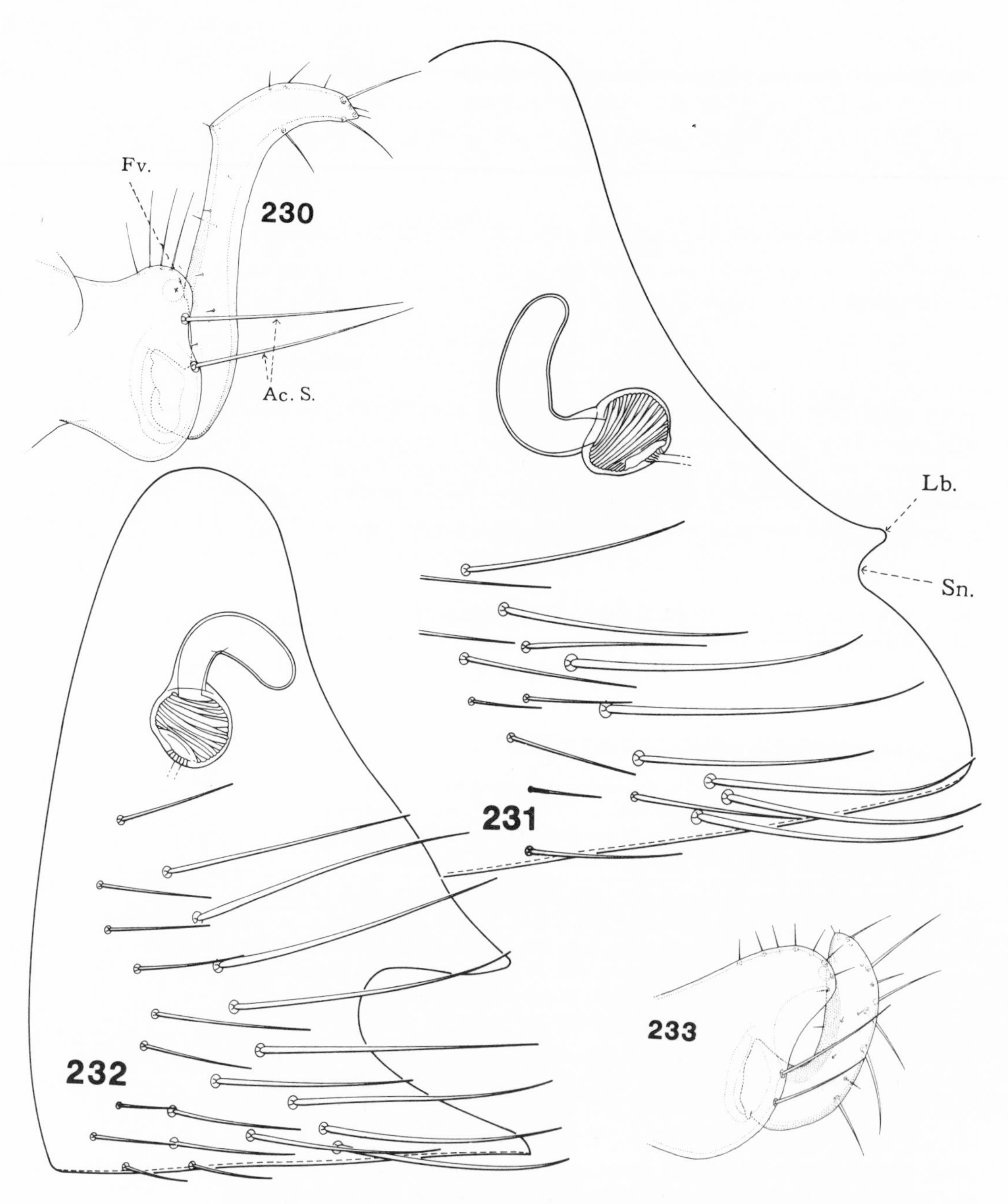

230-231 *Oropsylla (Opisocrostis) labis* (J. & R.). **230** Male clasper. **231** Female sternum VII and spermatheca. After Holland, 1985.

232-233 *Oropsylla (Opisocrostis) tuberculata tuberculata* (Baker). **232** Female sternum VII and spermatheca. **233** Male clasper. After Holland, 1985.

belonging to this subgenus, ten of which are known from the Pacific Northwest. The males are difficult to recognize and the females are even more so. The following key is a modification of that of Stark (1970).

KEY TO THE NORTHWESTERN TAXA OF THE SUBGENUS *THRASSIS*

1 Males **2**

Females **13**

2 Marginal spinelets on metanotum and abdominal segments heavily pigmented; a parasite of *Dipodomys* spp. .... **aridis hoffmani**

Marginal spinelets lightly pigmented, not typically found on *Dipodomys*...... **3**

3 Fewer than 11 setae on ventral half of t. VIII; marginal spinelets on first three abdominal terga; parasites of *Ammospermophilus, Lagurus* and other small rodents such as *Neotoma* and *Onychomys* **4**

More than twelve setae on ventral half of t. VIII; marginal spinelets on first four abdominal terga; parasites of *Spermophilus* and *Marmota* species **5**

4 Manubrium narrowly rounded; apex of crochet pointed posteriorly; ventral lobe of st. IX broad, blunt and prominent; parasites of *Lagurus curtatus* .... **bacchi johnsoni**

Manubrium blunt and flared apically; apex of crochet rounded dorsally and hooked posteriorly; ventral lobe of st. IX not prominent; not found on *Lagurus curtatus* .... **bacchi gladiolis**

5 Setae on ventral lobe of st. IX not highly modified; dorsal membranous lobe separated from ventral lobe, its surface turned partly posterior; st. VIII with three or fewer setae; apex of manubrium pointed; setae on posterior half of t. VIII not distinctly divided **6**

Setae on ventral lobe of st. IX modified into a prominent short apical seta and paired lateral setae which may be long and sharp or blade-shaped; dorsal membranous lobe broad, overlapping ventral lobe, its surface facing laterally; st. VIII with three to 22 setae; apex of manubrium tapered or small and rounded; setae on posterior half of t. VIII divided into two distinct groups **8**

6 Apex of manubrium small, pointed or rounded, not markedly turned dorsally; movable process long, narrow, 2.1 to 3.2 times longer than wide; dorsal apical portion of border of st. VIII angular; median dorsal lobe of aedeagus sclerotized, arched forward **7**

Apex of manubrium usually turned dorsally; movable process very long, narrow, 3.6 to 4.3 times longer than wide; dorsal apical portion of border of st. VIII rounded; median dorsal lobe filamentous and difficult to see .... **petiolata**

7 Movable process triangular, apex pointed; angle at dorsal apical border of st. VIII forming a small process .... **pandorae pandorae**

Movable process oblong, apex blunt; angle at dorsal apical border of st. VIII forming a large process .... **pandorae jellisoni**

8 Portion of posterior border of t. VIII above ventral apical angle concave; host: *Marmota* species **9**

Ventral apical angle not developed, or, portion of posterior border of t. VIII above ventral apical angle straight or convex **11**

9 Total setae in ventral half of tergum VIII 13-25; lateral lobe of aedeagus rounded ventrally, neck of aedeagus without a collar .... **spenceri spenceri**

Total setae in ventral half of tergum VIII 25-60; lateral lobe of aedeagus prominent, sharply pointed, neck of aedeagus with collar **10**

10 Portion of posterior border of t. VIII above ventral apical angle deeply concave; upper group of long setae some distance from ventral apical angle; neck of aedeagus with a hump .... **acamantis howelli**

Portion of posterior border of t. VIII above ventral apical angle only slightly concave; upper group of long setae near ventral apical angle; neck of aedeagus slender .... **acamantis media**

11 Neck of aedeagus with collar .... **acamantis acamantis**

Neck of aedeagus lacking a collar **12**

12 Modified lateral setae of st. IX moderately broad, curved only at tips; apical seta long, equal to lateral; two most ventral setae on movable process thick .... **francisi francisi**

Modified lateral setae of st. IX long and arched dorsally; apical seta short, about one-third the length of lateral; two most ventral setae on movable process thin .... **francisi rockwoodi**

13 Marginal spinelets on metanotum and abdominal segments heavily pigmented; a parasite of *Dipodomys* .... **aridis hoffmani**

Marginal spinelets lightly pigmented; not typically found on *Dipodomys* spp. **14**

14 First segment of hind tarsus long, 38-42 percent total length of tarsus; marginal spinelets on first two or three abdominal terga; hilla of spermatheca short, 1.1 to 1.7 times length of bulga **15**

First segment of hind tarsus shorter, 33-38 percent total length of tarsus; marginal spinelets on first three or four abdominal terga; hilla of spermatheca longer, 1.5 to 3.2 times length of bulga **16**

15 Host: *Lagurus curtatus* .... **bacchi johnsoni**

Host: *Ammospermophilus leucurus* (and occasionally other small rodents) .... **bacchi gladiolis**

16 Labial palpi short, less than 0.35 mm, or less than 15 percent of total length of insect **17**

Labial palpi long, more than 0.43 mm, or more than 17 percent of total length of insect **20**

17 Antesensilial setae usually three; bulga of spermatheca short and wide, length only 0.63 to 0.83 times width; posterior margin of st. VII usually angular, rarely with a concavity or deep sinus **18**

Antesensilial setae usually two; bulga of spermatheca rounded, length 0.73 to 1.4 times width; posterior margin of st. VII often angular, usually with concavity or deep sinus **19**

18 Western Great Basin and Cascades, on *Spermophilus beldingi* .... **francisi rockwoodi**

Western Great Basin, on *Spermophilus townsendi* .... **francisi francisi**

19 Posterior margin of st. VII angular with a deep sinus immediately below angle .... **pandorae jellisoni**

Posterior margin of st. VII straight or concave below angle. .... **pandorae pandorae**

20 Antesensilial setae two; posterior border of st. VII variable, usually not angular .... **petiolata**

Antesensilial setae three .... **spenceri & acamantis** ssp.

Oropsylla (Thrassis) spenceri spenceri (Wagner)

[93632]

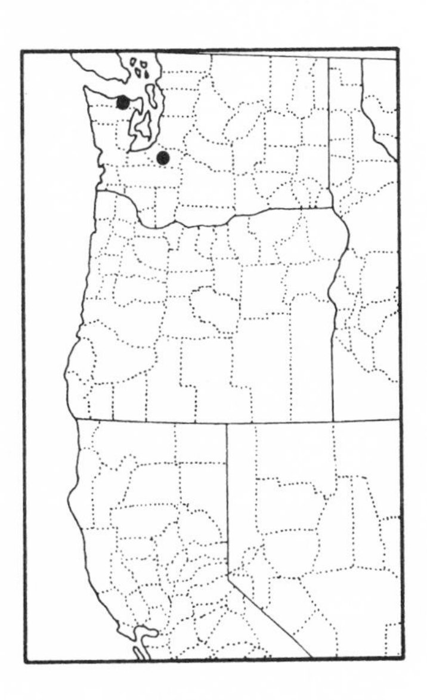

Thrassis acamantis Rothschild. Spencer, 1936, Proc. Ent. Soc. B. C. 32: 13 (Vancouver Island marmot, no locality given).
Thrassis spenceri. Wagner, 1936, Z. Parasitenk. 8: 654-655 (Granite Mountain, near Birch Island, British Columbia, from *Marmota caligata*).
Thrassis fousti. Hubbard, 1954, Ent. News 65: 173 (Olympic National Park, Washington, from *Marmota olympus*).
Thrassis spenceri vancouverensis. Stark, 1957, J. Parasitol. 43: 334-335 (Vancouver Island, British Columbia, from *Marmota vancouverensis*).
Thrassis spenceri spenceri Wagner. Stark, 1970, Univ. Calif. Publ. Ent. 53: 39 (synonymizes *fousti* and *vancouverensis*).
Oropsylla (Thrassis) s. spenceri (Wagner, 1936). Smit, in: Traub et al., 1983, p. 28.

An ectoparasite of various species of marmots, this species occurs in the state of Washington, northward to British Columbia and Alberta.

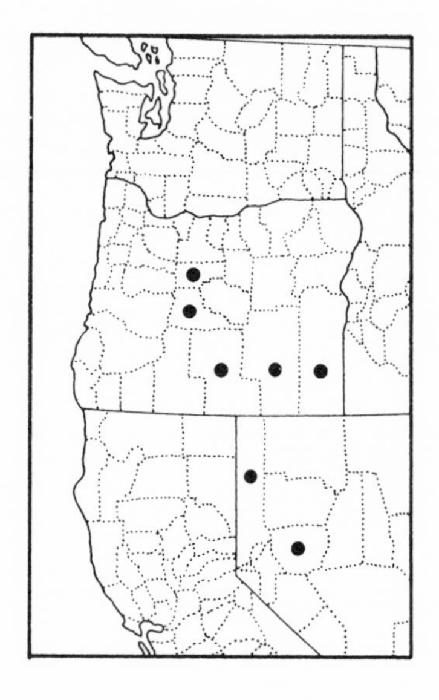

Oropsylla (Thrassis) bacchi gladiolis (Jordan)

[92530]

(Figs. 234, 235)

Ceratophyllus gladiolis. Jordan, 1925, Novit. Zool. 32: 108 (San Diego, California, from *Spermophilus turdicaudus* [= *tereticaudus*] and from San Francisco, from *Perognathus* and *Tamias* caged together).

Ceratophyllus gladiolis caducus. Jordan, 1930, Novit. Zool. 35: 268-269 (Vernal, Utah, from *Ammospermophilus leucurus cinnamomeus*).

Thrassis gladiolis Jordan, 1925. Jordan, 1933, Novit. Zool. 39: 73 (*Thrassis* erected).

Thrassis bacchi gladiolis (Jordan, 1925). Stark, 1955, Syst. Zool. 4: 46.

Thrassis bacchi consimilis Stark. Augustson and Durham, 1961, Bull. S. Calif. Acad. Sci. 60: 103.

Thrassis setosis Prince. Augustson and Durham, 1961, *op. cit.*

Thrassis (Nomadopsylla) bacchi gladiolis (Jordan). Stark, 1970, Univ. Calif. Publ. Ent. 53: 107-110.

Oropsylla (Thrassis) bacchi gladiolis Jordan, 1925a. Smit and Wright, 1978b, A catalogue of primary type-specimens..., p. 21.

Oropsylla (Thrassis) b. gladiolis (Jordan, 1925). Smit, in: Traub et al., 1983, p. 28.

Oropsylla (Thrassis) bacchi gladiolis (Jordan, 1925). Haddow et al., in: Traub et al., 1983, p. 145.

This subspecies has a rather broad distribution that includes parts of Oregon, Idaho, California, Nevada, Utah, and Arizona. It is primarily a parasite of *Ammospermophilus leucurus* but also has been taken on wood rats and other small rodents.

Stark (1970) discusses the intergradation of this taxon with *O. (T.) b. caduca* (Jordan, 1930), pointing out that it involves a gradual transition of several characters over a broad area from northeastern Utah to southwestern California.

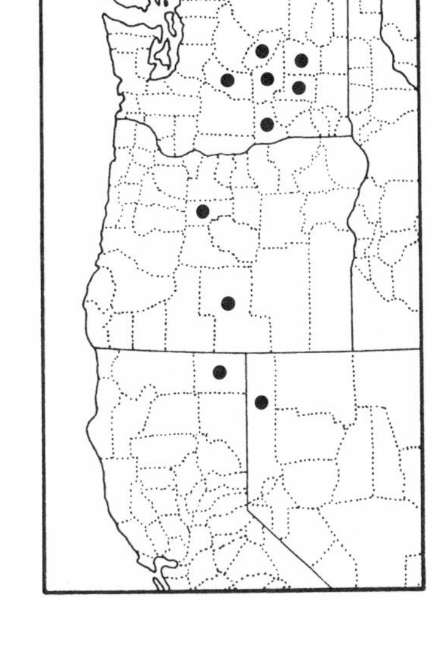

Oropsylla (Thrassis) bacchi johnsoni (Hubbard)

[94921]

(Figs. 236, 237)

Thrassis gladiolis johnsoni. Hubbard, 1949, Ent. News 60: 143-144 (Davenport, Lincoln Co., Washington, from *Lagurus curtatus pauperrimus*).
Thrassis bacchi johnsoni (Hubbard). Anonymous, 1950, Publ. Hlth. Rep., Wash. 65: 614 (*fide* Stark).
Thrassis (Nomadopsylla) bacchi johnsoni (Hubbard). Stark, 1970, Univ. Calif. Publ. Ent. 53: 110-112.
Oropsylla (Thrassis) b. johnsoni (Hubbard, 1949). Smit, in: Traub et al., 1983, p. 28.
Oropsylla (Thrassis) bacchi johnsoni (Hubbard, 1949). Haddow et al., in: Traub et al., 1983, p. 145.

Unlike other members of this subgenus, this subspecies is a specific parasite of a microtine rather than of a sciurid rodent. The sage vole, *Lagurus curtatus,* has a rather broad distribution in the northwestern United States and southwestern Canada, but this flea is currently known only from Washington and Oregon.

The role of this species in the maintenance of plague in *Lagurus* colonies was addressed by Lewis in Maser et al. (1974).

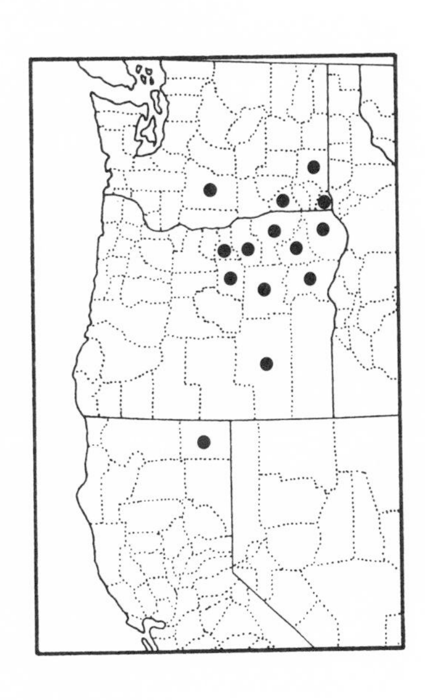

Oropsylla (Thrassis) pandorae pandorae (Jellison)

[93703]

(Figs. 238, 239)

Thrassis pandorae. Jellison, 1937, Publ. Hlth. Rep., Wash. 52: 726-729 (P. & O. Ranch, Beaver Head Co., Montana, from *Spermophilus elegans*. Paratypes also from Montana, Oregon, and Wyoming from *S. elegans, S. armatus* and *S. oregonus*).
Thrassis arizonensis ssp. Stanford, 1944, Proc. Utah Acad. Sci. Arts & Letters 19 & 20: 176 (*nec* Baker, 1898, misidentified).
Thrassis arizonensis arizonensis (Baker). Allred, 1952, Great Basin Nat. 12: 71, 74 (*nec* Baker, 1898, misidentified).
Thrassis petiolatus (Baker). Allred, 1952, *op. cit.* (*nec* Baker, 1904, misidentified).
Thrassis (Pandoropsylla) pandorae pandorae Jellison. Stark, 1970, Univ. Calif. Publ. Ent. 53: 84-89.
Oropsylla (Thrassis) p. pandorae (Jellison, 1937). Smit, in: Traub et al., 1983, p. 28.
Oropsylla (Thrassis) pandorae pandorae (Jellison, 1937). Haddow et al., in: Traub et al., 1983, p. 147.

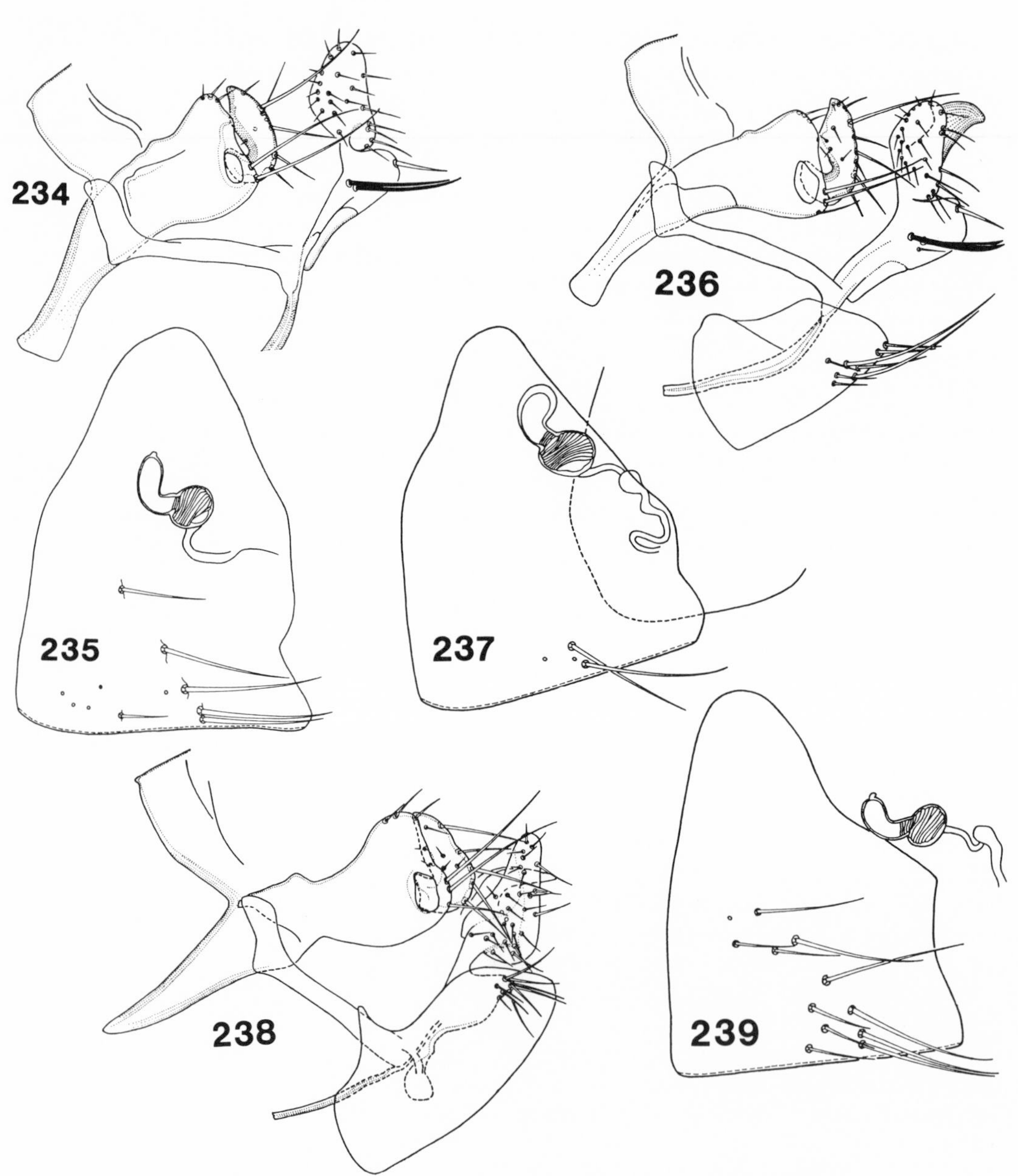

234-235 *Oropsylla (Thrassis) bacchi gladiolus* (Jordan). **234** Male clasper and sternum IX. **235** Female sternum VII and spermatheca. Original.
236-237 *Oropsylla (Thrassis) bacchi johnsoni* (Hubbard). **236** Male clasper and sternum IX. **237** Female sternum VII and spermatheca. Original.
238-239 *Oropsylla (Thrassis) pandorae pandorae* (Jellison). **238** Male clasper and sternite IX. **239** Female sternite VII and spermatheca. Original.

This, and the following subspecies, are distributed over the eastern two-thirds of the area under consideration as parasites of *Spermophilus beldingi* and, to a lesser extent, *S. columbianus*. Total range for the species includes parts of California, Washington, Nevada, Idaho, Montana, Wyoming, Utah, and Colorado. Within this area it also parasitizes *S. richardsoni* and *S. armatus*. However, there are numerous accidental records of this taxon from many small rodents, as well as lagomorphs and predators, and the gross distribution of the flea seems independent of the distribution of the three preferred host species of ground squirrels (*S. beldingi, armatus,* and *richardsoni*), as pointed out by Stark (1970).

Oropsylla (Thrassis) pandorae jellisoni (Hubbard)

[94052]

(Figs. 240, 241)

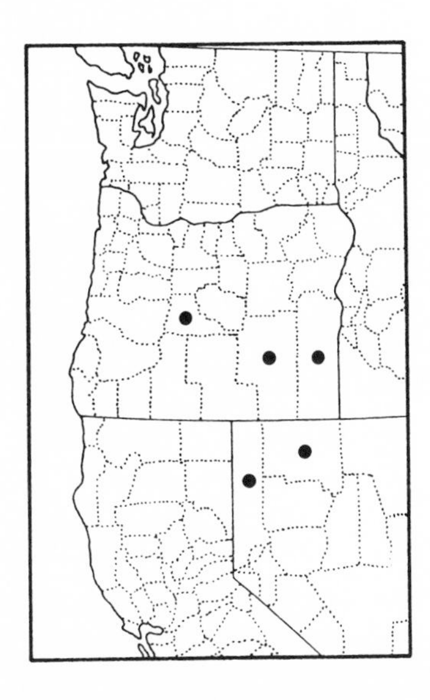

Thrassis jellisoni. Hubbard, 1940, Pacific Univ. Bull. 37: 1-4 (Franklin's Ranch, northwest Nevada, 15 mi. S. Denio, Oregon, from *Spermophilus beldingi oregonus*).
Thrassis petiolatus Baker, 1904. Hubbard, 1941, Pacific Univ. Bull. 37(9A): 1-4 (*partim,* Nevada specimens misidentified).
Thrassis (Pandoropsylla) pandorae jellisoni Hubbard. Stark, 1970, Univ. Calif. Publ. Ent. 53: 90.
Oropsylla (Thrassis) pandorae jellisoni (Hubbard, 1940). Smit, in: Traub et al., 1983, p. 28.
Oropsylla (Thrassis) pandorae jellisoni (Hubbard, 1940). Haddow et al., in: Traub et al., 1983, p. 147.

The distribution of this subspecies includes southeastern Oregon, northern Nevada, southern Idaho, and extreme northwestern Utah. Where *Spermophilus richardsoni nevadensis* occurs, this is the preferred host. Elsewhere it may be collected on *S. beldingi*. According to Stark (1970), it is suggested that the subspecies evolved on the former host species and transferred to the latter as the range of *S. r. nevadensis* contracted to that of a relict, out of contact with other subspecies of *S. richardsoni*.

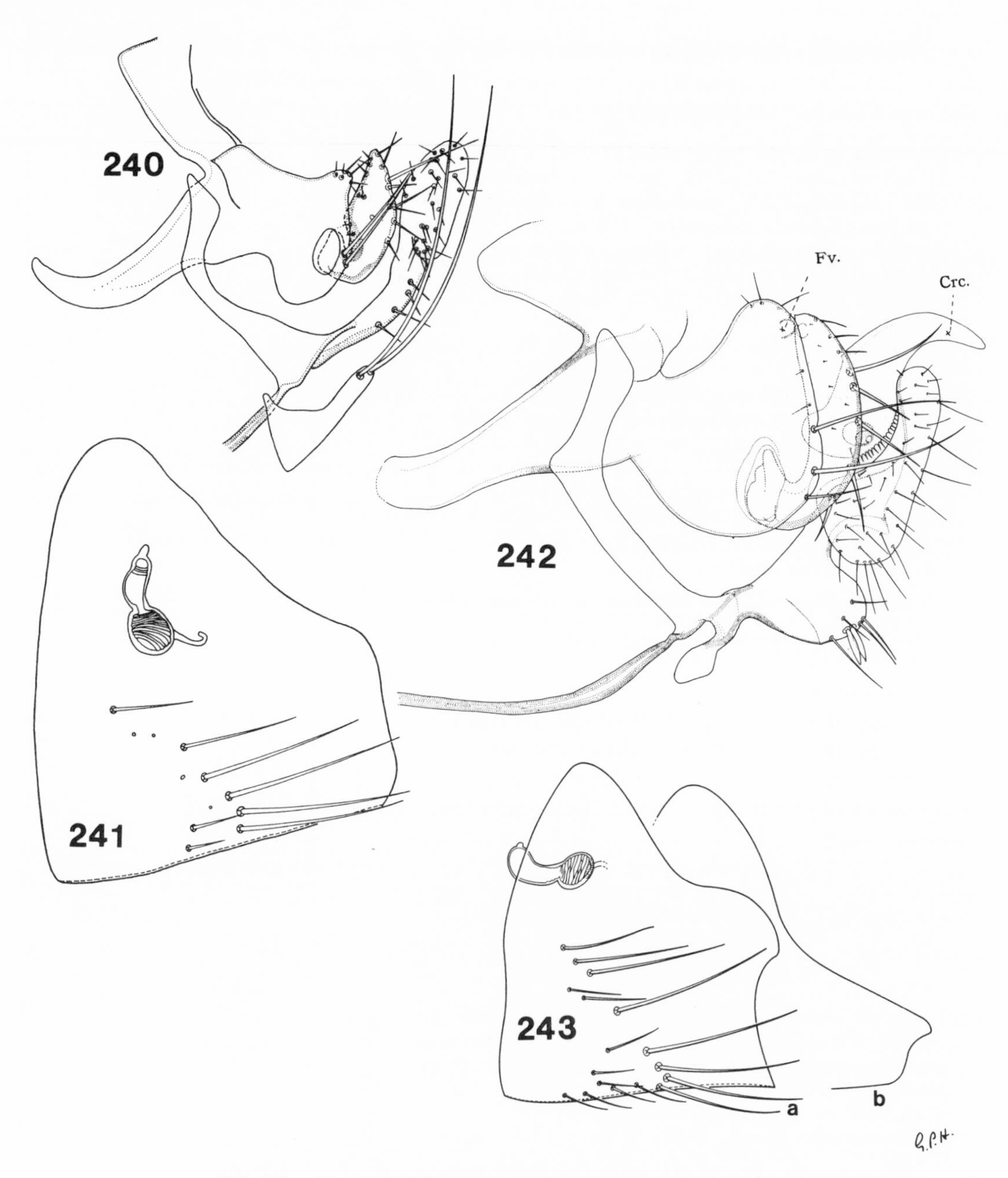

240-241 *Oropsylla (Thrassis) pandorae jellisoni* (Hubbard). **240** Male clasper and sterna VIII and IX. **241** Female sternum VII and spermatheca. Original.
242-243 *Oropsylla (Thrassis) petiolata* (Baker). **242** Male clasper, sternum IX and crochet. **243** Female sternum VII and spermatheca. After Holland, 1985.

Oropsylla (Thrassis) petiolata (Baker)

[90431]

(Figs. 242, 243)

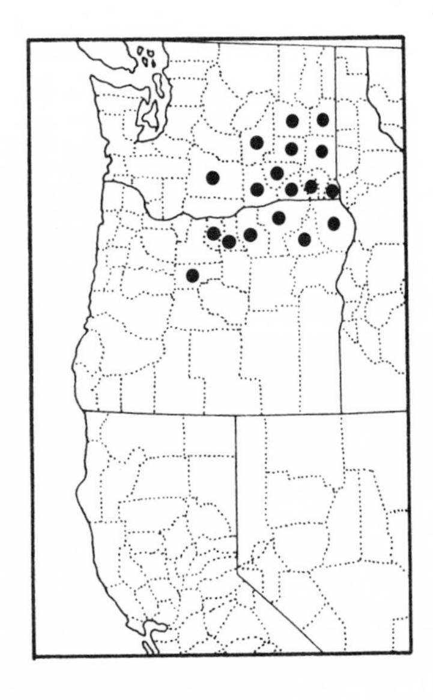

Ceratophyllus petiolatus. Baker, 1904, Proc. U. S. Nat. Mus. 27: 388, 415-416, 446, pl. 18, figs. 7-11 (Moscow, Idaho, from *Lynx canadensis*). (According to Jordan, 1929, fig. 7, pl. 18 is *Oropsylla idahoensis*).

Ceratophyllus idahoensis Baker. Dunn and Parker, 1923, Publ. Hlth. Rep., Wash. 38: 2763-2775 (misidentified).

Ceratophyllus petiolatus Baker (1904). Jordan, 1929, Novit. Zool. 35: 31.

Thrassis petiolatus Baker, 1904. Jordan, 1933, Novit. Zool. 39: 72-73.

Thrassis (Pandoropsylla) petiolatus (Baker). Stark, 1970, Univ. Calif. Publ. Ent. 53: 90-95.

Oropsylla (Thrassis) petiolata (Baker, 1904). Smit, in: Traub et al., 1983, p. 28.

Oropsylla (Thrassis) petiolata (Baker, 1904). Haddow et al., in: Traub et al., 1983, p. 148.

Thrassis petiolatus (Baker). Holland, 1985, Mem. Ent. Soc. Canada 130, pp. 452-454.

This species ranges from northeastern Oregon, eastern Washington, northern Idaho, and western Montana into southeastern British Columbia and western Alberta. This distribution coincides closely with the central portion of the distribution of its preferred host, *Spermophilus columbianus.* It is also known from *S. washingtoni* and *S. townsendi* and, rarely, on *Tamiasciurus* and *Onychomys* species.

Oropsylla (Thrassis) acamantis acamantis (Rothschild) [90503]

(Figs. 244, 245)

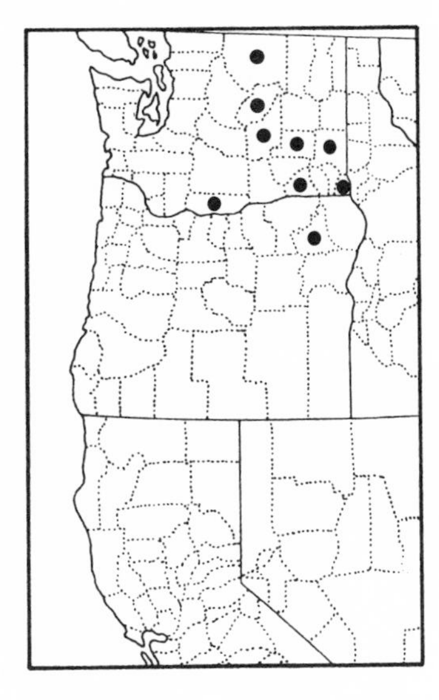

Ceratophyllus acamantis. Rothschild, 1905, Novit. Zool. 12: 156-158 (Okanagan, British Columbia, from *Mephitis mephitis spissigrada* and Eagle River, Sicamous, from *Marmota flaviventris avara,* and Sumas, from *Mustela vison energumenos* and *Canis latrans*).

Oropsylla acamantis (Roths., 1905). Wagner, 1929, Konowia 8: 315.

Thrassis acamantis Roths., 1905. Jordan, 1933, Novit. Zool. 39: 72-73.

Thrassis howelli (Jordan). Svihla, 1941, Univ. Wash. Publ. Biol. 12: 16.

Oropsylla (Thrassis) acamantis acamantis Rothschild, 1905a. Smit and Wright, 1978b, A catalogue of primary type-specimens..., p. 3 (selection of lectotype male).
Oropsylla (Thrassis) acamantis acamantis (Rothschild, 1905). Smit, in: Traub et al., 1983, p. 27.
Oropsylla (Thrassis) acamantis acamantis (Rothschild, 1905). Haddow et al., in: Traub et al., 1983, p. 142.
Thrassis acamantis acamantis (Rothschild). Holland, 1985, Mem. Ent. Soc. Canada 130, pp. 447-449.

We report this subspecies from a single collection locality in Union County, Oregon, from *Marmota flaviventris,* its usual host, and *Spermophilus columbianus,* plus literature records from Washington. Most specimens of this species collected in Oregon belonged to the following subspecies.

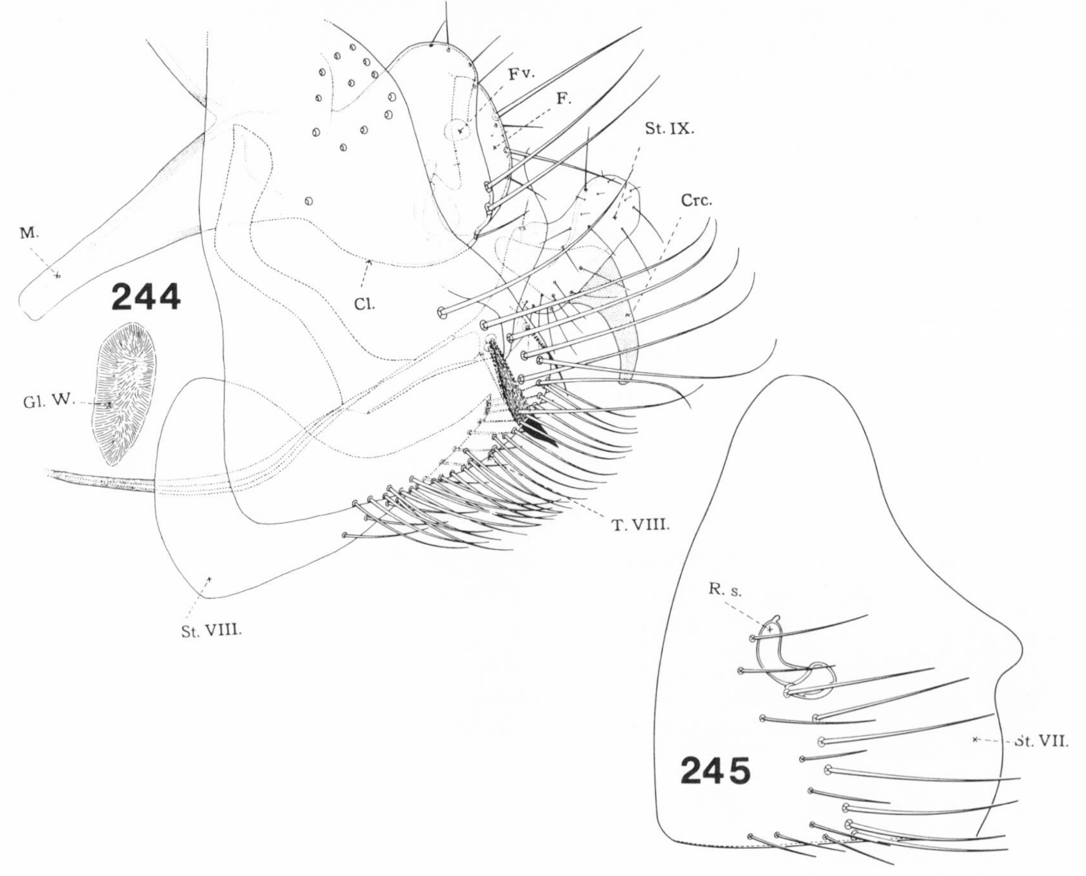

244-245 *Oropsylla (Thrassis) acamantis acamantis* (Rothschild). **244** Male genitalia. **245** Female sternum VII and spermatheca. After Holland, 1985.

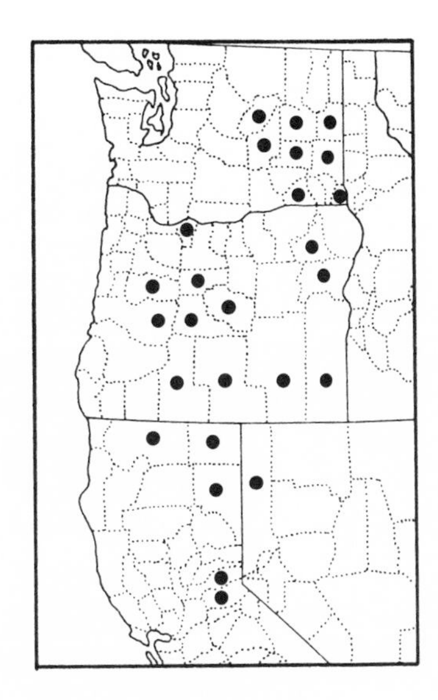

Oropsylla (Thrassis) acamantis howelli (Jordan)

[92532]

(Figs. 246, 247)

Ceratophyllus howelli. Jordan, 1925, Novit. Zool. 32: 109 (Pine City, Mono Co., California, from *Mustela frenata* ssp. and *Marmota flaviventris sierrae*).

Thrassis howelli Jord., 1925. Jordan, 1933, Novit. Zool. 39: 73.

Thrassis howelli howelli Jord., 1925. Wagner, 1936, Z. Parasitenk. 8: 340-342.

Thrassis acamantis howelli (Jordan). Eskey and Haas, 1939, Publ. Hlth. Rep., Wash. 54: 1472.

Thrassis (Thrassis) acamantis howelli (Jordan). Stark, 1970, Univ. Calif. Publ. Ent. 53: 54-57.

Oropsylla (Thrassis) acamantis howelli Jordan, 1925a. Smit and Wright, 1978b, A catalogue of primary type-specimens ..., p. 23 (selection of lectotype male).

Oropsylla (Thrassis) a. howelli (Jordan, 1925). Smit, in: Traub et al., 1983, pp. 27-28.

Oropsylla (Thrassis) acamantis howelli (Jordan, 1925). Haddow et al., in: Traub et al., 1983, p. 142.

This subspecies occurs throughout most of the eastern two-thirds of Oregon except for the small area in the southeast that is occupied by *O. (T.) a. media.* According to Stark (1970), the zone of intergradation between this and the nominate form is relatively narrow; the Columbia River serves as at least a partial barrier between the two subspecies.

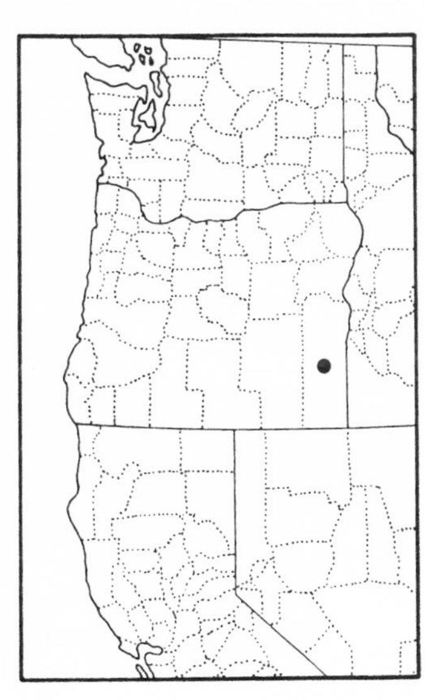

Oropsylla (Thrassis) acamantis media (Stark)

[97001]

(Figs. 248, 249)

Ceratophyllus acamantis Rothschild. Dunn and Parker, 1923, Publ. Hlth. Rep., Wash. 38: 2771, 2775 (*nec* Rothschild, 1905).

Oropsylla acamantis. Stanford, 1931, Proc. Utah Acad. Sci. Arts and Letters 8: 154 (*partim; nec* Rothschild, 1905).

Thrassis howelli utahensis. Wagner, 1936, Z. Parasitenk. 8: 340 (Salina and Logan, Utah, from *Marmota flaviventris engelhardti* and *M. f. nosophora*).

Thrassis acamantis medius. Stark, 1970, Univ. Calif. Publ. Ent. 53: 57-60 (16 mi. N.W. of West Yellowstone, Gallatin Co., Montana, from *Marmota flaviventris* ssp.).

Oropsylla (Thrassis) a. media (Stark, 1970). Smit, in: Traub et al., 1983, p. 28.

Oropsylla (Thrassis) acamantis media (Stark, 1970). Haddow et al., in: Traub et al., 1983, p. 142.

The bulk of the range of this subspecies is in Idaho, Montana, and Wyoming, but it does extend slightly into northern Utah and extreme southeastern Oregon. *Marmota flaviventris* is the normal host.

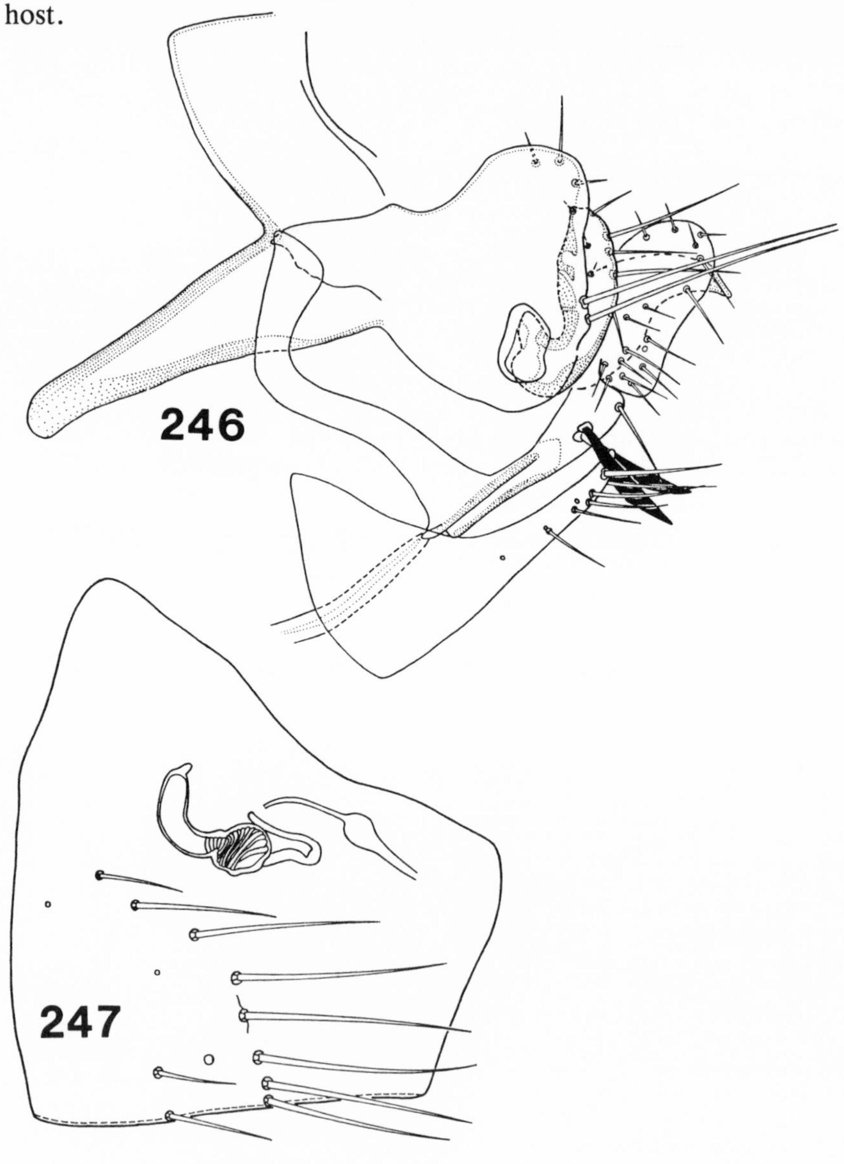

246-247 *Oropsylla (Thrassis) acamantis howelli* (Jordan). **246** Male clasper and sterna VIII and IX. **247** Female sternite VII and spermatheca. Original.

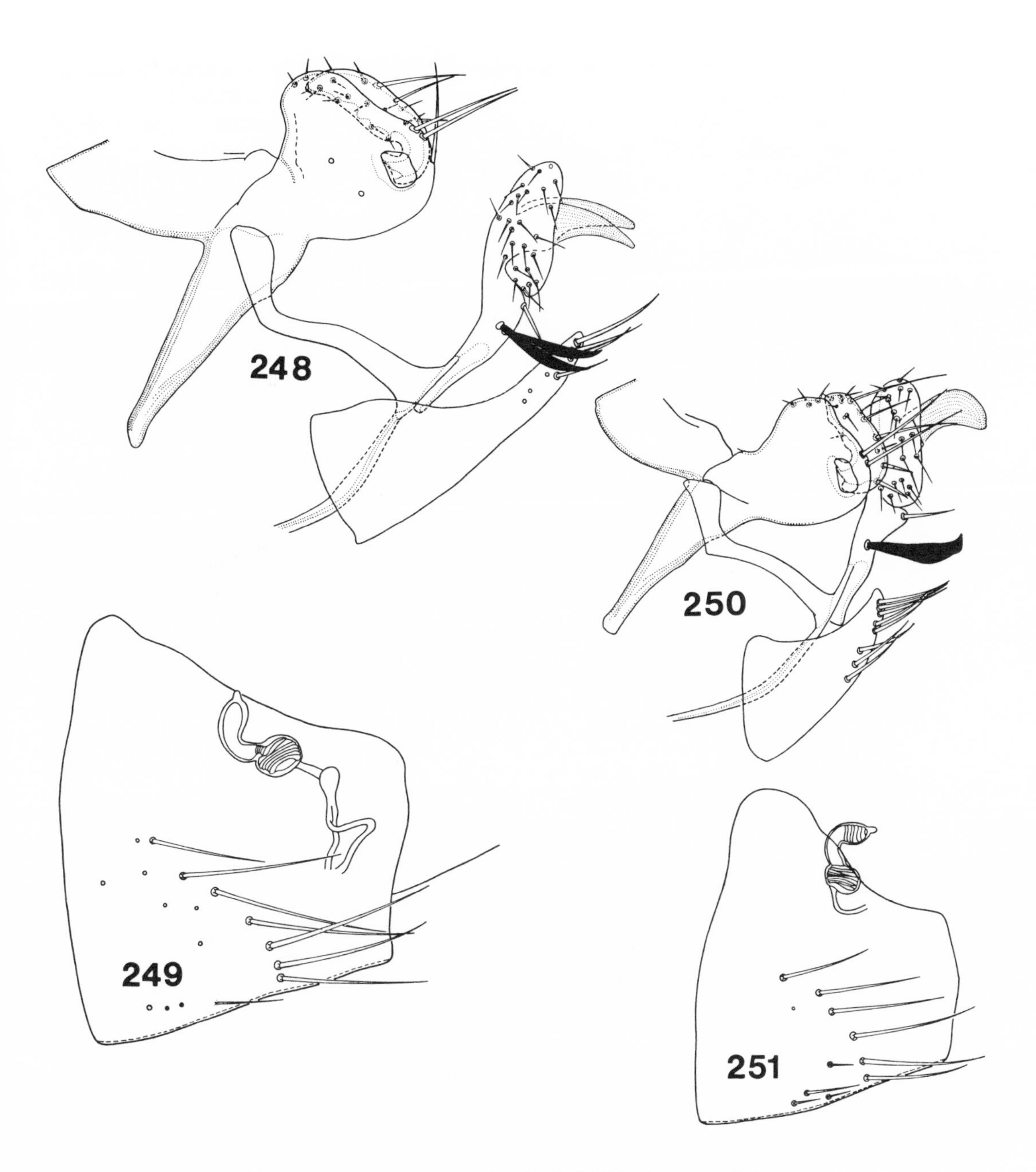

248-249 *Oropsylla (Thrassis) acamantis media* (Stark). **248** Male clasper and sterna VIII and IX. **249** Female sternum VII and spermatheca. Original.
250-251 *Oropsylla (Thrassis) francisi francisi* (C. Fox). **250** Male clasper and sterna VIII and IX. **251** Female sternum VII and spermatheca. Original.

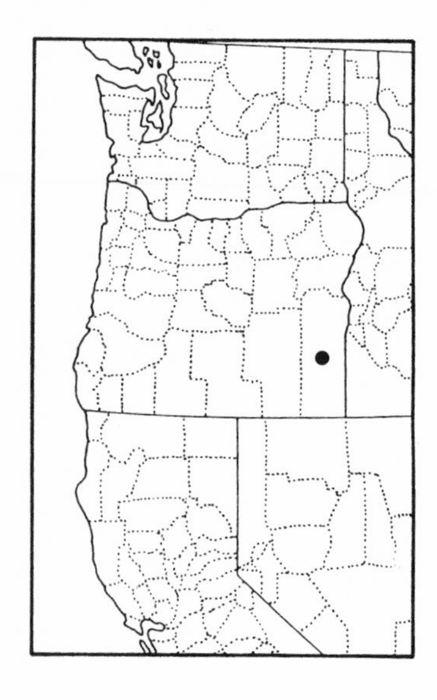

Oropsylla (Thrassis) francisi francisi (C. Fox)

[92707]

(Figs. 250, 251)

Ceratophyllus francisi. C. Fox, 1927, Trans. Amer. ent. Soc. 53: 210-211 (Utah, from *Citellus townsendi molis*).
Thrassis francisi Fox, 1924. Jordan, 1933, Novit. Zool. 39: 73.
Thrassis (Thrassis) francisi francisi (Fox). Stark, 1970, Univ. Calif. Publ. Ent. 53: 73-74.
Oropsylla (Thrassis) f. francisi (Fox, 1927). Smit, in: Traub et al., 1983, p. 28.
Oropsylla (Thrassis) francisi francisi (Fox, 1927). Haddow et al., in: Traub et al., 1983, p. 146.

This subspecies ranges from central Wyoming west through southern Montana, Idaho and extreme southeastern Oregon, south through most of Utah and the eastern half of Nevada. Although it has been reported from a broad range of hosts, in Oregon it is associated with *Spermophilus townsendi,* which isolates it from *O. (T.) f. rockwoodi,* which is found on various subspecies of *S. beldingi.*

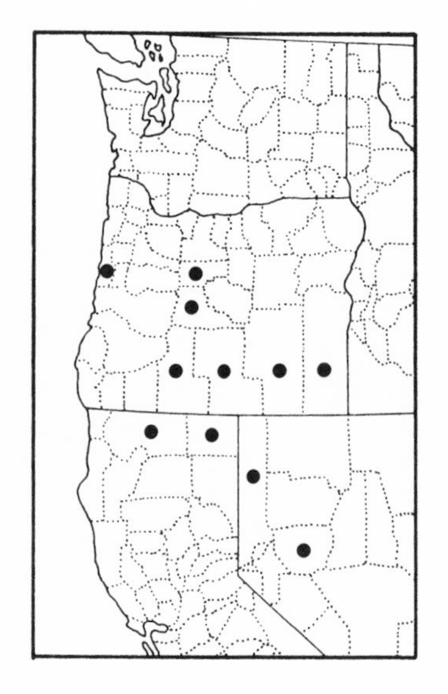

Oropsylla (Thrassis) francisi rockwoodi (Hubbard)

[94210]

(Figs. 252, 253)

Thrassis francisi Fox, 1927. Hubbard, 1940, Pacific Univ. Bull. 37: 2.
Thrassis rockwoodi. Hubbard, 1942, Pacific Univ. Bull. 38: 1-3 (20 mi. S. of Blitzen, Harney Co., Oregon, host unspecified. Later in same paper *Citellus beldingi oregonus, C. mollis mollis* and *C. m. canus* are listed as hosts).
Thrassis (Thrassis) francisi rockwoodi Hubbard (**New Status**). Stark, 1970, Univ. Calif. Publ. Ent. 53: 78-81.
Oropsylla (Thrassis) f. rockwoodi (Hubbard, 1942). Smit, in: Traub et al., 1983, p. 28.
Oropsylla (Thrassis) francisi rockwoodi (Hubbard, 1942). Haddow et al., in: Traub et al., 1983, p. 147.

This subspecies ranges over most of the eastern two-thirds of Oregon, except that small area occupied by the nominate form, extending into northeastern California and northwestern Nevada. Most collections are from *Spermophilus beldingi,* although a few have been taken from *S. townsendi.*

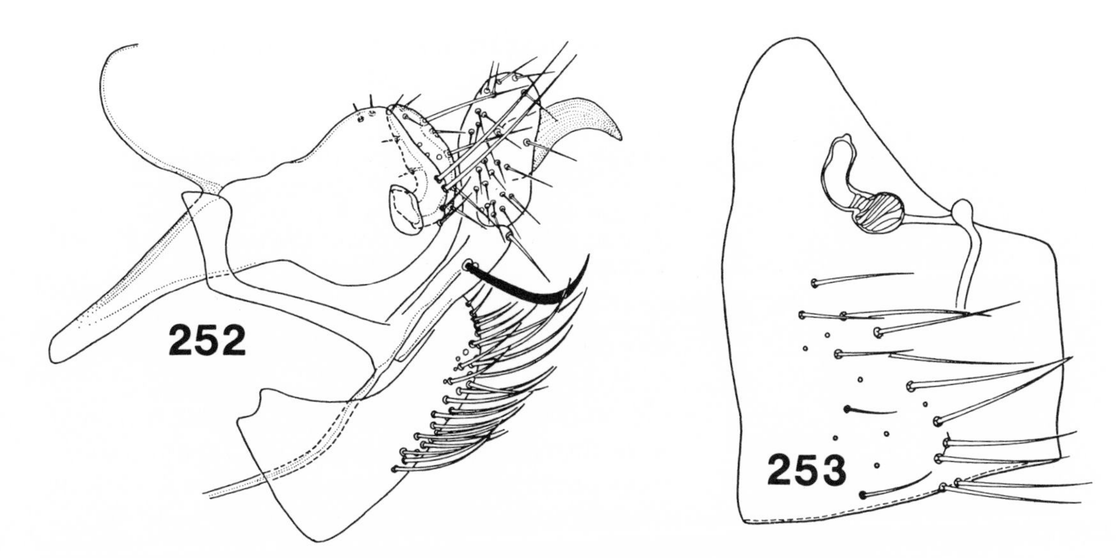

252-253 *Oropsylla (Thrassis) francisi rockwoodi* (Hubbard). **252** Male clasper, sterna VIII and IX and crochet. **253** Female sternum VII and spermatheca. Original.

Oropsylla (Thrassis) aridis hoffmani (Hubbard)

[94928]

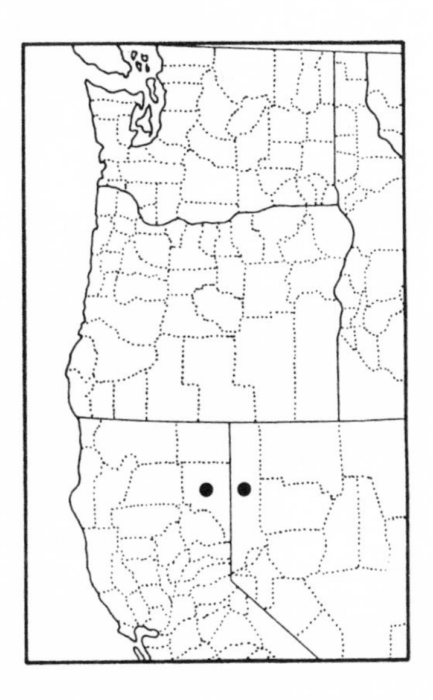

Thrassoides hoffmani. Hubbard, 1949, Bull. S. Calif. Acad. Sci. 48: 51-52 (Beaty, Nye Co., Nevada, from *Dipodomys deserti deserti*).

Thrassis hoffmani (Hubbard, 1949). Tipton and Allred, 1952, Great Basin Nat. 11: 108.

Thrassis aridis hoffmani Hubbard. Stark, 1959, The Siphonaptera of Utah, U.S.D.H.E.W., Publ. Hlth. Ser., CDC, pp. 138-139.

Thrassis aridis hoffmani Hubbard. Stark, 1970, Univ. Calif. Publ. Ent. 33: 144-147.

Oropsylla (Thrassis) aridis hoffmani (Hubbard, 1949). Smit, in: Traub et al., 1983, p. 28.

The normal hosts of this flea are species of *Dipodomys,* and its range extends into the area under consideration only along the California-Nevada border.

GENUS **Pleochaetis** Jordan

Pleochaetis. Jordan, 1933, Novit. Zool. 39: 77 (type species by designation: *P. mundus* J. & R., 1922).
Pleochaetis Jordan, 1933. Smit, in: Traub et al., 1983, p. 33.
Pleochaetis Jordan, 1933. Haddow et al., in: Traub et al., 1983, p. 151.

Following the Smit (1983) revision of the Ceratophyllidae, this genus has been reduced to three species, one of which is listed from eastern Oregon by Haddow et al. (1983). The other two species are restricted to Mexico. The normal hosts are cricetid rodents but all of the species have been reported from a number of other small rodent hosts.

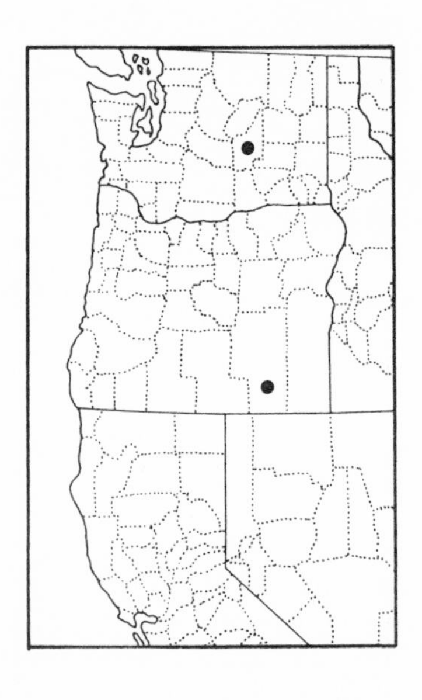

Pleochaetis exilis (Jordan)

[93706]

(Figs. 254, 255)

Megabothris exilis. Jordan, 1937, Novit. Zool. 40: 264 (Powderville, Powder River Co., Montana, from *Onychomys leucogaster*).
Monopsyllus exilis Jordan, 1937. Jordan, 1938, Novit. Zool. 41: 120.
Monopsyllus exilis opadus. Jordan, 1938, *op. cit.*, p. 121 (Yavapai, Arizona, from *Dipodomys ordii* and *Onychomys leucogaster capitulatus*).
Monopsyllus exilis triptus. Jordan, 1938, *op. cit.*, p. 122 (Roggen, Weld Co., Colorado, from *Dipodomys* or *Onychomys*).
Monopsyllus exilis kansensis. Hubbard, 1943, Pacific Univ. Bull. 39: 1 (Meade Co., Kansas, from *Onychomys leucogaster articeps*).
Monopsyllus exilis (Jordan). Johnson, 1961, U. S. Dept. Agric. Tech. Bull. 1227: 13-19 (subspecies synonymized).
Pleochaetis exilis (Jordan, 1937). Smit, in: Traub et al., 1983, p. 33.
Pleochaetis exilis (Jordan, 1937). Haddow et al., in: Traub et al., 1983, p. 151.
Monopsyllus exilis (Jordan). Holland, 1985, Mem. Ent. Soc. Canada 130, p. 335.

This species, recently removed from *Monopsyllus* by Smit (1983), has been collected on a broad range of small rodents, and even a bird, but its preferred host is evidently the grasshopper mouse, *Onychomys leucogaster*. It was not collected during our surveys but is included here based on the Hubbard (1947) record from Harney County, Oregon, plus the reference in Bacon (1953).

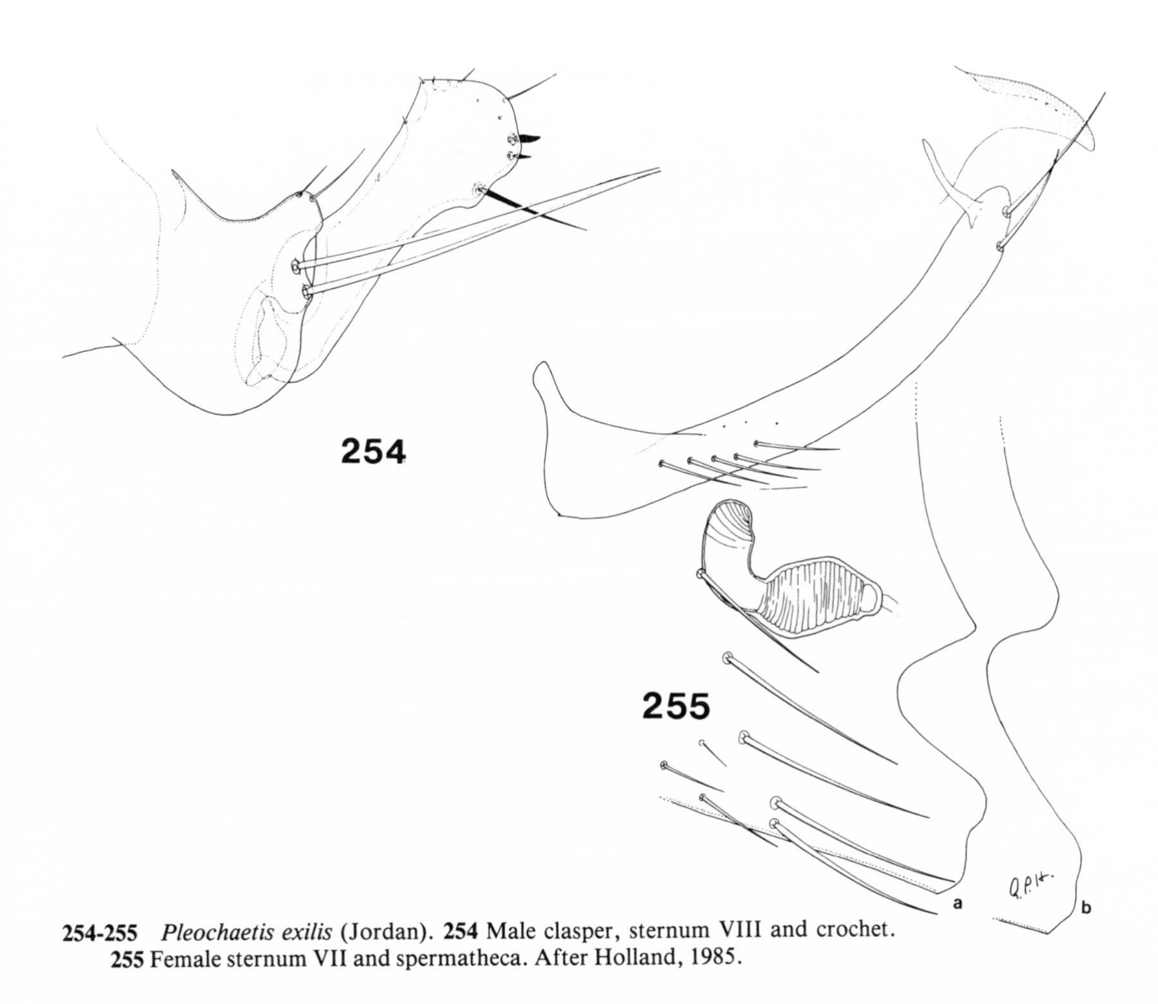

254-255 *Pleochaetis exilis* (Jordan). **254** Male clasper, sternum VIII and crochet. **255** Female sternum VII and spermatheca. After Holland, 1985.

GENUS **Tarsopsylla** Wagner

Tarsopsylla. Wagner, 1927, Konowia 6: 108 (type species: *Ctenonotus octodecimdentatus* Kolenati, 1862).
Tarsopsylla Wagner, 1927. Smit, in: Traub et al., 1983, p. 20.
Tarsopsylla Wagner, 1927. Haddow et al., in: Traub et al., 1983, p. 160.

This is a monotypic Holarctic genus known from one subspecies from Eurasia and another from western North America, south to New Mexico, and east to Ontario, Canada. Both taxa are parasites of tree squirrels and their predators.

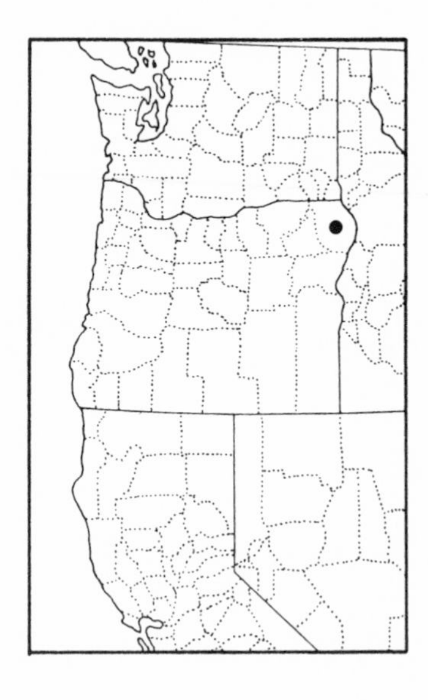

Tarsopsylla octodecimdentata coloradensis (Baker)

[89505]

(Figs. 256, 257)

Pulex coloradensis. Baker, 1895, Canad. Ent. 27: 111 (Georgetown, Colorado, from Fremont's chickaree [*Tamiasciurus hudsonicus fremonti*]).

Ceratophyllus coloradensis (Baker). Baker, 1904, Proc. U. S. Nat. Mus. 27: 388, 417, 441.

Tarsopsylla coloradensis (Baker). Jordan, 1933, Novit. Zool. 39: 72.

Opisodasys jellisoni I. Fox, 1941. Holland, 1949, The Siphonaptera of Canada, p. 43 (states that the male is *T. coloradensis* and the female is *Opisodasys vesperalis.* Redescribes *T. coloradensis*).

T.[arsopsylla] o.[ctodecimdentata] coloradensis (Baker). Smit, 1957, Siphonaptera. Handbk. Ident. Brit. Insects 1(16): 59.

Tarsopsylla octodecimdentata coloradensis (Baker, 1895). Smit, in: Traub et al., 1983, p. 20.

Tarsopsylla octodecimdentata coloradensis (Baker, 1895). Haddow et al., in: Traub et al., 1983, p. 160.

Tarsopsylla octodecimdentata coloradensis (Baker). Holland, 1985, Mem. Ent. Soc. Canada 130, pp. 428-431.

This taxon typically is associated with tree squirrels and their nests, especially *Glaucomys sabrinus* and *Tamiasciurus hudsonicus,* and to a lesser extent their predators and ecological associates. Only a single Oregon record is reported here from Wallowa County on the flying squirrel.

GENUS **Amphalius** Jordan

Amphalius. Jordan, 1933, Novit. Zool. 39: 74 (type species by designation: *C. runatus* Jordan and Rothschild, 1923).

Amphalius Jordan. Holland, 1985, Mem. Ent. Soc. Canada 130, p. 465.

This Holarctic genus currently consists of four species divided into ten subspecies. Nine of these are distributed in Central Asia, south to the Himalayas and east through parts of the People's Republic of China. A single subspecies is known from central and southeastern Alaska, south through the Rocky Mountains to northern New Mexico where it is a parasite of *Ochotona collaris* and *O. princeps.*

Amphalius runatus necopinus (Jordan)

[92533]

(Figs. 258, 259, 260)

Ceratophyllus necopinus. Jordan, 1925, Novit. Zool. 32: 110 (Pine City, Mono Co., California, from *Ochotona [princeps] muiri*).
Amphalius necopinus (Jordan). Jordan, 1933, Novit. Zool. 39: 74.
Amphalius necopinus (Jordan). Hubbard, 1947, Fleas of Western North America, pp. 170-172.
Amphalius runatus necopinus (Jordan). Holland, 1958, Proc. Tenth Int. Congr. Ent. 1: 652-654.
Amphalius runatus necopinus (Jordan, 1925). Haddow et al., in Traub et al., 1983, p. 49.
Amphalius runatus necopinus (Jordan). Holland, 1985, Mem. Ent. Soc. Canada 130, pp. 466-467.

This is another taxon that has yet to be reported from the area. Holland (1958) shows the species from southeastern British Columbia and southwestern Alberta, Canada, and Wyoming, Colorado, Utah, and New Mexico in the United States. All of these records came from *Ochotona princeps,* a pika distributed throughout most of the western states at high elevations in the Rocky Mountains and associated ranges. Hall and Kelson (1958) show four subspecies of this pika occurring in Oregon, one in the Blue Mountains in the northeast, the other three in the Cascade, Warner, and Steens mountains. Hubbard (1947) annotated his illustrations of this species with "Range—Rocky Mtns., Sierra Mtns., but not the Cascade Mtns." Presumably Hubbard had tried to collect this flea in Oregon without success. However, this still leaves the inland Blue Mountain pikas, which seem likely hosts. Additional collecting should be done in this area. As indicated earlier, *Ctenophyllus armatus* might also be expected.

GENUS **Dactylopsylla** Jordan

Dactylopsylla. Jordan, 1929, Novit. Zool. 35: 37-38 (type species by designation: *Odontopsylla bluei* C. Fox, 1909).
Dactylopsylla Jordan, 1929. Jordan, 1933, Novit. Zool. 39: 75.
Dactylopsylla Jordan. Ewing and Fox, 1943, U. S. Dept. Agric. Miscl. Publ. 500: 39-43.
Dactylopsylla Jordan, 1929. Smit, in: Traub et al., 1983, p. 22.
Dactylopsylla Jordan, 1929. Haddow et al., in: Traub et al., 1983, p. 82.

The species of this and the related genera *Spicata* and *Foxella* recently were rearranged by Smit (1983). Under his system, no species assigned to this genus have yet been collected in the area. However, the main hosts for many of the species are species of *Thomomys* that

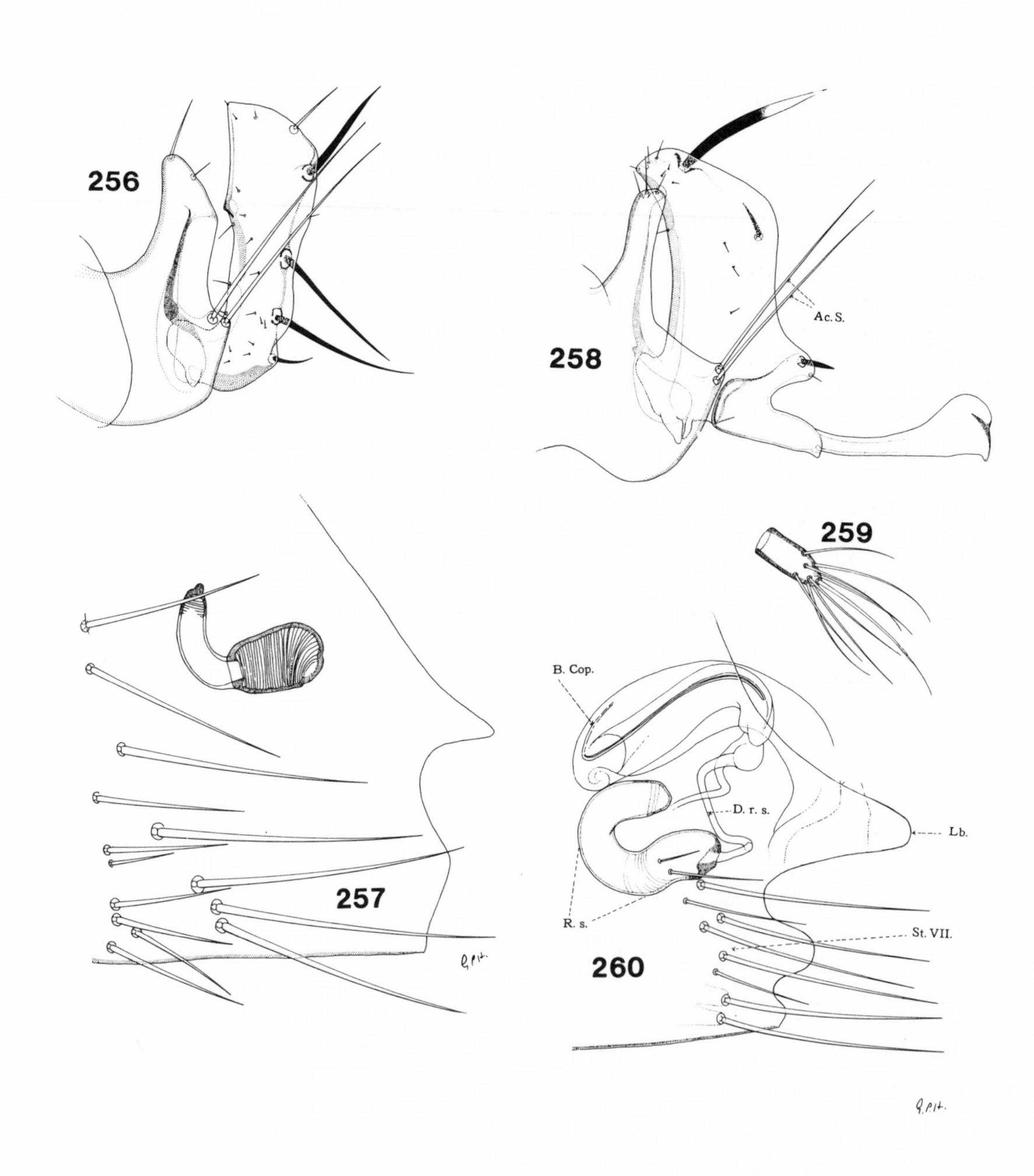

256-257 *Tarsopsylla octodecimdentata coloradensis* (Baker). **256** Male clasper. **257** Female sternum VII and spermatheca. After Holland, 1985.
258-260 *Amphalius runatus necopinus* (Jordan). **258** Male clasper. **259** Female anal stylet. **260** Female sternum VII and spermatheca. After Holland, 1985.

occur over most of the area. Consequently, collectors should be on the lookout for large, hairy fleas when pocket gophers are trapped.

According to some workers, these three genera may be grouped in the subfamily Foxellinae, a strictly Nearctic group which is exclusively associated with gophers as their true hosts. Unfortunately, the status of these fleas is poorly known. While specimens of *Foxella* may be taken in numbers from some hosts, specimens belonging to the other two genera are seldom encountered, and then only one or two specimens may be collected at a time. Thus they tend to be rare in collections. Add to this the fact that most of the descriptions and illustrations are of poor quality and it becomes obvious that the genera are in need of taxonomic attention.

It has been suggested that perhaps species of *Dactylopsylla* and *Spicata* are nest fleas. If this is the case, students of pocket gopher ecology are in an excellent position to contribute to our knowledge of gopher ectoparasites by collecting gopher nests for examination.

As will be seen, keys to the species belonging to these two genera have not been provided. Reasons for this omission are: 1) difficulty of identification, and 2) lack of sufficient material to construct keys. Persons collecting fleas from pocket gophers should seek the assistance of a specialist for identifications.

GENUS **Foxella (Foxella)** Wagner

Foxella. Wagner, 1929, Konowia 8: 314 (type species by designation: *Pulex ignotus* Baker, 1895).
Foxella Wagner. Jordan, 1933, Novit. Zool. 39: 75.
Foxella Wagner. Ewing and Fox, 1943, U. S. Dept. Agric. Miscl. Publ. 500: 41 (as a subgenus of *Dactylopsylla*).
Foxella (Foxella) Wagner, 1929. Smit, in: Traub et al., 1983, p. 23.
Foxella (Foxella) Wagner, 1929. Haddow et al., in: Traub et al., 1983, p. 90.
Foxella Wagner. Holland, 1985, Mem. Ent. Soc. Canada 130, pp. 467-470.

This is a genus and subgenus of gopher fleas, currently known from a single species which has been split into eleven subspecies. The nominate taxon occurs from Indiana west to Nebraska and Kansas on *Geomys bursarius.* All other subspecies are reported from *Thomomys* species. Three of these are known from the area and a fourth might be expected in the eastern portion. The only common subspecies is *F. (F.) i. recula* (J. & R., 1915).

As often happens with species which have a large range embracing considerable ecological diversity, taxonomists have tended to describe a number of subspecies to account for the variation observed in the various local populations. While this has not been taken to the excess in *Foxella,* as it has been with their hosts, there are certainly more named subspecies in the genus than there are in nature. The subspecies erected by Hubbard, Prince, and Augustson are essentially unrecognizable, having been poorly described and even more poorly illustrated. The group is in dire need of the taxonomic revision that is underway.

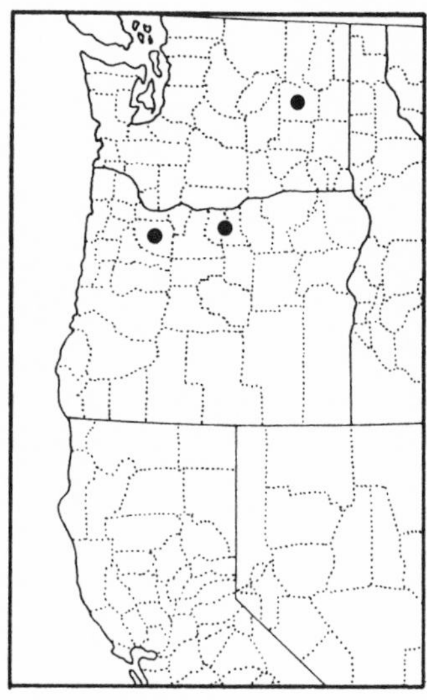

Foxella (Foxella) ignota clantoni Hubbard

[94925]

Foxella ignota clantoni. Hubbard, 1949, Bull. S. Calif. Acad. Sci. 48: 48 (4 mi. E. of Odessa, Lincoln Co., Washington, from *Rattus norvegicus*).

Foxella (Foxella) ignota clantoni Hubbard, 1949. Smit, in: Traub et al., 1983, p. 23.

Foxella (Foxella) ignota clantoni Hubbard, 1949. Haddow et al., in: Traub et al., 1983, p. 91.

Although this subspecies was not collected during the survey, Haddow et al. (1983) list Oregon as part of its range and show it from the north-central part of the state on map 57. No additional information is provided. We also show a record from east-central Washington.

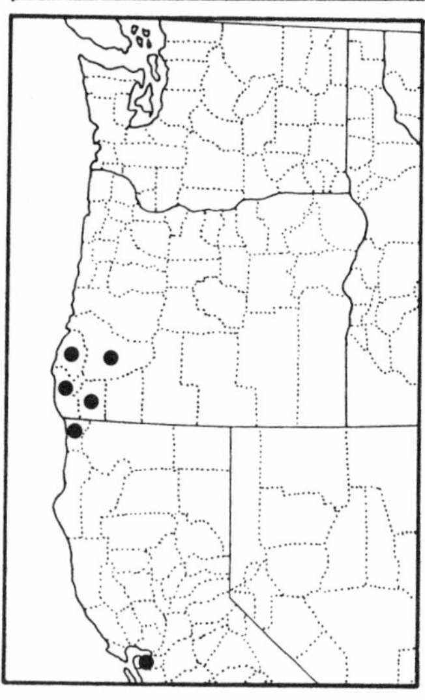

Foxella (Foxella) ignota franciscana (Rothschild)

[91002]

(Figs. 261, 262)

Ceratophyllus franciscanus. Rothschild, 1910, Ent. mon. Mag. 21: 88 (San Francisco, California, from *Thomomys bottae*).

Ceratophyllus ignotus franciscanus Roths., (1910). Jordan and Rothschild, 1915, Ectoparasites I: 58.

Foxella franciscanus Roth. Wagner, 1929, Konowia 8: 314-315.

Foxella ignota franciscana Rothschild, 1910. Hubbard, 1947, Fleas of Western North America, pp. 173-175.

Foxella ignotus acutus. Stewart, 1940, Pan-Pacific Ent. 16: 20-21 (Hastings Natural History Reservation, Monterey Co., California, from *Thomomys bottae* ssp.).

Foxella ignota (Baker). Linsdale and Davis, 1956, Univ. Calif. Publ. Zool. 54: 34 (*acutus* synonymized).

Foxella ignota franciscana Rothschild, 1910. Smit and Wright, 1978b, A catalogue of primary type-specimens..., p. 20 (selection of lectotype male).

Foxella (Foxella) i. franciscana (Rothschild, 1910). Smit, in: Traub et al., 1983, p. 23.

Foxella (Foxella) ignota franciscana (Rothschild, 1910). Haddow et al., in: Traub et al., 1983, p. 91.

This subspecies was not taken during our Oregon surveys; however, Haddow et al. (1983) include southwestern Oregon and northwestern California in its range. A few female specimens of the following species collected in the southwestern counties show a marginal lobe on st. VII characteristic of some individuals of this subspecies, but the character is variable and no males from these collections show any sign of intergradation with this subspecies.

Foxella (Foxella) ignota recula (Jordan and Rothschild)

[91535]

(Figs. 263, 264)

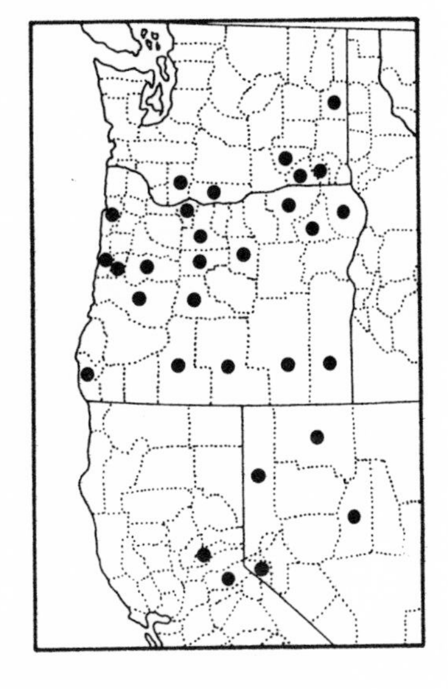

Ceratophyllus ignotus recula. Jordan and Rothschild, 1915, Ectoparasites I: 58 (Okanagan Landing, British Columbia, from *Mustela frenata* ssp.).
Foxella ignotus recula (Jordan and Rothschild). Wagner, 1936, Canad. Ent. 68: 198.
Dactylopsylla (Foxella) ignota recula (Jordan and Rothschild). Ewing and Fox, 1943, U. S. Dept. Agric. Miscl. Publ. 500: 42.
Foxella ignota recula (Jordan and Rothschild). Holland, 1949, The Siphonaptera of Canada, p. 126.
Foxella (Foxella) i. recula (Jordan and Rothschild, 1915). Smit, in: Traub et al., 1983, p. 23.
Foxella (Foxella) ignota recula (Jordan and Rothschild, 1915). Haddow et al., in: Traub et al., 1983, p. 91.
Foxella ignota recula (Jordan and Rothschild). Holland, 1985, Mem. Ent. Soc. Canada 130, pp. 470-471.

This was the only subspecies of *Foxella ignota* collected during our Oregon surveys. Three hundred and twenty-seven specimens (146 males and 181 females) were collected from various localities. All were taken on *Thomomys talpoides* and *T. mazama* except seven males and twelve females which came from an assortment of accidental hosts, including predator records. Seasonal distribution was as follows.

Seasonal distribution of *Foxella (F.) i. recula* in Oregon

	J	F	M	A	M	J	J	A	S	O	N	D
Males	1	0	3	0	8	15	63	8	9	39	0	0
Females	0	0	5	0	7	14	82	14	23	36	0	0

Absence during certain months is likely an indication of collecting activity as much as seasonal variation, although numbers certainly increase during the summer and early fall.

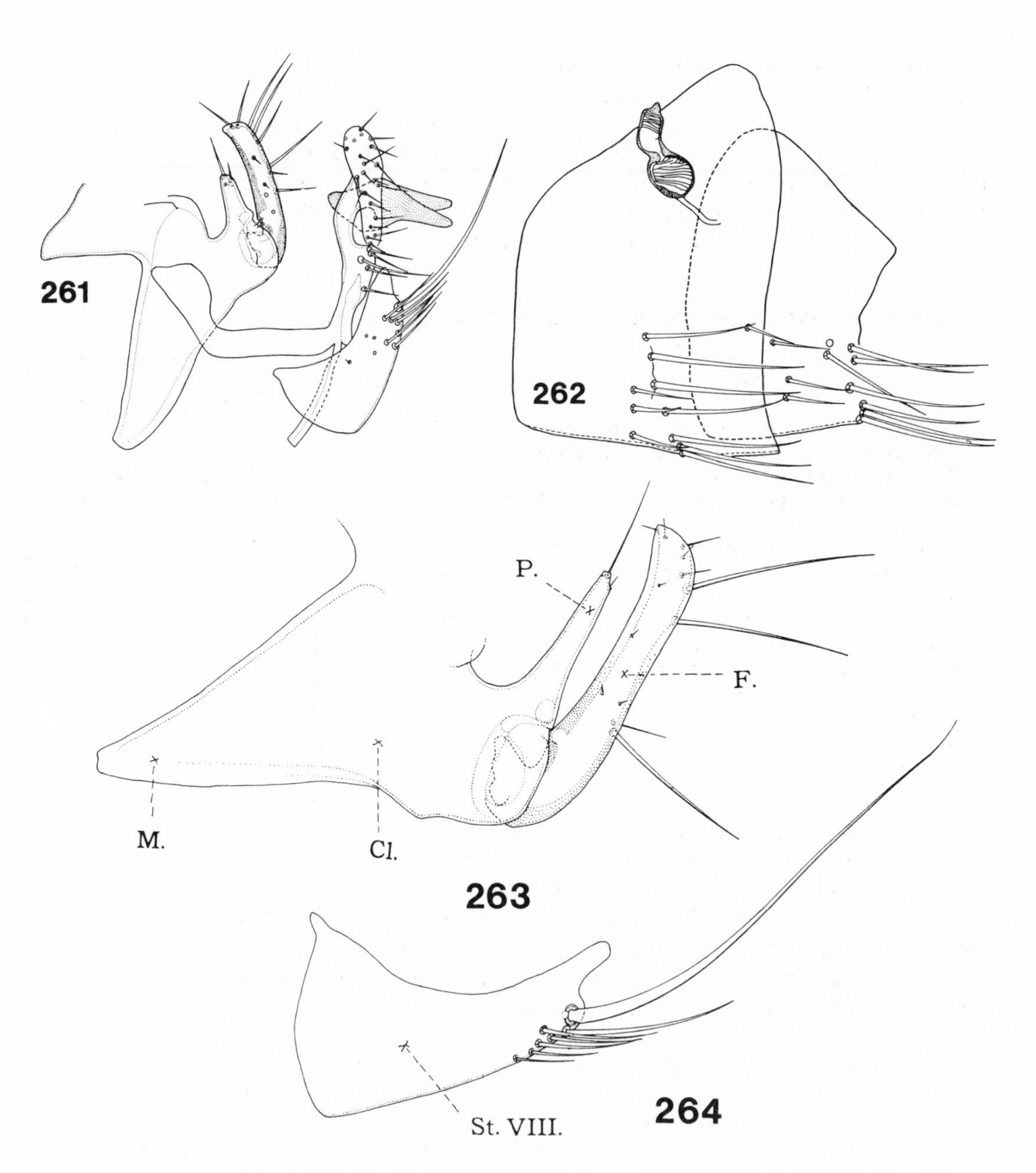

261-262 *Foxella (Foxella) ignota franciscana* (Rothschild). **261** Male clasper, sterna VIII and IX and crochet. **262** Female sterna VII and VIII and spermatheca. Original.

263-264 *Foxella (Foxella) ignota recula* (J. & R.). **263** Male clasper. **264** Male sternum VIII. After Holland, 1985.

Foxella (Foxella) ignota utahensis Wagner

[93633]

(Figs. 265, 266)

Foxella ignotus utahensis. Wagner, 1936, Z. Parasitenk. 8: 655 (Wellswille [Wellsville], Utah, *Thomomys* sp. (probably *T. uinta*).
Foxella (Foxella) ignota utahensis Wagner, 1936. Smit, in: Traub et al., 1983, p. 23.
Foxella (Foxella) ignota utahensis Wagner, 1936. Haddow et al., in: Traub et al., 1983, p. 91.

This subspecies has not been reported from the area but might be expected to occur in the extreme southeastern portion. It is immediately separable from the other subspecies of *Foxella ignota* in having three antepygidial bristles in the male and four in the female. The normal numbers in the other subspecies are one (or two) in the male and three in the female. This taxon is now known from Idaho, Montana, Wyoming, Colorado, and Utah.

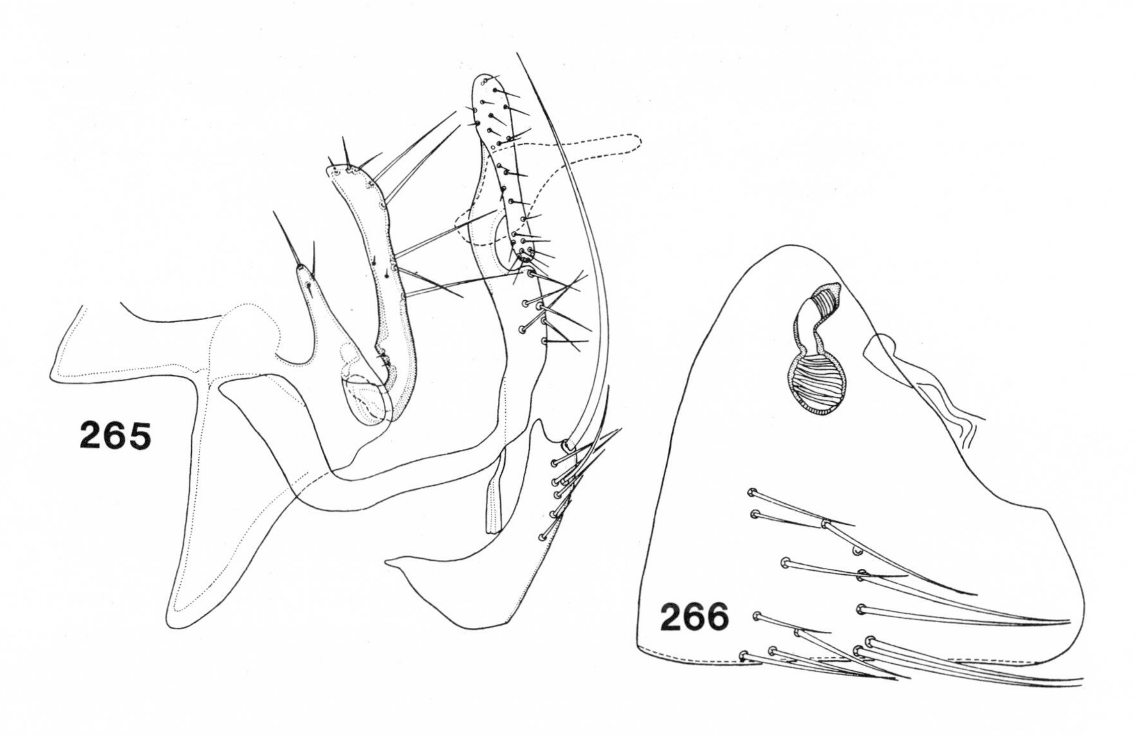

265-266 *Foxella (Foxella) ignota utahensis* Wagner. **265** Male clasper, sterna VIII and IX and crochet. **266** Female sternum VII and spermatheca. Original.

GENUS **Spicata** I. Fox

Spicata n. subgen. I. Fox, 1940, J. Wash. Acad. Sci. 30: 275 (type species by monotypy: *D. rara* I. Fox, 1940).
Spicata Fox, 1940. Smit, in: Traub et al., 1983, pp. 22-23 (elevated to full generic status).
Spicata I. Fox, 1940. Haddow et al., in: Traub et al., 1983, p. 157.

Until recently this generic name has been treated as a junior synonym of *Foxella* or *Dactylopsylla,* depending upon how the foxelline genera were treated. It is missing from the lists of genera in the Hopkins and Rothschild catalogues (1953, 1956, 1962, 1966) until volume V (1971) in which it appears as a genus or subgenus in the Foxellinae. Its elevation to full generic status was intimated in Smit and Wright (1978b) and formally published in Smit (1983).

There are eight currently recognized species in this genus, all limited to the western Nearctic Region where they parasitize species of *Thomomys.* The status of these taxa is as uncertain as those assigned to *Foxella ignota* and the group is in serious need of taxonomic revision. Females unaccompanied by males are impossible to identify with any degree of certainty and so little material is available that it has not been practical to attempt to write a key to the nine taxa said to occur in the area. Only *S. c. comis* was taken with certainty during the surveys, and then in very limited numbers.

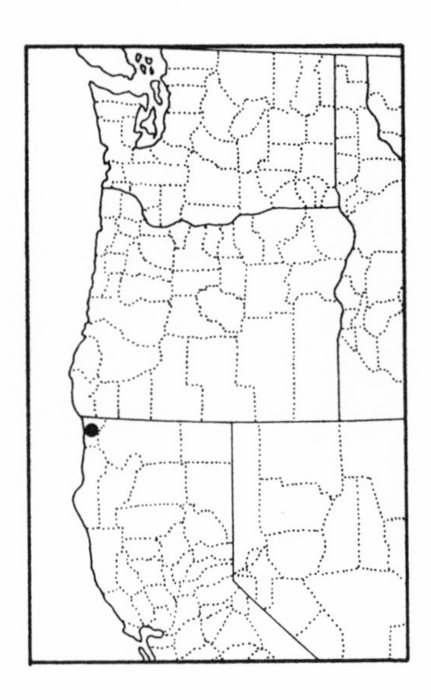

Spicata bottaceps (Hubbard)

[94308]

Dactylopsylla (Foxelloides) bottaceps. Hubbard, 1943, Pacific Univ. Bull. 40: 5 (Fort Dick, Del Norte Co., California, from *Thomomys bottae laticeps*).
Spicata bottaceps (Hubbard, 1943). Smit, in: Traub et al., 1983, p. 22.

As far as is known, this species has not been collected subsequent to the two females in the type series. The holotype was deposited in the U. S. National Museum, Washington, D. C. and the only paratype was sent to the British Museum (Natural History) collection, originally at Tring, Herts., and now in London.

Spicata comis comis (Jordan)

[92918]

(Figs. 267, 268)

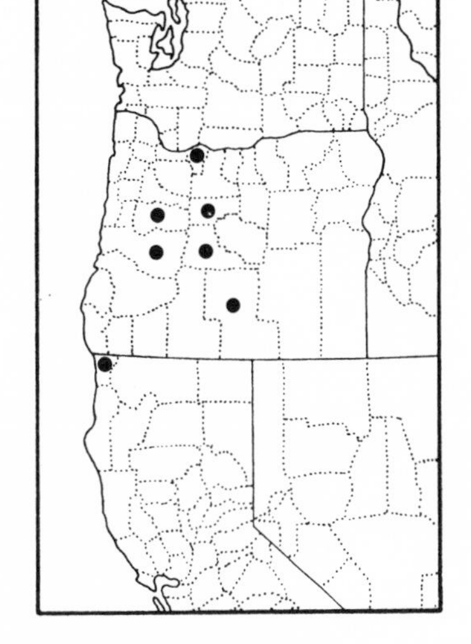

Dactylopsylla comis. Jordan, 1929, Novit. Zool. 35: 38 (Okanagan Landing, British Columbia, from *Thomomys [talpoides] fuscus*).
Dactylopsylla (Foxelloides) comis Jordan, 1929. Hubbard, 1943, Pacific Univ. Bull. 40: 3-4 (from the Siskiyou and Cascade Mountains of Oregon. Male described and figured).
Dactylopsylla comis Jordan, 1929. Hubbard, 1947, Fleas of Western North America, pp. 186-188.
Dactylopsylla comis Jordan. Holland, 1949, The Siphonaptera of Canada, p. 125.
Spicata comis comis (Jordan, 1929). Smit, in: Traub et al., 1983, p. 22.
Spicata comis comis (Jordan, 1929). Haddow et al., in: Traub et al., 1983, p. 157.
Dactylopsylla comis Jordan. Holland, 1985, Mem. Ent. Soc. Canada 130, pp. 476-477.

This would seem to be the commonest member of this genus in the area under consideration, although only two males and five females were taken during all of the surveys. Holland (1985) comments that although the type locality is in British Columbia, the species has not been collected in Canada subsequently, in spite of extensive collections of pocket gophers from the Okanagan Valley. These are large, dark fleas and might be mistaken for one of the smaller species of *Hystrichopsylla* except that they lack a genal ctenidium.

Spicata comis scapoosei (Hubbard)

[95414]

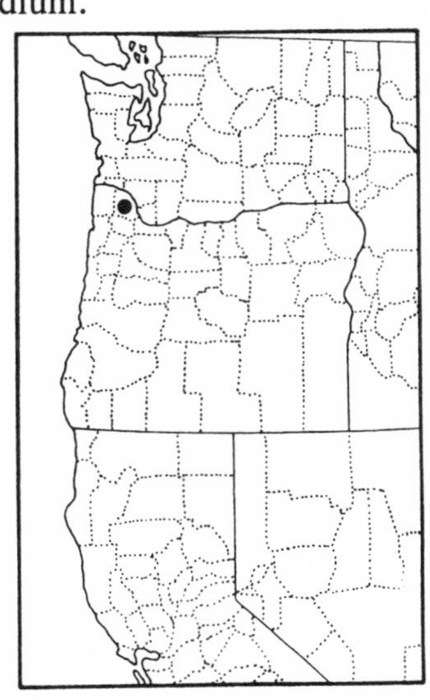

Dactylopsylla comis scapoosei. Hubbard, 1954, Ent. News 65: 169 (Scapoose, Columbia County, Oregon, from *Thomomys d. douglasi*).
Spicata c. scapoosei (Hubbard, 1954). Smit, in: Traub et al., 1983, p. 22.
Spicata comis scapoosei (Hubbard, 1954). Haddow et al., in: Traub et al., 1983, p. 158.

Except for the six specimens listed in Haddow et al. (1983), we know of no additional material beyond that on which the original description was based. The content of the type series was not stated.

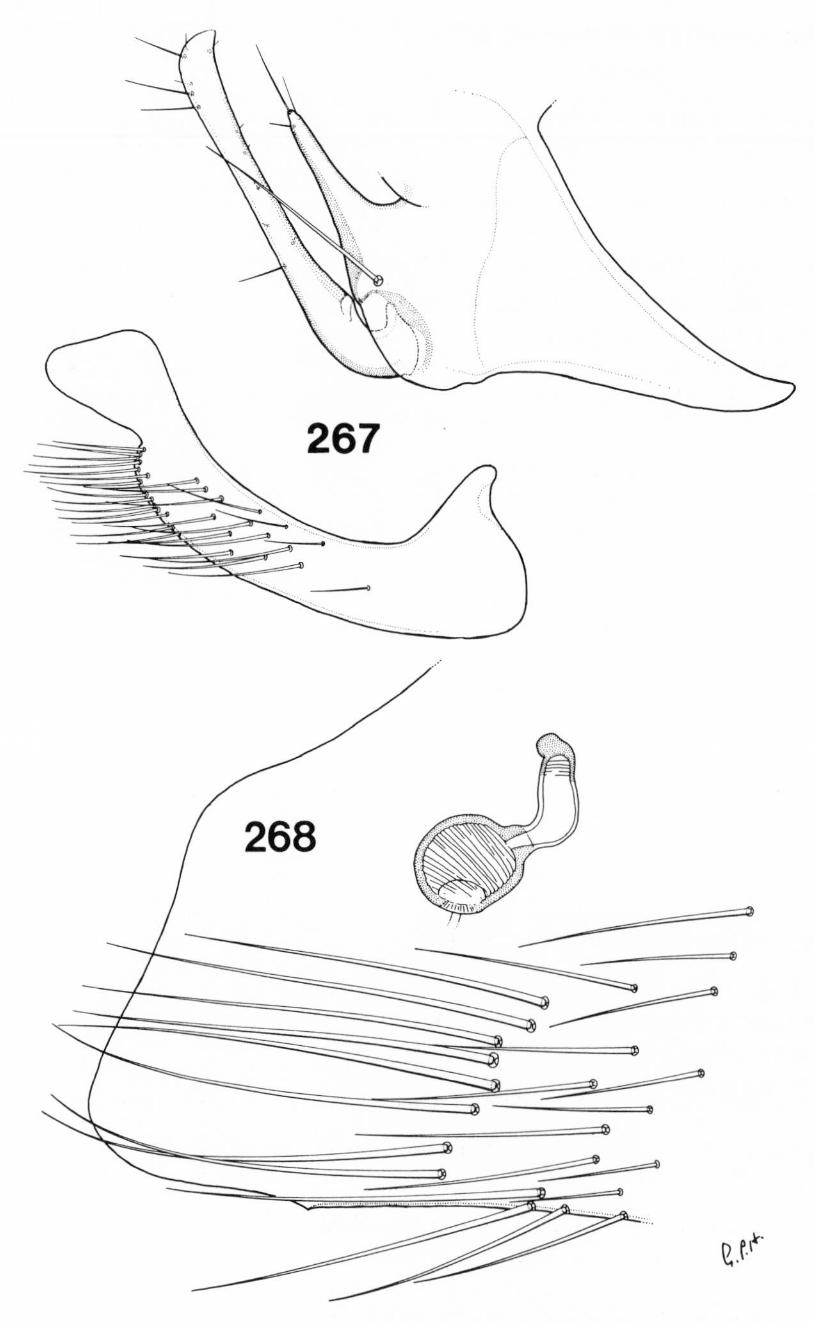

267-268 *Spicata comis comis* (Jordan). **267** Male clasper and sternum VIII. **268** Female sternum VII and spermatheca. After Holland, 1985.

Spicata comis tacomae (Hubbard)

[95415]

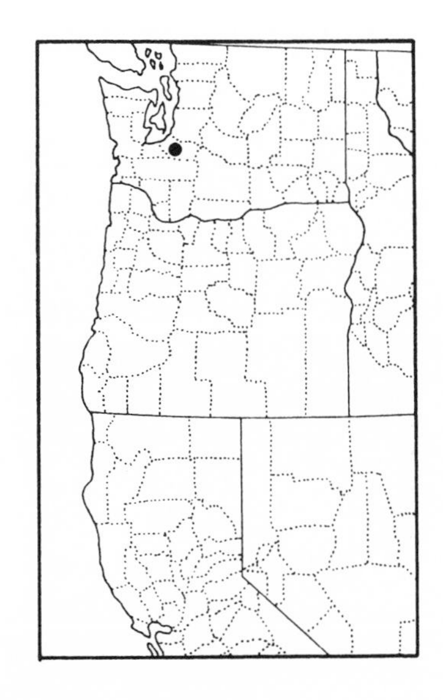

Dactylopsylla comis tacomae. Hubbard, 1954, Ent. News 65: 170 (Tacoma, Washington, from "Tacoma pocket gopher").

Spicate c. tacomae (Hubbard, 1954). Smit, in: Traub et al., 1983, p. 22.

As far as we know, this taxon is known only from the type series, and has not been subsequently collected or reported in the literature.

Spicata comis walkeri (Hubbard)

[95416]

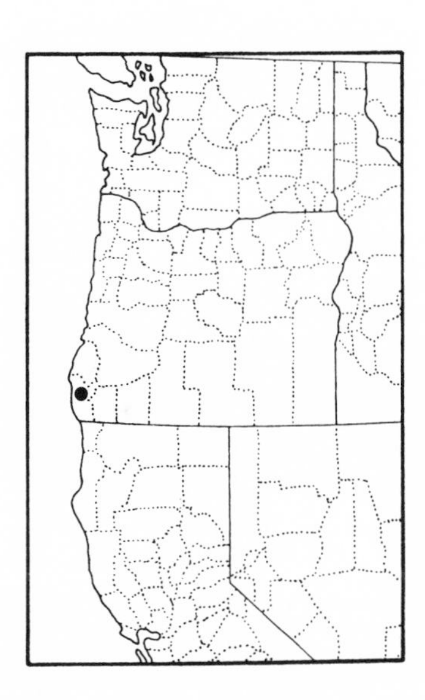

Dactylopsylla comis walkeri. Hubbard, 1954, Ent. News 65: 170 (Wedderburn, Curry Co., Oregon, from *Thomomys m.[onticola] helleri*).

Spicata c. walkeri (Hubbard, 1954). Smit, in: Traub et al., 1983, p. 22.

Spicata comis walkeri (Hubbard, 1954). Haddow et al., in: Traub et al., 1983, p. 158.

Beyond the three specimens listed in Haddow et al. (1983), we know of no additional specimens except those belonging to the type series. The content of the type series was not stated.

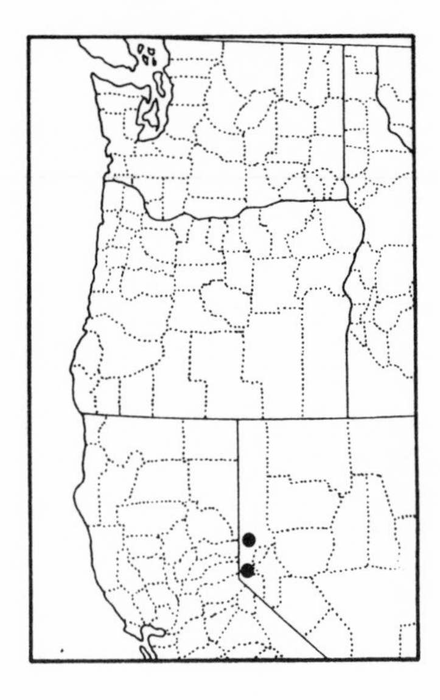

Spicata monticola (Prince)

[94503]

Dactylopsylla monticola. Prince, 1945, Canad. Ent. 77: 17 (13 mi. W. of Carson City, Nevada, from *Thomomys monticola*).
Spicata monticola (Prince, 1945). Smit, in: Traub et al., 1983, p. 22.

This is another species that does not appear to have been reported subsequent to the type series. However, Haddow et al., (1983) list two mounted specimens in the British Museum (Natural History) collection. Prince (1945) states that the holotype female was deposited in the U. S. Public Health Service Plague Investigation Station, San Francisco, California, and the single female paratype was deposited in the U. S. National Museum, Washington, D. C. The former collection now resides in the Division of Vector-Borne Viral Diseases, Centers for Disease Control, Fort Collins, Colorado, and contains a single female of this species from the type locality and bearing the word HOLOTYPE in Prince's printing. The provenance of the BM(NH) specimen is unknown to us.

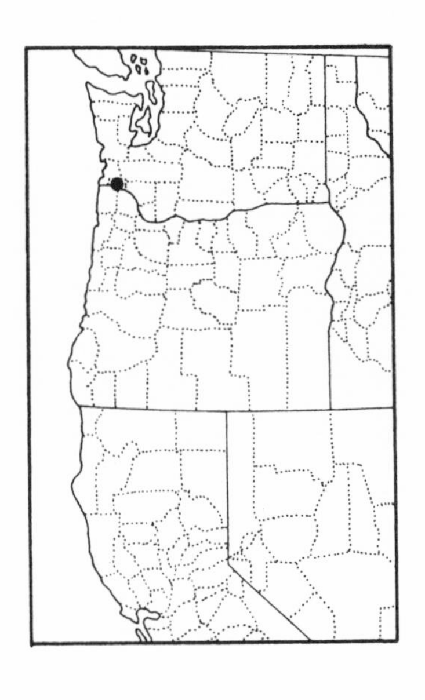

Spicata moorei moorei (Hubbard)

[94924]

Dactylopsylla moorei. Hubbard, 1949, Bull. S. Calif. Acad. Sci. 48: 47-48 (12 mi. N.E. of Cathlamet, Wahkiakum Co., Washington, on *Thomomys talpoides*).
Spicata moorei moorei (Hubbard, 1949). Smit, in: Traub et al., 1983, p. 23.

Although we know of no additional records, Haddow et al. (1983) list four mounted specimens in the British Museum (Natural History) collection. Only one pair of paratypes was said to have been sent to this collection by the author of the species and the origin of the other two specimens is unknown to us. Map 147 in Haddow et al. (1983) shows a symbol for the nominate subspecies within the boundaries of Oregon, but we find no confirmation for this record in the literature, and Oregon is not mentioned under distribution.

Spicata moorei oregona (Hubbard)

[95417]

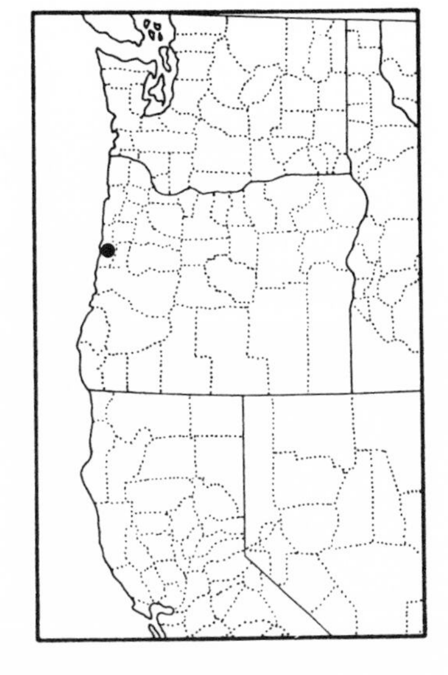

Dactylopsylla moorei oregona. Hubbard, 1954, Ent. News 65: 170 (Devil's Lake, Lincoln Co., Oregon, from *Thomomys [monticola] hesperus*).

Spicata m. oregona (Hubbard, 1954). Smit, in: Traub et al., 1983, p. 23.

Spicata moorei oregona (Hubbard, 1954). Haddow et al., in: Traub et al., 1983, p. 158.

With respect to this taxon, Haddow et al. (1983) mention four mounted specimens in the collection of the British Museum (Natural History). To our knowledge, these and the type series are the only specimens extant. The contents of the type series were not stated by the author.

Spicata pacifica (Hubbard)

[94307]

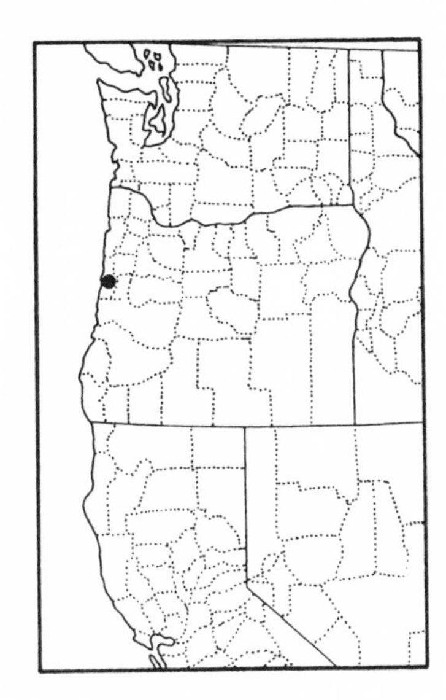

Dactylopsylla (Foxelloides) pacifica. Hubbard, 1943, Pacific Univ. Bull. 40: 4-5 (Devil's Lake, Lincoln Co., Oregon, from *Thomomys [monticola] hesperus*).

Spicata pacifica (Hubbard, 1943). Smit, in: Traub et al., 1983, p. 23.

Spicata pacifica (Hubbard, 1943). Haddow et al., in: Traub et al., 1983, p. 159.

As far as can be determined, this species is known only from the type series of two males and one female. Hubbard (1947) indicated that the holotype and allotype were deposited in the United States National Museum. The remaining male paratype was sent to Dr. Karl Jordan at Tring (U.K.). This is evidently the single mounted specimen listed by Haddow et al. (1983).

Literature Cited

American Ornithologists' Union. 1983. *Check-list of North American Birds,* 6th edition, pp. 1-877.

Amrine, J. W., Jr. and M. A. Jerabek. 1983. Possible ultrasonic receptors on fleas. *Ann. Ent. Soc. Amer.* 76: 395-399.

Amrine, J. W., Jr. and R. E. Lewis. 1978. The topography of the exoskeleton of *Cediopsylla simplex* (Baker, 1895) (Siphonaptera: Pulicidae). I. The head and its appendages. *J. Parasitol.* 64: 343-358.

Amrine, J. W., Jr. and R. E. Lewis. 1986. The topography of the exoskeleton of *Cediopsylla simplex* (Baker, 1895) (Siphonaptera: Pulicidae): The thorax, abdomen and associated appendages. *J. Parasitol.* 72: 71-87.

Anonymous. 1950a. Plague infection in Grant County, Washington. *Publ. Hlth. Rep., Wash.* 65: 575.

Anonymous. 1950b. Plague infection in the State of Washington. *Publ. Hlth. Rep., Wash.* 65: 614.

Anonymous. 1950c. Plague infection in Kittatas County, Washington. *Publ. Hlth. Rep., Wash.* 65: 901.

Bacon, M. 1953. A study of the arthropods of medical and veterinary importance in the Columbia River basin. Wash. Agric. Exp. Sta. Tech. Bull. 11: 1-40.

Bacon, M. and R. F. Bacon. 1964. Notes on the distribution of Siphonaptera on wild rabbits of eastern and central Washington. *Northwest Sci.* 38: 25-32.

Bacot, A. W. and C. J. Martin. 1914. LXVII. Observations on the mechanism of the transmission of plague by fleas. *J. Hyg.,* Plague Suppl. 3: 423-439.

Baker, C. F. 1904. A revision of the American Siphonaptera, or fleas, together with a complete list and bibliography of the group. *Proc. U. S. Nat. Mus.* 27: 365-469.

Barbour, R. W. and W. H. Davis. 1969. *Bats of America.* The University Press of Kentucky, Lexington. pp. 1-286, 24 color plates, 131 figs.

Barnes, A. M. and L. Kartman. 1960. Control of plague vectors on diurnal rodents in the Sierra Nevada of California by use of insecticide bait-boxes. *J. Hyg.,* Camb. 58: 347-355.

Barnes, A. M., V. J. Tipton and J. A. Wildie. 1977. The subfamily Anomiopsyllinae (Hystrichopsyllidae: Siphonaptera). I. A revision of the genus *Anomiopsyllus* Baker. *Great Basin Nat.* 37: 138-206.

Benton, A. H. 1955. The taxonomy and biology of *Epitedia wenmanni* (Rothschild, 1904) and *E. testor* (Rothschild, 1915) (Hystrichopsyllidae: Siphonaptera). *J. Parasitol.* 41: 491-495.

Benton, A. H. 1980. An atlas of the fleas of the eastern United States. Marginal Media, P. O. Box 241, Fredonia, NY, pp. 1-177 (Mimeographed).

Browning, B. M. and E. M. Lauppe. 1964. A deer study on Redwood-Douglas fir forest type. *Calif. Fish & Game* 50: 132-147.

Bruce, W. N. 1948. Studies on the biological requirements of the cat flea. *Ann. Ent. Soc. Amer.* 41: 346-352.

Campos, E. G. and H. E. Stark. 1979. A revaluation of the *Hystrichopsylla occidentalis* group, with description of a new subspecies (Siphonaptera: Hystrichopsyllidae). *J. Med. Entomol.* 15: 431-444.

Clifford, C. M., C. E. Yunker, E. R. Easton and J. E. Keirans. 1970. Ectoparasites and other arthropods from coastal Oregon. *J. Med. Entomol.* 7: 438-445.

Corbet, G. B. and J. E. Hill. 1980. *A world list of mammalian species.* Brit. Mus. (Nat. Hist.) & Comstock Publ. Assoc., London & Ithaca, pp. 1-226.

Costa Lima, A. da and C. R. Hathaway. 1946. Pulgas: Bibliografia, catálogo e animais por elas sugados. Monogr. Inst. Oswaldo Cruz No. 4, pp. 1-522.

Dunnet, G. M. and D. K. Mardon. 1974. A monograph of Australian fleas (Siphonaptera). Aust. J. Zool., Suppl. Ser. No. 30, pp. 1-273.

Easton, E. R. 1983. Ectoparasites in two diverse habitats in western Oregon. III. Interrelationship of fleas (Siphonaptera) and their hosts. *J. Med. Entomol.* 20: 216-219.

Egoscue, H. J. 1968. A new species of the genus *Stenistomera* (Siphonaptera: Hystrichopsyllidae). *Bull. So. Calif. Acad. Sci.* 67: 138-142.

Egoscue, H. J. 1980. Wood rat nest flea *Anomiopsyllus amphibolus* in southeastern Oregon. *Great Basin Nat.* 40: 361.

Eskey, C. R. and V. H. Haas. 1940. Plague in the western part of the United States. Part II. Flea investigations. Observations on the ecology of fleas. U. S. Public Health Bull. No. 254, pp. 29-74.

Franklin, J. F. and C. T. Dyrness. 1988. *Natural vegetation of Oregon and Washington.* Oregon State University Press, Corvallis, pp. 1-464.

Gresbrink, R. A. and D. D. Hopkins. 1982. Siphonaptera: Host records from Crater Lake National Park, Oregon. *Northwest Sci.* 56: 176-179.

Haas, G. E. 1982. Siphonaptera: The William M. Wallace collection of fleas from Oregon and Washington. *Wasmann J. Biol.* 40: 45-47.

Haddow, J., R. Traub and M. Rothschild. 1983. Distribution of ceratophyllid fleas and notes on their hosts. Material in the collection, with additional notes and maps of the genera. *In:* Traub, R., et al., 1983, pp. 42-163, 151 maps.

Hall, E. R. and K. R. Kelson. 1959. *The mammals of North America.* The Ronald Press. Vol. I, i-xxx, 1-546, plus 79 pp. index; Vol. II, i-viii, 547-1083, plus 79 pp index.

Hansen, C. G. 1964. Ectoparasites of mammals from Oregon. *Great Basin Nat.* 24: 75-81.

Holland, G. P. 1949. *The Siphonaptera of Canada.* Canada Dept. Agric. Tech. Bull. 70, pp. 1-306.

Holland, G. P. 1957. Notes on the genus *Hystrichopsylla* Rothschild in the New World, with descriptions of one new species and two new subspecies (Siphonaptera: Hystrichopsyllidae). *Canad. Entomol.* 89: 309-324.

Holland, G. P. 1958. Distribution patterns of northern fleas (Siphonaptera). Proc. Tenth Int. Congr. Entomol. 1: 645-658.

Holland, G. P. 1985. *The fleas of Canada, Alaska and Greenland (Siphonaptera).* Mem. Ent. Soc. Canada No. 130, pp. 1-630.

Holland, G. P. and E. W. Jameson, Jr. 1950. Notes on some Nearctopsyllinae, with descriptions of two new species of *Nearctopsylla* from California (Siphonaptera). *Canad. Entomol.* 81: 249-253.

Honaki, J. H., K. E. Kinman and J. W. Koeppl (Eds.). 1982. *Mammal species of the world.* Allen Press, Inc. and Association of Systematic Collections, Lawrence, Kansas, pp. i-ix, 1-694.

Hopkins, D. 1980. Ectoparasites of the Virginia opossum (*Didelphis virginiana*) in an urban environment. *Northwest Sci.* 54: 199-201.

Hopkins, D. 1985. Host associations of Siphonaptera from Central Oregon. *Northwest Sci.* 59: 108-114.

Hopkins, G. H. E. and M. Rothschild. 1953. *An illustrated catalogue of the Rothschild collection of fleas (Siphonaptera) in the British Museum (Natural History).* Vol. I. Tungidae and Pulicidae. University Press, Cambridge. pp. i-xv, 1-361, 45 pls.

Hopkins, G. H. E. and M. Rothschild. 1956. *An illustrated catalogue of the Rothschild collection of fleas (Siphonaptera) in the British Museum (Natural History).* Vol. II. Coptopsyllidae, Vermipsyllidae, Stephanocircidae, Ischnopsyllidae, Hypsophthalmidae and Xiphiopsyllidae [Macropsyllidae]. University Press, Cambridge. pp. i-xi, 1-445, 32 pls.

Hopkins, G. H. E. and M. Rothschild. 1962. *An illustrated catalogue of the Rothschild collection of fleas (Siphonaptera) in the British Museum (Natural History).* Vol. III. Hystrichopsyllidae (*partim*). University Press, Cambridge. pp. i-ix, 1-560, 10 pls.

Hopkins, G. H. E. and M. Rothschild. 1966. *An illustrated catalogue of the Rothschild collection of fleas (Siphonaptera) in the British Museum (Natural History).* Vol. IV. Hystrichopsyllidae (concluded). University Press, Cambridge. pp. i-viii, 1-549, 12 pls.

Hopkins, G. H. E. and M. Rothschild. 1971. *An illustrated catalogue of the Rothschild collection of fleas (Siphonaptera) in the British Museum (Natural History).* Vol. V. Leptopsyllidae & Ancistropsyllidae. University Press, Cambridge. pp. i-viii, 1-530, 30 pls.

Hopla, C. E. 1965. Alaskan hematophagous insects, their feeding habits and potential as vectors of pathogenic organisms. I. The Siphonaptera of Alaska. Arctic Aeromedical Laboratory, Ft. Wainwright, AK. AAL-TR-64-12. Vol. I. pp. i-xiii, 1-344.

Hubbard, C. A. 1940a. American mole and shrew fleas (a new genus, three new species). *Pacific Univ. Bull.* 37(2): 2-12.

Hubbard, C. A. 1940b. A review of the western fleas of the genus *Malaraeus* with one new species and the description of a new *Thrassis* from Nevada. *Pacific Univ. Bull.* 37(6): 1-4.

Hubbard, C. A. 1943. The fleas of California. *Pacific Univ. Bull.* 39(8): 1-12.

Hubbard, C. A. 1947. *Fleas of western North America.* Iowa State College Press, Ames, pp. 1-533.

Hubbard, C. A. 1949a. New fleas and records from the western states. *Bull. So. Calif. Acad. Sci.* 48(2): 47-54.

Hubbard, C. A. 1949b. Fleas of the State of Nevada. *Bull. So. Calif. Acad. Sci.* 48(3): 115-128.

Hubbard, C. A. 1949c. Fleas of the Sagebrush Meadow Mouse. *Ent. News* 60: 141-144.

Hubbard, C. A. 1961a. Host specificity of fleas from Kangaroo rats. *Ent. News* 72: 25-27.

Hubbard, C. A. 1961b. Fleas from the Kangaroo rats of Northern California. *Ent. News.* 72: 133-139.

Jameson, E. W., Jr. and J. M. Brennan. 1957. An environmental analysis of some ectoparasites of small forest mammals in the Sierra Nevada, California. *Ecol. Monogr.* 27: 45-54.

Jellison, W. L. 1947. Siphonaptera: Host distribution of the genus *Opisocrostis* Jordan. *Trans. Amer. Microsc. Soc.* 66: 64-69.

Jellison, W. L. and L. Glesne. 1967. *Index to the literature of Siphonaptera of North America.* Suppl. 2, 1951-1960, pp. 1-406.

Jellison, W. L. and C. Senger. 1973. Fleas of Montana. Montana Agric. Exp. Sta. Res. Rep. 29: 1-75 (Mimeographed).

Jellison, W. L. and C. Senger. 1976. Fleas of western North America except Montana in the Rocky Mountain Laboratory Collection. *In: Papers in honor of Jerry Flora.* H. C. Taylor, Jr. and J. Clark, Eds., West. Wash. State Coll., Bellingham, WA. pp. 55-136.
Johnson, P. T. 1961. A revision of the species of *Monopsyllus* Kolenati in North America (Siphonaptera, Ceratophyllidae). *U. S. Dept. Agric. Tech. Bull.* 1227, pp. 69.
Jordan, K. 1933. A survey of the classification of the American species of *Ceratophyllus s. lat. Novit. Zool.* 39: 70-79.
Jordan, K. 1937. On some North American Siphonaptera. *Novit. Zool.* 40: 262-271.
Kartman, L. and F. M. Prince. 1956. Studies on *Pasteurella pestis* in fleas. V. The experimental plague-vector efficiency of wild rodent fleas compared with *Xenopsylla cheopis,* together with observations on the influence of temperature. *Amer. J. Trop. Med. Hyg.* 5: 1058-1070.
Kohls, G. M. 1940. Siphonaptera. A study of the species infesting wild hares and rabbits of North America north of Mexico. *Natl. Inst. Hlth. Bull.* 175, pp. 1-34.
Lewis, R. E. 1967. Contributions to a taxonomic revision of the genus *Nosopsyllus* Jordan, 1933 (Siphonaptera: Ceratophyllidae). *J. Med. Entomol.* 4: 123-142.
Lewis, R. E. 1972. Notes on the geographical distribution and host preferences in the order Siphonaptera. Part 1. Pulicidae. *J. Med. Entomol.* 9: 511-520.
Lewis, R. E. 1974. Ectoparasites. *In:* Maser, C. (Ed.) The sage vole, *Lagurus curtatus* (Cope, 1868), in the Crooked River National Grassland, Jefferson County, Oregon. A contribution to its life history and ecology. *Saugetierk. Mitt.* 22: 215-218.
Lewis, R. E. and C. Maser. 1978. *Phalacropsylla oregonensis* sp. n., with a key to the species of *Phalacropsylla* Rothschild, 1915 (Siphonaptera: Hystrichopsyllidae). *J. Parasitol.* 64: 147-150.
Lewis, R. E. and C. Maser. 1981. Invertebrates of the H. J. Andrews Experimental Forest, western Cascades, Oregon. I. An annotated checklist of fleas. USDA For. Serv. PNW Res. Note 378, pp. 1-10.
Lewis, R. E. and N. Wilson. 1982. A new species of *Nycteridopsylla* (Siphonaptera: Ischnopsyllidae) from southwestern United States, with a key to the North American species. *J. Med. Entomol.* 19: 605-614.
Link, V. B. 1955. Plague in the United States. *Publ. Hlth. Rep., Wash.* 70: 335-336.
Marshall, A. G. and B. C. Nelson. 1967. Bird ectoparasites from South Farallon Island, California. *J. Med. Entomol.* 4: 335-338.
Maser, C., E. W. Hammer, C. Brown, R. E. Lewis, R. L. Rausch and M. L. Johnson. 1974. The sage vole, *Lagurus curtatus* (Cope, 1868), in the Crooked River National Grassland, Jefferson County, Oregon. A contribution to its life history and ecology. *Saugetierk. Mitt.* 22: 193-222.
Maser, C., B. R. Mate, J. F. Franklin and C. T. Dyrness. 1981. Natural history of Oregon coast mammals. USDA For. Serv. Gen. Tech. Rep. PNW-133, pp. 1-496. Pacific Northwest Forest and Range Experiment Station, Portland, Oregon.
McCammon, R. B. and G. Wenninger. 1970. The dendrograph. State Geol. Surv. Comput. Contrib. pp. 1-28.
McCoy, G. W. and M. B. Mitzmain. 1910. Fleas collected from squirrels from various parts of California. *Publ. Hlth. Rep. Wash.* 25: 737-738.
Mead, R. A. 1963. Some aspects of parasitism in skunks of the Sacramento Valley of California. *Amer. Midl. Natur.* 70: 164-167.
Méndez, E. 1956. A revision of the genus *Megarthroglossus* Jordan and Rothschild, 1915 (Siphonaptera: Hystrichopsyllidae). *Univ. Calif. Publ. Entomol.* 11: 159-192.

Merriam, C. H. 1894. Laws of temperature control of the geographic distribution of terrestrial animals and plants. *Natl. Geog. Mag.* 6: 229-238.

Miller, N. G. and C. H. Drake. 1954. Infectious diseases in native wild animals of the Columbia Basin, Washington. *Northwest Sci.* 28: 135-156.

Nelson, B. C. 1972. Fleas from the archaeological site at Lovelock Cave, Nevada (Siphonaptera). *J. Med. Entomol.* 9: 211-214.

Nelson, B. C. and C. R. Smith. 1976. Ecological effects of a plague epizootic on the activities of rodents inhabiting caves at Lava Beds National Monument, California. *J. Med. Entomol.* 13: 51-61.

Nelson, B. C., C. A. Wolf and B. A. Sorrie. 1979. The natural introduction of *Hectopsylla psittaci,* a neotropical sticktight flea (Siphonaptera: Pulicidae) on Cliff swallows in California, USA. *J. Med. Entomol.* 16: 548-549.

O'Farrell, T. P. 1975. Small mammals, their parasites and pathologic lesions on the Arid Lands Ecology Reserve, Benton County, Washington. *Amer. Midl. Nat.* 93: 377-387.

Olsen, P. F. 1981. Sylvatic Plague. *In:* Davis, J. W., L. H. Karstad and D. O. Trainer, *Infectious diseases of wild mammals.* 2nd Ed., Iowa State University Press, pp. 232-243.

Parman, D. C. 1923. Biological notes on the hen flea *Echidnophaga gallinacea. J. Agric. Res.* 23: 1007-1009.

Perdue, J. C. 1980. Taxonomic review of the genus *Opisocrostis* Jordan, 1933. Unpublished M.S. Thesis, Iowa State University, Ames, Iowa 50011. pp. 1-87.

Poinar, G. O., Jr. and B. C. Nelson. 1973. *Psyllotylenchus viviparus,* n. gen., n. sp. (Nematoda: Tylenchida: Allantonematidae) parasitizing fleas (Siphonaptera) in California. *J. Med. Entomol.* 10: 349-354.

Pollitzer, R. 1954. *Plague.* WHO Organ. Monogr. Ser. No. 22, pp.1-698.

Poole, V. V. and R. A. Underhill. 1953. Biology and life history of *Megabothris clantoni clantoni* (Siphonaptera: Dolichopsyllidae). *Walla Walla Coll. Publ. Dept. Biol. & Biol. Sta.* 9: 1-19.

Prince, F. M. 1943. Species of fleas on rats collected in states west of the 102nd meridian and their relation to the dissemination of plague. *Publ. Hlth. Rep., Wash.* 58: 700-708.

Prince, F. M. 1945. Descriptions of three new species of *Dactylopsylla* Jordan and one new species of *Foxella* Wagner, with records of other species in the genera (Siphonaptera). *Canad. Entomol.* 77: 15-20.

Rothschild, M. 1975. Recent advances in our knowledge of the order Siphonaptera. *Ann. Rev. Entomol.* 20: 241-259.

Rothschild, M. and B. Ford. 1973. Factors influencing the breeding of the rabbit flea (*Spilopsyllus cuniculi*): A spring-time accelerator and a kairomone in nestling rabbit urine with notes on *Cediopsylla simplex,* another "hormone bound" species . *J. Zool.,* Lond. 170: 87-137.

Sapegina, V. F. and N. N. Kharitonova. 1969. On the ability of bird fleas to transmit Omsk hemorrhagic fever to white mice under experimental conditions. *In:* Cherepanov, A. I. et al. (Eds.) *Pereletnye ptitsy i ikh rol'v rasprostranenii arbovirusov.* Sibirsk. Otd. Akad. Nauk SSSR, Biol. Inst., Akad. Med. Nauk SSSR, Inst. Polio. Virus. Entsef., Minist. Zdravsokhr. RSFSR, Omsk Inst. Prirod. Ochag. Infekts., Novosibirsk. pp. 363-367.

Schwan, T. G., M. L. Higgins and B. C. Nelson. 1983. *Hectopsylla psittaci,* a South American sticktight flea (Siphonaptera: Pulicidae), established in Cliff swallow nests in California, USA. *J. Med. Entomol.* 20: 690-692.

Senger, C. M. 1967. Description of the female and notes on the male of *Nearctopsylla martyoungi* (Siphonaptera). *J. Kansas Ent. Soc.* 40: 190-191.

Smit, F. G. A. M. 1958. A preliminary note on the occurrence of *Pulex irritans* L. and *Pulex simulans* Baker in North America. *J. Parasitol.* 44: 523-526.

Smit, F. G. A. M. 1973. Siphonaptera (Fleas). *In:* K. G. V. Smith (Ed.). *Insects and other arthropods of medical importance.* British Museum (Natural History), London. pp. 325-371.

Smit, F. G. A. M. 1975. *Aconothobius anthobius* n. gen., n. sp., a pika flea found in a flower in Nepal (Siphonaptera). *Senckenb. Biol.* 56: 247-256.

Smit, F. G. A. M. 1983. Key to the genera and subgenera of Ceratophyllidae. *In:* Traub, R. et al., 1983, pp. 1-36.

Smit, F. G. A. M. and A. M. Wright. 1978a. A list of code numbers of species and subspecies of Siphonaptera. British Museum (Natural History), pp. 1-49. (Mimeographed).

Smit, F. G. A. M. and A. M. Wright. 1978b. A catalogue of primary type-specimens of Siphonaptera in the British Museum (Natural History), pp. 1-71. (Mimeographed).

Snodgrass, R. E. 1946. The skeletal anatomy of fleas (Siphonaptera). *Smithsonian Miscl. Coll.* 104: 1-89, + 21 pls.

Stark, H. E. 1970. A revision of the flea genus *Thrassis* Jordan, 1933 (Siphonaptera: Ceratophyllidae), with observations on ecology and relationship to plague. *Univ. Calif. Publ. Entomol.* 53. i-vii, pp. 1-184.

Stark, H. E. and A. R. Kinney. 1969. Abundance of rodents and fleas as related to plague in Lava Beds National Monument, California. *J. Med. Entomol.* 6: 287-294.

Svihla, R. D. 1941. A list of the fleas of Washington. *Univ. Wash. Publ. Biol.* 12: 9-20.

Tevis, L., Jr. 1955. Observations on chipmunks and mantled squirrels in northeastern California. *Amer. Midl. Nat.* 53: 71-78.

Tipton, V. J. and C. E. Machado-Allison. 1972. *Fleas of Venezuela.* Brigham Young Univ. Sci. Bull., Biol. Ser. 17: 1-115, + 91 figs.

Tipton, V. J., H. E. Stark and J. A. Wildie. 1979. Anomiopsyllinae. (Siphonaptera: Hystrichopsyllidae), II. The genera *Callistopsyllus, Conorhinopsylla, Megarthroglossus* and *Stenistomera. Great Basin Nat.* 39: 351-418.

Traub, R. 1950. Siphonaptera from Central America and Mexico, a morphological study of the aedeagus with description of new genera and species. *Fieldiana: Zool. Mem.* 1: 1-127, + 54 pls.

Traub, R. 1983. Medical importance of the Ceratophyllidae. *In:* Traub, R. et al., 1983, pp. 202-228.

Traub, R. and W. L. Jellison. 1981. Evolutionary and biogeographic history and the phylogeny of vectors and reservoirs as factors in the transmission of diseases from other animals to man. *In:* Burgdorfer, W. and R. L. Anacker (Eds.). *Rickettsiae and Rickettsial Diseases,* Academic Press. pp. 517-546.

Traub, R. and M. Rothschild. 1983. Evolution of the Ceratophyllidae. *In:* Traub, R., et al., 1983, pp. 188-201.

Traub, R., M. Rothschild and J. F. Haddow. 1983. *The Rothschild collection of fleas. The Ceratophyllidae: Key to the genera and host relationships with notes on their evolution, zoogeography and medical importance.* (Published privately. Printed by the University Press, Cambridge, U.K. Distributed by Academic Press, Inc. (London) Ltd.) pp. i-xv, 1-288.

Trembley, H. L. and F. C. Bishopp. 1940. Distribution and hosts of some fleas of economic importance. *J. Econ. Entomol.* 33: 701-703.

Whitaker, J. O., Jr. and D. A. Easterla. 1975. Ectoparasites of bats from Big Bend National Park, Texas. *Southwest Nat.* 20: 241-254.

Whitaker, J. O., Jr., C. Maser and R. E. Lewis. 1985. Ectoparasitic mites (excluding chiggers), fleas and lice from pocket gophers, *Thomomys,* in Oregon. *Northwest Sci.* 59: 33-39.

APPENDIX I

Host / Flea Index

Following is a list of the host species and the species of fleas reported from them. Where possible, the mammal classification of Honaki et al. (1982) has been followed for the mammals, and the A.O.U. Checklist (1983) has been followed for the birds.

Class **MAMMALIA**

Order **MARSUPIALIA**

Family **Didelphidae**

Didelphis marsupialis (Virginia opossum)
- *Catallagia sculleni sculleni*
- *Cediopsylla inaequalis interrupta*
- *Odontopsyllus dentatus*
- *Orchopeas sexdentatus cascadensis*

Order **INSECTIVORA**

Family **Soricidae**

Sorex bendirei (Marsh shrew)
- *Catallagia sculleni sculleni*
- *Corypsylla kohlsi*
- *Dasypsyllus (Dasypsyllus) gallinulae perpinnatus*
- *Epitedia scapani*
- *Hystrichopsylla occidentalis occidentalis*
- *Nearctopsylla martyoungi*
- *Opisodasys keeni*

Sorex obscurus (Dusky shrew)
- *Catallagia decipiens*

Sorex pacificus (Pacific shrew)
- *Catallagia sculleni chamberlini*
- *Corrodopsylla curvata obtusata*
- *Corypsylla kohlsi*
- *Epitedia scapani*
- *Epitedia stewarti*
- *Hystrichopsylla occidentalis occidentalis*
- *Leptopsylla segnis*
- *Megabothris (Amegabothris) abantis*
- *Opisodasys keeni*
- *Peromyscopsylla selenis*
- *Trichopsylloides oregonensis*

Sorex palustris (American water shrew)
- *Malaraeus telchinus*
- *Corypsylla ornata*
- *Delotelis hollandi*
- *Nearctopsylla princei*

Sorex trowbridgei (Trowbridge shrew)
- *Aetheca wagneri*
- *Atyphloceras multidentatus multidentatus*
- *Catallagia sculleni sculleni*
- *Catallagia sculleni rutherfordi*
- *Ceratophyllus (Amonopsyllus) ciliatus protinus*
- *Corrodopsylla curvata obtusata*
- *Corypsylla jordani*
- *Corypsylla kohlsi*
- *Corypsylla ornata*
- *Delotelis hollandi*
- *Epitedia scapani*
- *Epitedia stewarti*
- *Hystrichopsylla occidentalis occidentalis*
- *Malaraeus telchinus*
- *Nearctopsylla jordani*
- *Nearctopsylla martyoungi*
- *Nearctopsylla princei*
- *Neopsylla inopina*
- *Opisodasys keeni*
- *Peromyscopsylla selenis*
- *Rhadinopsylla sectilis goodi*

Sorex vagrans (Vagrant shrew)
- *Catallagia decipiens*
- *Catallagia sculleni sculleni*
- *Corrodopsylla curvata obtusata*
- *Corypsylla jordani*
- *Corypsylla kohlsi*
- *Corypsylla ornata*
- *Epitedia scapani*
- *Hystrichopsylla occidentalis occidentalis*
- *Malaraeus telchinus*
- *Megabothris (Amegabothris) abantis*
- *Nearctopsylla hyrtaci*
- *Nearctopsylla martyoungi*
- *Peromyscopsylla selenis*

Sorex sp.
- *Corrodopsylla curvata obtusata*
- *Rhadinopsylla sectilis sectilis*
- *Spicata comis tacomae*

Family **Talpidae**

Neurotrichus gibbsi gibbsi (Gibbs shrew-mole)
- *Catallagia mathesoni*
- *Catallagia sculleni sculleni*
- *Catallagia sculleni rutherfordi*
- *Corypsylla jordani*
- *Corypsylla kohlsi*
- *Corypsylla ornata*
- *Delotelis hollandi*
- *Epitedia scapani*
- *Epitedia stewarti*
- *Hystrichopsylla occidentalis occidentalis*
- *Malaraeus telchinus*
- *Nearctopsylla hamata*

Scapanus latimanus (Broad-footed mole)
- *Corypsylla ornata*
- *Nearctopsylla hamata*

Scapanus orarius (Pacific mole)
- *Catallagia sculleni sculleni*
- *Corypsylla jordani*
- *Corypsylla kohlsi*
- *Corypsylla ornata*
- *Epitedia scapani*
- *Epitedia stewarti*
- *Nearctopsylla jordani*
- *Nearctopsylla traubi*

Scapanus townsendi (Townsend mole)
- *Catallagia charlottensis*
- *Catallagia sculleni sculleni*
- *Corypsylla ornata*
- *Epitedia scapani*
- *Foxella ignota recula*
- *Megabothris (Amegabothris) abantis*
- *Nearctopsylla genalis hygini*
- *Nearctopsylla jordani*
- *Nearctopsylla martyoungi*

Scapanus sp.
- *Rhadinopsylla sectilis sectilis*
- *Spicata comis tacomae*

Order **CHIROPTERA**

Family **Vespertilionidae**

Antrozous pallidus (Pallid bat)
- *Cediopsylla inaequalis interrupta*
- *Myodopsylla palposa*
- *Opisodasys keeni*

Eptesicus fuscus (Big brown bat)
- *Myodopsylla borealis*
- *Myodopsylla gentilis*
- *Myodopsylla palposa*

Lasionycteris noctivagans (Silver-haired bat)
- *Myodopsylla gentilis*

Macrotis californicus (California leaf-nosed bat)
- *Myodopsylla gentilis*

Myotis californicus (California myotis)
- *Myodopsylla gentilis*
- *Nycteridopsylla vancouverensis*
- *Orchopeas sexdentatus cascadensis*

Myotis evotis (Long-eared myotis)
- *Myodopsylla gentilis*

Myotis lucifugus (Little brown bat)
- *Myodopsylla gentilis*

Myotis volans (Long-legged myotis)
- *Myodopsylla gentilis*

Myotis yumanensis (Yuma myotis)
- *Myodopsylla gentilis*
- *Orchopeas sexdentatus cascadensis*

Plecotus rafinesquii (Rafinesque's big-eared bat)
- *Myodopsylla gentilis*

Order **CARNIVORA**

Family **Canidae**

Canis latrans (Coyote)
- *Cediopsylla inaequalis inaequalis*
- *Cediopsylla inaequalis interrupta*
- *Ceratophyllus (Amonopsyllus) ciliatus protinus*
- *Ctenocephalides canis*
- *Odontopsyllus dentatus*
- *Orchopeas nepos*
- *Pulex irritans*

Urocyon cinereoargenteus (Gray fox)
- *Cediopsylla inaequalis interrupta*
- *Ctenocephalides canis*
- *Odontopsyllus dentatus*
- *Pulex irritans*

Urocyon sp.
- *Pulex irritans*

Vulpes vulpes (Red fox)
- *Cediopsylla inaequalis interrupta*
- *Ctenocephalides canis*
- *Odontopsyllus dentatus*
- *Pulex irritans*

Family **Mustelidae**
Martes americana (American marten)
Orchopeas sexdentatus agilis
Rhadinopsylla difficilis
Mephitis mephitis (Striped skunk)
Ceratophyllus (Amonopsyllus) ciliatus protinus
Ctenocephalides felis felis
Echidnophaga gallinacea
Foxella ignota ssp.
Hoplopsyllus anomalus
Orchopeas sexdentatus cascadensis
Oropsylla (Diamanus) montana
Oropsylla (Thrassis) acamantis howelli
Pulex irritans
Mustela erminea (Short-tailed weasel)
Atyphloceras multidentatus multidentatus
Catallagia decipiens
Catallagia sculleni sculleni
Ceratophyllus (Amonopsyllus) ciliatus protinus
Epitedia scapani
Foxella ignota recula
Malaraeus telchinus
Megabothris (Amegabothris) abantis
Nearctopsylla martyoungi
Orchopeas sexdentatus cascadensis
Mustela frenata (Long-tailed weasel)
Aetheca wagneri
Foxella ignota recula
Foxella ignota ssp.
Malaraeus telchinus
Megabothris (Amegabothris) abantis
Nearctopsylla hyrtaci
Orchopeas leucopus
Oropsylla (Thrassis) francisi francisi
Rhadinopsylla difficilis
Mustela vison (Mink)
Ceratophyllus (Amonopsyllus) ciliatus protinus
Dolichopsyllus stylosus
Hystrichopsylla schefferi
Megabothris (Amegabothris) abantis
Opisodasys keeni
Orchopeas sexdentatus cascadensis
Spilogale putorius (Spotted skunk)
Atyphloceras multidentatus multidentatus
Catallagia sculleni sculleni
Cediopsylla inaequalis interrupta
Ceratophyllus (Amonopsyllus) ciliatus protinus
Ctenocephalides felis felis
Dolichopsyllus stylosus
Echidnophaga gallinacea
Epitedia scapani
Foxella ignota ssp.
Hoplopsyllus anomalus
Hystrichopsylla dippiei spinata
Hystrichopsylla dippiei ssp.
Hystrichopsylla schefferi
Megabothris (Amegabothris) quirini
Opisodasys keeni
Orchopeas sexdentatus cascadensis
Oropsylla (Diamanus) montana
Pulex irritans
Rhadinopsylla sectilis goodi
Trichopsylloides oregonensis
Spilogale sp.
Ceratophyllus (Amonopsyllus) ciliatus ciliatus

Family **Felidae**
Felis rufus (Bobcat)
Cediopsylla inaequalis interrupta
Dolichopsyllus stylosus
Epitedia scapani
Odontopsyllus dentatus
Orchopeas sexdentatus cascadensis

Order **ARTIODACTYLA**

Family **Cervidae**
Odocoileus hemionus (Mule deer)
Pulex irritans

Order **RODENTIA**

Family **Aplodontidae**
Aplodontia rufa (Mountain beaver)
Ceratophyllus (Amonopsyllus) ciliatus protinus
Dolichopsyllus stylosus
Epitedia scapani
Hystrichopsylla schefferi
Opisodasys keeni
Oropsylla (Diamanus) montana
Paratyphloceras oregonensis
Trichopsylloides oregonensis

Family **Sciuridae**
Ammospermophilus leucurus (Antelope ground squirrel)
Euhoplopsyllus glacialis foxi
Hoplopsyllus anomalus
Oropsylla (Thrassis) bacchi gladiolis
Rhadinopsylla heiseri

Eutamias amoenus (Yellow-pine chipmunk)
- *Catallagia sculleni rutherfordi*
- *Ceratophyllus (Amonopsyllus) ciliatus protinus*
- *Eumolpianus cyrturus*
- *Eumolpianus eumolpi eumolpi*
- *Malaraeus telchinus*
- *Oropsylla (Diamanus) montana*

Eutamias minimus (Least chipmunk)
- *Aetheca wagneri*
- *Eumolpianus cyrturus*
- *Eumolpianus eumolpi eumolpi*
- *Megabothris clantoni*

Eutamias quadrimaculatus (Long-eared chipmunk)
- *Catallagia sculleni rutherfordi*
- *Eumolpianus eumolpi eumolpi*
- *Oropsylla (Diamanus) montana*

Eutamias speciosus (Lodgepole chipmunk)
- *Aetheca wagneri*
- *Catallagia sculleni rutherfordi*
- *Ceratophyllus (Amonopsyllus) ciliatus mononis*
- *Eumolpianus cyrturus*
- *Eumolpianus eumolpi eumolpi*
- *Oropsylla (Diamanus) montana*

Eutamias townsendi (Townsend chipmunk)
- *Aetheca wagneri*
- *Catallagia charlottensis*
- *Catallagia sculleni sculleni*
- *Ceratophyllus (Amonopsyllus) ciliatus ciliatus*
- *Ceratophyllus (Amonopsyllus) ciliatus mononis*
- *Ceratophyllus (Amonopsyllus) ciliatus protinus*
- *Ceratophyllus (Monopsyllus) vison*
- *Ctenocephalides felis felis*
- *Dasypsyllus (Dasypsyllus) gallinulae perpinnatus*
- *Epitedia scapani*
- *Epitedia wenmanni*
- *Eumolpianus cyrturus*
- *Eumolpianus eumolpi eumolpi*
- *Hystrichopsylla dippiei spinata*
- *Malaraeus telchinus*
- *Megabothris (Amegabothris) abantis*
- *Megarthroglossus procus*
- *Opisodasys keeni*
- *Oropsylla (Diamanus) montana*
- *Peromyscopsylla hesperomys pacifica*
- *Rhadinopsylla sectilis goodi*

Eutamias sp.
- *Aetheca wagneri*
- *Catallagia sculleni rutherfordi*
- *Ceratophyllus (Amonopsyllus) ciliatus mononis*
- *Ceratophyllus (Amonopsyllus) ciliatus protinus*
- *Eumolpianus cyrturus*
- *Eumolpianus eumolpi eumolpi*
- *Megabothris (Amegabothris) abantis*
- *Orchopeas nepos*
- *Oropsylla (Oropsylla) idahoensis*
- *Oropsylla (Diamanus) montana*

Glaucomys sabrinus (Northern flying squirrel)
- *Aetheca wagneri*
- *Callistopsyllus terinus*
- *Ceratophyllus (Amonopsyllus) ciliatus protinus*
- *Delotelis telegoni*
- *Eumolpianus cyrturus*
- *Hystrichopsylla occidentalis occidentalis*
- *Megarthroglossus divisus*
- *Megarthroglossus procus*
- *Opisodasys keeni*
- *Opisodasys vesperalis*
- *Orchopeas caedens caedens*
- *Orchopeas caedens durus*
- *Orchopeas nepos*
- *Peromyscopsylla selenis*
- *Tarsopsylla octodecimdentata coloradensis*

Marmota caligata (Hoary marmot)
- *Oropsylla (Thrassis) spenceri spenceri*

Marmota flaviventris (Yellow-bellied marmot)
- *Neopsylla inopina*
- *Oropsylla (Oropsylla) idahoensis*
- *Oropsylla (Diamanus) montana*
- *Oropsylla (Hubbardipsylla) oregonensis*
- *Oropsylla (Opisocrostis) tuberculata tuberculata*
- *Oropsylla (Thrassis) acamantis acamantis*
- *Oropsylla (Thrassis) acamantis howelli*

Marmota olympus (Olympic marmot)
- *Oropsylla (Oropsylla) eatoni*
- *Oropsylla (Thrassis) spenceri spenceri*

Tamiasciurus douglasi (Chickaree)
Atyphloceras multidentatus multidentatus
Catallagia decipiens
Catallagia sculleni sculleni
Ceratophyllus (Amonopsyllus) ciliatus protinus
Ceratophyllus (Amonopsyllus) ciliatus mononis
Eumolpianus eumolpi eumolpi
Foxella ignota recula
Megarthroglossus procus
Opisodasys keeni
Orchopeas caedens durus
Orchopeas nepos
Oropsylla (Diamanus) montana
Tamiasciurus hudsonicus (Red squirrel)
Ceratophyllus (Monopsyllus) vison
Eumolpianus eumolpi eumolpi
Orchopeas sexdentatus agilis
Oropsylla (Hubbardipsylla) oregonensis
Tamiasciurus sp.
Orchopeas nepos
Family **Geomyidae**
Thomomys bottae (Botta's pocket gopher)
Foxella ignota franciscana
Spicata bottaceps
Spicata comis comis
Thomomys monticola (Mountain pocket gopher)
Foxella ignota recula
Spicata monticola
Thomomys mazama (Mazama pocket gopher)
Foxella ignota recula
Spicata comis comis
Thomomys talpoides (Northern pocket gopher)
Epitedia scapani
Foxella ignota recula
Meringis hubbardi
Orchopeas sexdentatus cascadensis
Spicata comis comis
Spicata moorei moorei
Thomomys townsendi (Townsend pocket gopher)
Foxella ignota recula
Orchopeas sexdentatus cascadensis
Thomomys sp.
Corypsylla ornata
Megabothris (Amegabothris) clantoni
Foxella ignota recula
Oropsylla (Thrassis) francisi rockwoodi
Spicata comis tacomae
Family **Heteromyidae**
Dipodomys deserti (Desert kangaroo-rat)
Meringis parkeri
Oropsylla (Thrassis) aridis hoffmani
Dipodomys heermanni (Heermann's kangaroo-rat)
Eumolpianus eumolpi eumolpi
Meringis cummingi
Meringis parkeri
Opisodasys keeni
Dipodomys merriami (Merriam's kangaroo-rat)
Meringis dipodomys
Meringis parkeri
Oropsylla (Thrassis) aridis hoffmani
Dipodomys microps (Chisel-toothed kangaroo-rat)
Aetheca wagneri
Meringis cummingi
Meringis dipodomys
Dipodomys ordi (Ord kangaroo-rat)
Meringis dipodomys
Meringis parkeri
Oropsylla (Opisocrostis) tuberculata tuberculata
Dipodomys panamintensis (Panamint kangaroo-rat)
Meringis parkeri
Rhadinopsylla sectilis sectilis
Microdipodops megacephalus (Dark kangaroo-mouse)
Meringis hubbardi
Orchopeas leucopus
Perognathus longimembris (Little pocket mouse)
Aetheca wagneri
Meringis hubbardi
Perognathus parvus (Great Basin pocket mouse)
Aetheca wagneri
Eumolpianus eumolpi eumolpi
Meringis hubbardi
Meringis parkeri
Meringis shannoni

HOST/FLEA INDEX

Family **Cricetidae**

- *Neotoma cinerea* (Bush-tailed wood rat)
  - *Aetheca wagneri*
  - *Amaradix bitterrootensis bitterrootensis*
  - *Amaradix euphorbi*
  - *Anomiopsyllus amphibolus*
  - *Anomiopsyllus falsicalifornicus congruens*
  - *Catallagia decipiens*
  - *Catallagia mathesoni*
  - *Catallagia sculleni sculleni*
  - *Catallagia sculleni chamberlini*
  - *Cediopsylla inaequalis interrupta*
  - *Eumolpianus cyrturus*
  - *Malaraeus telchinus*
  - *Megarthroglossus jamesoni*
  - *Megarthroglossus spenceri spenceri*
  - *Opisodasys keeni*
  - *Orchopeas sexdentatus agilis*
  - *Orchopeas sexdentatus cascadensis*
  - *Orchopeas sexdentatus nevadensis*
  - *Oropsylla (Diamanus) montana*
  - *Oropsylla (Opisocrostis) tuberculata tuberculata*
  - *Oropsylla (Thrassis) francisi francisi*
  - *Oropsylla (Thrassis) petiolata*
  - *Phalacropsylla allos*
  - *Stenistomera alpina*
- *Neotoma fuscipes* (Dusky-footed wood rat)
  - *Aetheca wagneri*
  - *Anomiopsyllus falsicalifornicus congruens*
  - *Atyphloceras multidentatus multidentatus*
  - *Catallagia sculleni sculleni*
  - *Eumolpianus eumolpi eumolpi*
  - *Meringis hubbardi*
  - *Myodopsylla gentilis*
  - *Opisodasys keeni*
  - *Orchopeas nepos*
  - *Orchopeas sexdentatus agilis*
  - *Orchopeas sexdentatus cascadensis*
  - *Orchopeas sexdentatus nevadensis*
  - *Stenistomera alpina*
- *Neotoma lepida* (Desert wood rat)
  - *Orchopeas sexdentatus nevadensis*
  - *Phalacropsylla oregonensis*
- *Neotoma* sp.
  - *Anomiopsyllus falsicalifornicus congruens*
  - *Atyphloceras multidentatus multidentatus*
  - *Orchopeas sexdentatus nevadensis*
- *Onychomys leucogaster* (Northern grasshopper mouse)
  - *Aetheca wagneri*
  - *Catallagia decipiens*
  - *Foxella ignota recula*
  - *Malaraeus sinomus*
  - *Meringis hubbardi*
  - *Meringis parkeri*
  - *Meringis shannoni*
  - *Oropsylla (Thrassis) petiolata*
  - *Pleochaetis exilis*
- *Onychomys* sp.
  - *Oropsylla (Hubbardipsylla) washingtonensis*
- *Peromyscus boylii* (Brush mouse)
  - *Atyphloceras multidentatus multidentatus*
  - *Catallagia mathesoni*
  - *Corypsylla kohlsi*
  - *Eumolpianus eumolpi eumolpi*
  - *Malaraeus telchinus*
  - *Opisodasys keeni*
- *Peromyscus crinitus* (Canyon mouse)
  - *Aetheca wagneri*
  - *Callistopsyllus terinus deuterus*
  - *Catallagia decipiens*
  - *Catallagia sculleni chamberlini*
  - *Malaraeus sinomus*
  - *Malaraeus telchinus*
  - *Megarthroglossus divisus*
  - *Opisodasys keeni*
  - *Peromyscopsylla selenis*
  - *Phalacropsylla oregonensis*
  - *Stenistomera alpina*
  - *Stenistomera macrodactyla*
- *Peromyscus maniculatus* (Deer mouse)
  - *Aetheca wagneri*
  - *Amaradix bitterrootensis bitterrootensis*
  - *Amaradix euphorbi*
  - *Anomiopsyllus amphibolus*
  - *Atyphloceras multidentatus multidentatus*
  - *Callistopsyllus terinus deuterus*
  - *Catallagia charlottensis*
  - *Catallagia decipiens*
  - *Catallagia mathesoni*

Catallagia sculleni sculleni
Catallagia sculleni rutherfordi
Catallagia sculleni chamberlini
Ceratophyllus (Amonopsyllus) ciliatus protinus
Corrodopsylla curvata obtusata
Corypsylla kohlsi
Corypsylla ornata
Delotelis hollandi
Epitedia scapani
Epitedia stanfordi
Epitedia wenmanni
Eumolpianus cyrturus
Eumolpianus eumolpi eumolpi
Foxella ignota recula
Hystrichopsylla dippiei ssp.
Hystrichopsylla dippiei neotomae
Hystrichopsylla occidentalis occidentalis
Malaraeus sinomus
Malaraeus telchinus
Megabothris (Amegabothris) abantis
Megarthroglossus procus
Meringis hubbardi
Meringis parkeri
Meringis shannoni
Nearctopsylla hamata
Nearctopsylla princei
Odontopsyllus dentatus
Opisodasys keeni
Orchopeas caedens durus
Orchopeas leucopus
Orchopeas sexdentatus agilis
Orchopeas sexdentatus cascadensis
Oropsylla (Thrassis) bacchi johnsoni
Oropsylla (Thrassis) francisi francisi
Peromyscopsylla hesperomys pacifica
Peromyscopsylla selenis
Phalacropsylla allos
Phalacropsylla oregonensis
Rhadinopsylla fraterna
Rhadinopsylla sectilis sectilis
Rhadinopsylla sectilis goodi
Stenistomera alpina
Stenistomera hubbardi
Stenistomera macrodactyla

Peromyscus truei (Piñon mouse)
Atyphloceras multidentatus multidentatus
Catallagia mathesoni
Epitedia stanfordi
Meringis shannoni
Opisodasys keeni

Peromyscus sp.
Aetheca wagneri
Catallagia mathesoni
Ceratophyllus (Monopsyllus) vison
Delotelis hollandi
Delotelis telegoni
Meringis shannoni
Odontopsyllus dentatus
Opisodasys keeni
Oropsylla (Thrassis) acamantis acamantis

Reithrodontomys megalotus (Western harvest mouse)
Aetheca wagneri
Atyphloceras multidentatus multidentatus
Catallagia decipiens
Epitedia stanfordi
Eumolpianus cyrturus
Malaraeus telchinus
Orchopeas caedens durus

Reithrodontomys sp.
Orchopeas leucopus
Oropsylla (Thrassis) petiolata

Family **Arvicolidae**

Arborimus albipes (White-footed vole)
Catallagia sculleni sculleni
Epitedia scapani
Hystrichopsylla occidentalis occidentalis
Megabothris (Amegabothris) abantis
Rhadinopsylla sectilis goodi

Clethrionomys californicus (Western red-backed vole)
Atyphloceras multidentatus multidentatus
Catallagia mathesoni
Catallagia sculleni sculleni
Catallagia sculleni chamberlini
Catallagia sculleni rutherfordi
Ceratophyllus (Amonopsyllus) ciliatus protinus
Corypsylla kohlsi
Delotelis hollandi
Epitedia stewarti
Hystrichopsylla occidentalis occidentalis
Malaraeus telchinus
Megabothris (Amegabothris) abantis

Catallagia sculleni sculleni
Ceratophyllus (Amonopsyllus) ciliatus protinus
Corypsylla kohlsi
Delotelis hollandi
Epitedia scapani
Epitedia stanfordi
Epitedia stewarti
Epitedia wenmanni
Hystrichopsylla occidentalis occidentalis
Malaraeus telchinus
Megabothris (Amegabothris) abantis
Opisodasys keeni
Oropsylla (Diamanus) montana
Peromyscopsylla selenis
Rhadinopsylla sectilis goodi

Microtus richardsoni (Water vole)
Atyphloceras multidentatus multidentatus
Catallagia sculleni sculleni
Catallagia sculleni chamberlini
Foxella ignota recula
Hystrichopsylla dippiei ssp.
Hystrichopsylla occidentalis occidentalis
Megabothris (Amegabothris) abantis
Peromyscopsylla selenis

Microtus townsendi (Townsend vole)
Atyphloceras multidentatus multidentatus
Catallagia charlottensis
Catallagia sculleni sculleni
Catallagia sculleni chamberlini
Epitedia scapani
Epitedia stewarti
Hystrichopsylla dippiei ssp.
Hystrichopsylla occidentalis occidentalis
Megabothris (Amegabothris) abantis
Megabothris (Amegabothris) quirini
Peromyscopsylla selenis
Rhadinopsylla sectilis goodi

Microtus sp.
Aetheca wagneri
Atyphloceras multidentatus multidentatus
Hystrichopsylla occidentalis linsdalei
Peromyscopsylla selenis

Ondatra zibethicus (Muskrat)
Oropsylla (Opisocrostis) tuberculata tuberculata

Phenacomys intermedius (Heather vole)
Catallagia sculleni sculleni
Megabothris (Amegabothris) abantis
Peromyscopsylla selenis

Family **Muridae**

Mus musculus (House mouse)
Aetheca wagneri

Rattus norvegicus (Norway rat)
Atyphloceras multidentatus multidentatus
Catallagia sculleni sculleni
Foxella ignota clantoni
Leptopsylla segnis
Megabothris (Amegabothris) clantoni
Nosopsyllus (Nosopsyllus) fasciatus
Oropsylla (Thrassis) bacchi johnsoni
Xenopsylla cheopis

Rattus rattus (Black rat)
Nosopsyllus (Nosopsyllus) fasciatus

Rattus sp.
Aetheca wagneri
Echidnophaga gallinacea
Megabothris (Amegabothris) clantoni
Nosopsyllus (Nosopsyllus) fasciatus
Xenopsylla cheopis

Family **Zapodidae**

Zapus trinotatus (Pacific jumping mouse)
Aetheca wagneri
Epitedia scapani
Epitedia stewarti
Megabothris (Amegabothris) abantis
Opisodasys keeni

Zapus sp.
Megabothris (Amegabothris) abantis

Family **Erethizontidae**

Erethizon dorsatum (North American porcupine)
Ceratophyllus (Amonopsyllus) ciliatus protinus

Order **LAGOMORPHA**

Family **Ochotonidae**

Ochotona princeps (Pika)
Amaradix bitterrootensis bitterrootensis
Ctenophyllus armatus
Megabothris (Amegabothris) abantis
Orchopeas nepos
Oropsylla (Oropsylla) idahoensis

Family **Leporidae**

Lepus americanus (Snowshoe hare)
Cediopsylla inaequalis interrupta

Class **AVES**

Miscellaneous

Redwood litter

Hystrichopsylla occidentalis linsdalei

APPENDIX II

The Collection and Preservation of Fleas

As is true for many other groups of insects, the collection and preservation of fleas requires rather specialized techniques. Since fleas are blood-sucking ectoparasites of some birds and most mammals, their collection usually involves the collection and examination of the host animal. Following is a description of the methods most commonly employed in the collection of fleas.

COLLECTING FLEAS FROM THEIR HOSTS

With the exception of the species of three or four genera, adult fleas do not attach themselves permanently to the body of their host although a temporary attachment is necessary during feeding. Even the slightest disturbance will cause them to withdraw their mouthparts and escape, either by jumping off the host or by hiding in the pelage. For this reason host animals must be enclosed in some type of container until they can be examined.

Any of the standard collecting techniques employed by mammalogists or ornithologists are satisfactory for obtaining host animals. These fall into two general categories: methods such as shooting, trapping, etc., which usually result in the death of the host, or live trapping, netting, etc., methods that produce living hosts for examination. Dead host animals should be retrieved as soon after death as possible and placed, individually, in cloth or plastic bags. As the body cools, many of the fleas will jump off or, if more than one animal is present, will transfer from one body to another. For this reason two host animals of different species or from different localities should *never* be placed together in the same bag.

Experience has shown that fleas tend to leave the dead body of the host more readily in warm weather than in cold. Individual species differ in this respect, but generally speaking traps should be inspected more frequently during the summer than in winter. Since the bulk of the small mammals will be caught during the first few hours of darkness, it is usually sufficient to check traps about 2 hours after sunset and again 2 hours later. Traps may be left out all night in which case they should be checked and taken up as early in the morning as possible.

Slight mechanical disturbances of the body of the dead host will stimulate fleas to jump, even under ideal conditions. For this reason the body should be bagged as quickly as possible. The most satisfactory method is to place the mouth of the bag as close to the trapped host as possible and with one deft motion place both the trap and the animal in the bag. The animal can then be released from the trap and the trap shaken briskly in order to insure the removal of fleas from the trap itself before it is withdrawn from the bag. Bags must be tightly closed with a rubber band or tied with cord immediately to prevent escape of fleas.

It is necessary to immobilize living hosts with some type of anesthetic or by killing them. When using a gas anesthetic such as ether, care must be taken to inspect the trap thoroughly for fleas that have fallen off the host. This method unfortunately cannot be recommended since fleas frequently respond to fright in the host by abandoning it.

There are two schools of thought concerning the advisability of killing the host animal versus trapping it alive. In our experience the former method is most satisfactory for the following reasons. First, snap traps are much smaller and easier to carry than even small live traps. For this reason more traps can be carried by a single person and more traps can be set in a given period of time. Second, many mammals that are readily trapped in snap traps are reluctant to enter live traps. Third, a live host presents additional complications in that it must be immobilized before it can be examined for fleas and other ectoparasites. There is little evidence that live-trapped animals yield more ectoparasites than snap-trapped animals, if handled properly.

The simplest way to remove fleas and other ectoparasites from small animals is to immobilize them first. If living specimens are required, ether or carbon dioxide may be used as an anesthetic. A number of hosts in cloth bags may be placed in a wide-mouthed gallon jar or some other tight container and a few drops of chloroform sprinkled over them. Exposure to the fumes for 10 minutes is usually sufficient to anesthetize fleas although ticks and mites frequently require longer. Plastic bags require individual fumigation.

Each host must be inspected individually. The body should be removed from the bag and placed on a light background such as a large sheet of white paper or an enamel tray. Many fleas will be found adhering to the pelage, and these should be removed with the aid of a pair of forceps or a camel's hair brush moistened in 70 percent ethyl alcohol. When all of the visible fleas have been removed, the body of the host should be grasped by the feet and tail and tapped briskly on the surface to dislodge additional specimens which have become trapped in the pelage. The host also may be combed or washed in detergent, although this usually is not necessary in the collection of fleas. The body of the host should be examined until no further fleas can be found. *Be sure to examine the bag* for fleas which have left the body during anesthetization.

Ectoparasites from *each* host should be placed in a separate vial of 70-80 percent ethyl alcohol along with a slip of paper written in either lead pencil or India ink (never ball point or fountain pen) indicating either the field number of the host or collection data. Collection data must include name of host, locality of collection, date of collection, and name of collector. Specialized studies may require more information, but that listed above is the absolute minimum.

Host animals that cannot be identified with certainty should be preserved by standard methods for later identification by specialists.

Under no conditions, unless the fleas are needed for histological studies, should they be "fixed" in fluids containing formalin since they are impossible to clear properly when so treated. In the absence of preservatives, fleas may be stored quite satisfactorily in dry vials if care is taken to protect them from mechanical damage.

In the case of large hosts such as foxes, raccoons, jackals, hyenas, etc., it is better to omit the anesthetic and collect the living fleas from the body of the host with the aid of an aspirator. Since fleas respond positively to warmth and moisture, they may be attracted to the surface of the pelage by simply blowing one's breath into the mouth of the bag. Thus attracted they are easily aspirated and can then be immobilized in the aspirator prior to preservation.

Large animals should be examined also, especially around the eyes, lips, and ears for sticktight fleas which do not dislodge when anesthetized. Fleshy areas between the plates of animals such as pangolins and armadillos should be examined also.

COLLECTING FLEAS FROM OTHER SOURCES

Since fleas are holometabolous insects whose larval stages depend upon a different type of food from that of the adults, it is possible to obtain specimens from sources other than the host animals. Mating and oviposition usually occur in the nest or the lair of the host, and at any time there are likely to be more adults in the nesting material than on the host. In addition, certain species are "nest" fleas and seldom feed or occur on the host except when it is occupying its nest or retreat. This is particularly true of bird fleas. For these reasons, one of the most productive methods of obtaining adults is by sorting them from the nesting material of their hosts.

Nests may be obtained by various means. The nests of passerine birds and some arboreal mammals usually are easily collected. Those of hole-nesting birds, such as swallows, petrels, etc., and those of burrowing mammals must be excavated. Although this is one of the most onerous methods of collecting insects, it is likewise one of the most productive for fleas.

Nesting material should be placed in a cloth or plastic bag and sealed to prevent escape of the ectoparasites, although this is of less importance with nests than with host animals. The material should be examined as soon as possible after collection because nests in cloth bags dry rapidly and those in plastic bags are prone to mold. Small amounts of the nest should be removed from the bag and picked apart in a white enamel tray or on a sheet of white paper. Never fumigate nesting material prior to examination as dead fleas are practically impossible to detect amid the rubble in the nest. Blowing one's breath over the nesting material usually activates the fleas and they can then be picked up with an aspirator.

Once all adult fleas have been removed, the nesting material may be placed in a paper bag or other container, sprinkled lightly with water, and set aside for a week or two in a warm, moist location. Examination at a later date usually reveals additional adults that have emerged in the interim.

The Preparation of Fleas for Study

Preserved fleas generally are not suitable for study until they have been cleared and mounted on microscope slides. While the literature abounds with routines for preparing fleas for study, the following method is generally conceded to yield the best results. It also is applicable, with slight modifications, to other small insects.

CLEARING

Since fleas frequently are engorged with blood when they are collected, it usually is necessary to remove the contents of the body by means of a chemical clearing agent. A 10-15 percent aqueous solution of potassium hydroxide (KOH) is most commonly employed for this purpose. Specimens to be cleared may be placed in the KOH solution directly from preservative. At first they will float but after an hour or two they usually will sink to the bottom of the container. Floating specimens will clear but they require longer because only half of the body is in contact with the clearing solution so it is advisable to "scuttle" any individuals that do not sink of their own accord.

Specimens should be cleared until all traces of blood are removed, as well as all of the unsclerotized portions of the internal anatomy. No specific time can be predicted because some species clear quickly while others may require days. Twenty-four to 36 hours usually is sufficient at normal room temperatures. Never attempt to speed the process by heating or boiling the specimens in the KOH solution.

As clearing progresses, small droplets of oil will accumulate inside the specimens, but these will be removed by one of the reagents employed in a later step.

NEUTRALIZING

Once cleared, the specimens must be neutralized or acidified in order to stop the clearing process. It is absolutely essential that this be done or the specimens will continue to clear, even after they have been mounted. Specimens should be removed from the KOH solution and placed in distilled water to which a few drops of acetic or hydrochloric acid have been added. A stock solution of approximately 5 cc of acid to 500 cc of distilled water is adequate for this purpose. Neutralization requires a minimum of 1 hour but material can remain in the acid-water for days or weeks without being damaged.

SPREADING AND HARDENING

Although fleas are strongly laterally compressed, it is helpful to spread the legs and harden the specimens in a spread position before they are mounted. This is a stage in the process which is omitted by most workers and frequently results in specimens that have critical details of their anatomy obscured by the appendages.

For the beginner, it is best to treat each specimen on a separate slide in the following manner. A flea is removed from the acid-water and placed in the center of a clean microscope slide. Under a dissecting microscope the specimen should be oriented so that the legs are pointed in the direction of the operator. Excess water should be drawn off with the aid of a Kleenex or other absorbent tissue but sufficient liquid should remain to cause the specimen to adhere to the slide. Only experience can demonstrate how much liquid is necessary but in no case should the specimen be allowed to dry.

The legs are now drawn away from the body by means of a fine dissecting needle. Needles made from insect pins are most satisfactory for this purpose. Approximately 1 mm of the pin point should be bent at a 45 degree angle, and the pin should be mounted in a needle holder.

Once the legs are positioned, a cover-glass or small fragment of a broken slide should be placed on top of the specimen, and the preparation should be flooded with *absolute* ethyl alcohol (EtOH). Most specimens are sufficiently hardened in a few seconds although large species may require additional time. Alcohol should be replaced as it evaporates, and the specimens should never be allowed to dry. Refer to the drawing on page 283 for an example of proper positioning of the legs. When properly hardened, the glass fragment should be removed with forceps and the specimen *washed* off the slide with absolute alcohol. All individuals of a series should be accumulated in a small container of absolute EtOH. When the series is completed, the EtOH should be drawn off and replaced with fresh absolute EtOH.

RUNNING-UP

Material in alcohol cannot be mounted directly in resinous mounting media such as Canada Balsam. Specimens must be transferred to the resin solvent or some other compound that is miscible with it. Most resins are dissolved in xylene, which is miscible with *absolute* alcohol, and the process of transferring the material from alcohol to xylene is called "running-up" (at least by us). Since fleas and most other ectoparasites are not particularly delicate, plasmolysis is not a problem and graded series of reagents are not necessary as with more delicate materials. Immediately upon completion of the spreading process the material can be transferred from absolute EtOH to a solution of equal proportions of absolute EtOH and xylene. If, at this or any subsequent stage of the running-up, the solution appears milky, it is an indication that there is still some water present that is reacting with the xylene. In such cases, the material should be returned to absolute EtOH for an additional period of dehydration. If the condition persists, the EtOH is probably contaminated with water and should not be used.

Material can remain almost indefinitely in the xylene-alcohol solution, but it should not be progressed to pure xylene until it has soaked an hour or two at room temperature. The alcohol-xylene solution can then be drawn off and replaced with pure xylene, where the specimens should remain for at least one hour. After this period the xylene should be drawn off and replaced with fresh xylene after which the specimens are ready to mount.

MOUNTING

Fleas can be mounted in either natural or synthetic resin. Canada Balsam is the best known natural resin. It is soluble in either xylene or toluene, but xylene is preferred in the routine described above. Specimens should never be mounted in Hoyers or other water soluble mounting media.

Standard 1 inch by 3 inch microscope slides are used and No. 1, 12-15 mm, round cover slips are preferred. Slides should be cleaned in 70 percent alcohol, and cover slips may be cleaned dry or in alcohol, although the former method is easier.

Since two labels usually are required on slides of ectoparasites, it is best to mount specimens in the center of the slide in order to leave space for a label on each end. We prefer to mount specimens slightly closer to one edge than the other. Uniformity may be obtained by using a guide drawn on a 3 x 5 white index card as shown below.

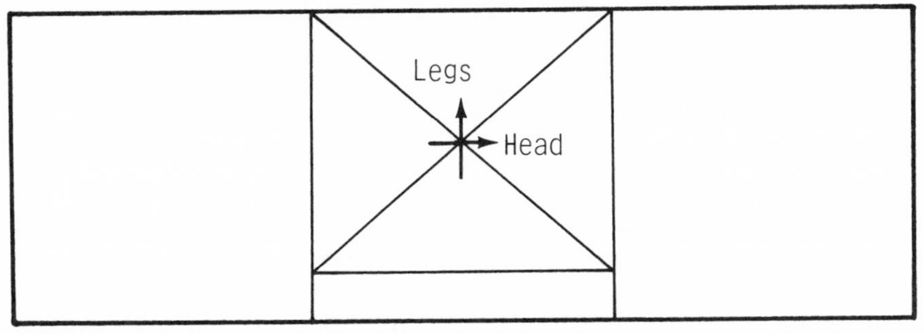

Because of inversion introduced by the compound microscope, individual fleas should be mounted with the head to the *right* and the legs toward the *top* of the slide as indicated by the arrows in the drawing. Thus, when examined under the microscope, all specimens face the *left,* and the legs extend downward. Most workers follow this convention but those who do not create considerable difficulty in using their specimens and drawings for comparative purposes.

Unless you are absolutely certain that the material consists of but one species or you are able to separate both sexes of all the species contained in the sample, *do not mount males and females on the same slide!* Actually, pairs are preferable, but only if both sexes belong to the same species. Males should be mounted above females when mounting pairs. The mounting routine is as follows.

Individual specimens or pairs should be removed from the pure xylene with the aid of a section lifter or, gently, with a pair of fine forceps. Once properly positioned on the slide, excess xylene may be drawn off with the forceps or with absorbent tissue. The cover slip is then placed on the top of the specimen(s), and a drop of thin mountant is placed on the slide along the *low* edge of the cover. It will flow under the cover slip slowly and surround the specimens without displacing them. Additional mountant may be added if needed. Air bubbles which are accidentally trapped usually will work their way to the edge of the cover slip in a few hours. Once mounted, the slides should be stored flat for a month or two in order to dry or, if urgently needed, may be dried sufficiently in a week in a slide drying oven.

Be sure to scratch the lot number which you have assigned to the material on each slide with a diamond pencil. This is the only reference you have to the collection data which ultimately must appear on the collecting label.

FINISHING

Although many workers simply label the dried slide without further attention, slides should be cleaned and ringed before they are labeled. Remember, any project requiring this much time should be an example of the worker's expertise, and the few additional minutes required to complete the job are well spent.

Slides are cleaned easily by scraping the excess mountant off with a scalpel. Ten or more slides should be scraped at one time. Any remaining traces of the resin may be wiped away with a soft cloth moistened in chloroform.

Ringing is accomplished easily on a slide ringing table with a revolving stage. Slides are centered on the stage, revolved rapidly, and the edge of the cover slip and the slide surrounding it are covered with a layer of ringing medium, applied with a No. 1 or 2 red sable watercolor brush. Several commercial ringing media are available, all at exhorbitant prices. Any quick-drying laquer or enamel may be used.

LABELLING

Ectoparasite slides should bear two labels. How these are placed on the slide depends upon how the slide is to be stored. Most people use boxes for storage and the usual practice is to place one label on each end of the slide with the top of the labels along the top of the slide as shown below.

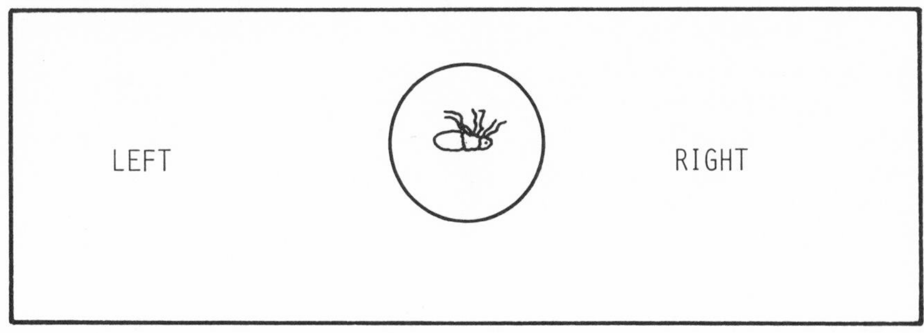

Although there is some variation in the data entered on the labels, the following description follows the system employed by most of the workers in the field and most of the major museums in the world. The left label is the determination label and bears the following information as illustrated below: number, sex(es), generic and specific name, describer and date of description, and the name of the person who identified the specimens. The right label is the collection label and bears the following information as illustrated: scientific name of host, collection locality, date of collection, and name of collector. Map coordinates and elevations are desirable also and should be included when available. Distances and elevations should follow the metric system.

LEWIS COLLECTION
15275
Actenopsyllus suavis J. & R., 1929
det. R. E. Lewis

LEWIS COLLECTION
Ptychoramphus aleuticus nest
USA: CA SE Farallon Isl. 17-III-1980
col. T. G. Schwan

HELPFUL HINTS

The following hints, based on experience gained from making over 50,000 slides of fleas, will save time and improve the quality of your mounts.

Specimens should be handled as little as possible in order to avoid breaking off setae and inflicting mechanical damage. Clearing, neutralization, and running-up all can be done in the same container. We prefer small stender dishes with ground glass tops. The collection number of each series may be written on the glass with wax pencil or in lead pencil on a small piece of paper glued to the lid of the dish.

All reagents may be stored easily in polyethylene squeeze bottles which are ideal stock bottles.

A waste bottle can be made by cutting the inner tube from a squeeze bottle. When changing solutions, the solution to be removed can be sucked up in the waste bottle and quickly and easily replaced by another from the stock squeeze bottles.

Remember: experience is the best teacher, so don't be too disappointed if your first few efforts do not live up to your expectations.

Index

This index contains all of the scientific names of the taxa cited in the text. Included are names and combinations that are no longer valid, but may be encountered in the literature. Junior synonyms are given in italics in order to differentiate them from currently valid names. The page number in bold-face indicates where the species account begins.

Anatomical terms are indexed and hosts are listed to family and genus. Additional information about the host/parasite associations may be obtained from Appendix 1, pages 265-276.